For permission to use material from this book please contact us:
Telephone 631-231-7269; Fax 631-231-8175
Web Site: http://www.novapublishers.com

Library of Congress Cataloging-in-Publication Data

ISBN: 978-1-63117-438-4

Published by Nova Science Publishers, Inc. † New York

CONTENTS

Preface   vii

Chapter 1   The Role of Soil Microbial Communities in Soil Carbon Processes and the Biogeochemical Carbon Cycle   1
O. Roger Anderson

Chapter 2   Land Use Effects on Soil Carbon and Nitrogen Stocks and Fluxes in the Pampas: Impact on Productivity   51
R. Alvarez, J. L. De Paepe, H. S. Steinbach, G. Berhongaray, M. M. Mendoza, A. A. Bono, N. F. Romano, R. J. C. Cantet and C. R. Alvarez

Chapter 3   State of Soil Carbon in Hungarian Sites: Loss, Pool and Management   91
C. Centeri, B. Szabó, G. Jakab, J. Kovács, B. Madarász, J. Szabó, A. Tóth, G. Gelencsér, Z. Szalai and M. Vona

Index   **119**

PREFACE

This book focuses mainly on soil carbon and its environmental benefits.

Chapter 1 - Carbon, a major constituent of structural and energy-releasing organic molecules, is essential to life. A substantial amount of C found in soils is available as soluble organic molecules (SOM) or particulate organic matter (POM) derived from plant primary production. Available atmospheric C as CO_2, used by plants during photosynthesis, is part of a larger biogeochemical cycle, including the cyclic release of CO_2 by respiration in living organisms and its fixation through photosynthesis. Carbon also is fixed through biological calcification and is eventually bound in carbonate-containing rocks, especially in aquatic and marine environments. This review is on soil microbial communities and their role in the carbon cycle, especially their ecological role in the utilization and decomposition of organic sources of carbon in a variety of terrestrial systems including deserts, grasslands, forests, and high latitude ecosystems. Special focus is given to the responses of microbial communities to global climate change and anthropogenic environmental alterations that effect carbon utilization and release as CO_2. The role of microbial respiratory CO_2 efflux to the atmosphere is also considered as a response to, and a contributory factor toward, the greenhouse effect and global warming, in addition to the positive ecosystem services to society by soil microbes.

Chapter 2 - The Pampas in Argentina is a vast plain of around 60 Mha that is considered as one of the most important grain producing regions in the World. Cultivation began by the last quarter of the 19th Century occupying 50% of the surface at present. Low input agriculture, in combination with livestock production, was performed till 1970, afterwards soybean (*Glicyne max* (L). Merr.) crop was introduced. This crop replaced pastures and at

present it accounts for 60% of the grain crop seeded surface. During the 2007-2008 growing season 382 sites, widespread over the region, were sampled up to 1 m depth and carbonate carbon, organic carbon, and total nitrogen stocks were determined in these soil samples. Land use had no effect on inorganic carbon content, which accounted for one third of total soil carbon. Aggregation of soil map information showed that the carbonate carbon stock of the region was estimated to be 1.9 Gt. Organic carbon was affected by land use. Cropped soils lost 16% of their organic carbon stock up to 50 cm depth. The overall organic carbon stock of the Pampas was estimated to be 4.2 Gt. This estimation was achieved combining an artificial neural network model for carbon estimation that used as inputs climate, textural, and land use information, and satellite image classification that referred to land use per pampean county. The net flux of C-CO$_2$ from soils to the atmosphere as the consequence of cultivation was estimated to be 0.33 Gt. Cropping effect on total nitrogen stock was similar to the effect on organic carbon. The average C/N ratio of the soils was 8.9, with only small differences between land use types and depth layers. When comparing present organic carbon and nitrogen stocks with those estimated from soil maps corresponding to surveys performed ca. 40 years ago, we observed that areas with high carbon and nitrogen contents tended to loose carbon and nitrogen and areas with low levels tended to maintain the levels or increase them. The maintenance or increase of carbon could be explained by increases in rainfall due to climate change and the increase of crop residue inputs of higher yielding crops, despite the soybean adoption in rotations. Additionally, the adoption of no-till as the most common tillage system favored carbon gain of many soils. The Pampean agroecosystems have a positive nitrogen balance with an annual gain of ca. 1 Mt N yr^{-1}. The main nitrogen input to soils is through biological nitrogen fixation, both by leguminous pasture and soybean, which represents ca. 5 times the fertilizer input. In sites with annual crop rotations the nitrogen balance is near neutral. Productivity of Pampean soils is regulated by available water holding capacity and organic carbon content. Changes in carbon levels produced by the agricultural use have little impact on productivity. The effects of different management practices on organic carbon and some labile fractions (microbial biomass, in vitro mineralized carbon, particulate carbon, light fraction carbon) were assessed in numerous field experiments, performed mainly in the humid portion of the Pampas. The labile fractions of organic matter are not good indicators of future changes in the carbon content of Pampean soils. These fractions change more pronounced than organic carbon only when carbon changes are greater than 25-50% but vary in equal measure

or even with an opposite trend below this threshold. Neither these fractions, nor the potentially mineralizable nitrogen or the organic matter/clay+silt ratio are good predictors of the nitrate level at planting or crop yields. Crop yield prediction could not be improved by using the mentioned labile fractions instead of organic carbon in models. Conversely, biological fractionations, determined by in vitro nitrogen mineralization, allowed a better soil nitrogen mineralization prediction under field conditions than using only carbon and nitrogen stocks.

An artificial neural network model was fitted that predicted field nitrogen mineralization using as inputs, land use information, soil texture, temperature, water content, total nitrogen stock, and the in vitro mineralization test results, with very good performance (R^2= 0.76). The impact of cultivation on soil nitrogen supplying capacity was assessed for by applying this network. In average, cultivation produced an 80% decrease in the nitrogen mineralization capacity of soils. Despite soil nitrogen was only moderately affected by cropping, the biological fractionation method, coupled with artificial intelligence modeling, showed great significant soil degradation in the Pampas. Present levels of nitrogen mineralization capacity of soils are not enough to sustain high to medium wheat (*Triticum aestivum* L.) and corn (*Zea mayz* L.) yield and fertilization has become a required practice. The author's results show that pampean soils loose only a small fraction of their organic matter content, but this fraction had a deep impact on soil fertility. Biological fractionation, not physical fractionation of organic matter, allowed to asses cropping effects on soil quality, and resulted in the development of suitable methods for agronomic purposes in the Pampas.

Chapter 3 - In the 19th century, in the Austrian-Hungarian Monarchy, Hungary has always been referred to as the food chamber. The reason for this is the huge areas having high agricultural potentials. Nowadays, 2/3 of the country's present area is suitable for agricultural production and 50 % has good potential for arable farming. The present study intend to present the state of soil carbon on some arable lands where it is lost due to unfavorable management and to present its state where it is protected in undisturbed, natural or semi-natural spots. Soil carbon was measured on various sloping areas of Hungary. Shallow drillings, up to 3.6 meters have been made where soil organic matter has been measured in order to present the amount of buried – and thus lost for agricultural production etc. – material. The above methodologies have been applied to areas with different agricultural management types.

This allow us to show the differences among the effects of land use on soil carbon loss and to formulate suggestions for soil carbon protecting and still, economically viable land use forms which might be called optimal. There are experimental areas where different land use forms and land management have been examined, providing information on soil carbon, too.

In: Soil Carbon
Editor: Aquila Margit

ISBN: 978-1-63117-438-4
© 2014 Nova Science Publishers, Inc.

Chapter 1

THE ROLE OF SOIL MICROBIAL COMMUNITIES IN SOIL CARBON PROCESSES AND THE BIOGEOCHEMICAL CARBON CYCLE

O. Roger Anderson[*]

Biology, Lamont-Doherty Earth Observatory of Columbia University,
Palisades, NY, US

ABSTRACT

Carbon, a major constituent of structural and energy-releasing organic molecules, is essential to life. A substantial amount of C found in soils is available as soluble organic molecules (SOM) or particulate organic matter (POM) derived from plant primary production. Available atmospheric C as CO_2, used by plants during photosynthesis, is part of a larger biogeochemical cycle, including the cyclic release of CO_2 by respiration in living organisms and its fixation through photosynthesis. Carbon also is fixed through biological calcification and is eventually bound in carbonate-containing rocks, especially in aquatic and marine environments. This review is on soil microbial communities and their role in the carbon cycle, especially their ecological role in the utilization and decomposition of organic sources of carbon in a variety of terrestrial systems including deserts, grasslands, forests, and high latitude

---

[*] E-mail: ora@LDEO.columbia.edu.

ecosystems. Special focus is given to the responses of microbial communities to global climate change and anthropogenic environmental alterations that effect carbon utilization and release as CO_2. The role of microbial respiratory CO_2 efflux to the atmosphere is also considered as a response to, and a contributory factor toward, the greenhouse effect and global warming, in addition to the positive ecosystem services to society by soil microbes.

INTRODUCTION

There is a substantial literature on the roles of individual microbial taxa in terrestrial C-related processes, including each of the major taxonomic groups to be considered in this review (bacteria, fungi and heterotrophic protists, previously referred to as protozoa). Only the free-living protists are considered. Due to the rapid decomposition and assimilation of organic compounds by terrestrial microbes (especially bacteria and fungi) a substantial amount of the available soil C is sequestered in particulate biomass. Consequently, to fully account for soil C content, it is important to assess the abundance, diversity, and distribution of terrestrial microbiota, particularly bacteria, fungi, and their eukaryotic microbial predators. Likewise, the amount and movement of available soluble organic carbon (SOC) in soils is significant in determining availability to microbial communities for metabolism and activity (e.g., Neff and Asner, 2001; Trumbore, 1993). Given the scope of this chapter, the focus will be on microbial ecology at the community-level, within an ecological context and the biogeochemical C-cycle. Less emphasis will be given to environmental and edaphic factors that determine the concentrations and availability of C to microbial communities. References to publications that address only one of the pertinent taxonomic groups will be cited when they contribute to elucidating particular community-level processes. An earnest effort has been made to be comprehensive and representative in reviewing the literature, while selectively choosing those contributions that make a coherent exposition of our current knowledge within the scope of the theme of this chapter.

The chapter sections are organized first within a biogeographic context of some major ecosystems, i.e. deserts, grasslands, forests, Arctic tundra and Antarctic, with some comments about the important ecosystem services provided by terrestrial microbial communities.

Finally, a more comprehensive perspective is taken on global ecosystem processes and the biogeochemical C-cycle with particular reference to climate

change and global warming. This chapter is intended to complement some current more broad-based reviews of terrestrial microbial community structure and function in relation to C dynamics and the biogeochemical cycle (e.g., Adl, 2003; Anderson, 2011, 2012a; De Bruijn, 2013; Jahn, et al., 2010; Raich and Schlesinger, 1992).

To maintain a reasonable scope, this chapter focuses on microbial C metabolism and the role of respiratory CO_2 in the biogeochemical C-cycle, however the role of microbial-produced methane is also of increasing importance in global warming. Methane is a more potent greenhouse gas. Additional information on its sources and effects is available elsewhere (e.g., Bridgham, et al., 2013).

BIOGEOGRAPHIC RESEARCH ON MICROBIAL ABUNDANCE, DIVERSITY AND DISTRIBUTION

The structure and function of soil microbial communities is reviewed in successive subsections for each major ecosystem with attention as appropriate to abundance, diversity and C biomass. Comparative research studies of two or more ecosystems are reviewed in transitions between subsections.

Research on microbial community organization has been substantially enhanced by modern techniques including marker molecules (e.g., phospholipid fatty acids, and sterols), enzyme assays, proteomics, and molecular genetics. Improved methods of laboratory culturing of some eukaryotic microbes have also contributed to better characterization of the morphological correlates with molecular genetics and community functions of various taxa.

However, relative to other soil microbial taxa, the eukaryotic microbes (protozoa) have received less attention. Some review papers highlight what is known across broad biogeographic locales for the role of these protists in soil fertility and ecosystem processes, including arid soils (e.g., Robinson, et al., 2002), temperate forest soils (e.g., Clarholm, 2002), tropical soils (Swift, 1996), Alaskan Tundra (Anderson, 2008, 2010a), and agroecosystems (Swift and Anderson, 1993). Application of rapid, high throughput molecular genetic analyses has supported more thorough large-scale comparative studies across major global biomes, especially for bacteria and to a lesser extent for eukaryotes (e.g., Fierer, et al., 2012).

Deserts

Desert ecosystems include diverse localized habitats varying in soil and rock composition, in vegetation type, and precipitation gradients across biogeographic ranges. To focus this review of desert microbial research, some examples of published reports on the following topics are presented: 1) microbial consortia associated with rocks (hypolith communities), 2) vegetation-related communities, and 3) the variations in microbial communities related to geographic range and precipitation gradients.

Edaphic and Vegetation Factors

A synoptic review of microorganisms associated with desert rocks has been published by Wierzchos et al. (2012) and includes an analysis of the unique habitats provided by rocks, such as thermal buffering, physical stability, protection against incident UV radiation, excessive photosynthetically active radiation, and freeze-thaw events. Makhalanyane, et al. (2013) reported that microbial communities found underneath translucent rocks in the Namib Desert were largely dominated by cyanobacteria affiliated to Pleurocapsales, whereas actinobacteria were prevalent in the surrounding soil. Molecular genetics coupled with multivariate statistical analysis indicated that there were six and nine indicator lineages for hypolith and soil microbial communities, respectively, based on 16S rRNA gene clone libraries. Their results are consistent with the concept of species sorting and suggest that the bottom of the quartz rocks provides conditions suitable for the development of discrete and demonstrably different microbial assemblages compared to surrounding habitats. However, they also found strong evidence for neutral assembly processes, as almost 90% of the taxa present in the hypoliths were also detected in the soil. These results suggest that hypolithons selectively recruit from local populations of microbial communities in the surrounding soil. At a more broad scale, Chan, et al. (2012) also confirmed that cyanobacteria are dominant in biofilms on the under surface of translucent rocks in desert pavement terrains and examined their role in geobiological processes and soil stabilization, especially focusing on global-scale trends in the ecology of hypoliths that are strongly related to climate. They particularly focused on climate-related shifts in cyanobacterial assemblages. A synthesis of available global data revealed a linear trend for colonization with regard to climate (i.e., decreasing colonization with increasing aridity). The authors

suggest a potential application for hypoliths as biomarkers of aridity on a landscape scale. They also discuss the potential to exploit the soil-stabilizing properties of hypolithic colonization in environmental engineering on dryland soils. Although, there are some clearly identifiable predictors for the composition of hypolith microbial communities, there is increasing evidence that stochastic and deterministic processes interact in the assembly of desert microbial communities on a global scale (e.g., Caruso, et al., 2011). Their study showed that multi-trophic microbial systems (including cyanobacteria and heterotrophs) may not be fully described by a single set of niche or neutral assembly rules and that stochasticity is likely a major determinant of such systems, with significant variation in the influence of these determinants on a global scale. The consistent presence of cyanobacteria in hypoliths represents a potentially significant C sink in an otherwise largely heterotrophic bacterial regime.

Sparse vegetation in desert habitats is expected to have some affect on soil microbial community composition based on the likely contribution of organic matter in the surrounding soil and especially secreted around the roots, among other factors. Some current research indicates that the effects are intricate, and varied, as may be expected in extreme environments, especially where abiotic and biotic factors interact in complex ways. Barness, et al. (2009) examined the differences in microbial communities associated with two desert shrubs, *Atriplex halimus* and *Hammada scoparia*; the former, a leaf-bearing halophyte with a much higher percentage of organic matter beneath its canopy compared to the bare-stemmed *H. scoparia*. Their results obtained from the field study demonstrated that plant ecophysiological adaptation of the associated microbes played an important role in the temporal and spatial distribution of microbial community levels such as microbial biomass, CO_2 evolution, and colony-forming units of both bacteria and fungi. For example, microbial biomass in soil samples collected in the vicinity of *H. scoparia* was fairly steady with values ranging between 108 and 128 μg C g^{-1} dry soil; whereas, biomass in soil samples collected in the vicinity of *A. halimus* was significantly higher, ranging between 150 to 200 μg C g^{-1} dry soil in the upper 20 cm of soil during summer. In a similar study, Saul-Tcherkas, et al. (2013) examined microbial composition and diversity in soil samples near canopies of two shrub species, *Artemisia sieberi* and *Noaea mucronata*. They found no obvious differences in the Shannon-Weaver index, evenness values, or total phylogenetic distances for the soil microbial communities. However, detailed denaturing gradient gel electrophoresis clustering as well as taxonomic diversity analyses indicated clear shifts in the soil microbial community composition. These shifts were

governed significantly by plant species type as well as seasonal variability in soil moisture.

Zhang, et al. (2012) examined bacteria associated with plant rhizosphere in the Tengger Desert using molecular genetic techniques and reported that Proteobacteria, consisting of the alpha, beta, and gamma subdivisions, were clearly the dominant group at all depths in rhizosphere soil. Moreover, bacterial community abundance closely correlated with soil enzyme activities in different soils. Among the few discovered studies that reported abundances of microbes, Norovsuren, et al. (2007) reported that density of actinomycetes in desert-steppe soil rhizosphere reached hundreds of thousands of colony-forming units g^{-1} of substrate. Jin, et al. (2010) found that rhizosphere bacteria associated with different desert shrubs in Xinjiang, China were dominant, accounting for 78.10% to 93.18% of the total soil microbes, while actinomyces were second dominant, and fungi accounted for only 0.03% to 0.12%. Studies of microbial communities associated with the desert sagebrush (*Artemisia tridentata*) using PCR-DGGE analysis (Gonzalez-Franco, et al., 2009) showed that actinomycete groups were less diverse in rhizosphere soils collected in winter than in bulk soils. However, rhizosphere soil of young *A. tridentata* collected in spring showed an enrichment of actinomycete diversity and/or selection of unique actinomycete populations, in contrast to actinomycete populations detected in the rhizosphere soil of old plants in the same season. In this latter case, a shift in the dominant detected taxa was observed between seasons. Overall, the enumeration analysis showed that actinomycete counts tended to be higher in spring, while total bacteria counts were higher in winter. Comparative studies of microbial communities associated with desert plant rhizosphere and surrounding barren soil (e.g., Andrew, et al., 2012) have shown that for cacti, microbial communities are shaped primarily by soil characteristics associated with geographic locations, while rhizosphere associations are secondary factors. Moreover, based on their molecular genetic analyses, operational taxonomic units (OTUs) common to both soil and rhizosphere samples comprised the bulk of raw sequence reads, suggesting that the shared community of soil and rhizosphere microbes constitute common and abundant taxa. These common taxa were particularly in the bacterial phyla Proteobacteria, Actinobacteria, Planctomycetes, Firmicutes, Bacteroidetes, Chloroflexi, and Acidobacteria. The vast majority of OTUs, however, were rare and unique to either soil or rhizosphere communities and differed among locations dozens of kilometers apart.

Bamforth (1984), focusing on edaphic and vegetation aspects, examined densities of microorganisms in litters, soils and cryptogamic crusts in Arizona

deserts and semi-arid woodlands, and reported numbers g^{-1} soil dry weight for bacteria of 1 to 5 x 10^8, and protozoa from 2 x10^3 to 1 x10^4 g^{-1}. In a further study (Bamforth, 1985) thirty-four species of protozoa were found in 27 soil collections from two desert sites in northern Utah. Most of the protozoa were highly edaphic types, and 74% of the species also occurred in the warm desert of Arizona. Free-living amoebae associated with shrubs in the Negev Desert in Israel numbered as high as 8,000 g^{-1} soil dry weight (Rodriguez-Zaragoza, et al., 2005). The amoeba population in the deeper layers (40 to 50 cm) was found to be as active as that in the upper layers, indicating that, in the desert, soil columns below 20 cm are fertile and worth further study. Among the few studies of protozoa in arid lands, Robinson, et al. (2002) examined the density and diversity of protozoa in Australian soils. Amoebae ranged from 1,000 – 5,000 g^{-1} soil, and were two orders of magnitude more abundant than ciliates. Testacea ranged 900 – 5,000 g^{-1} with similar species richness throughout the sampled vegetation. Cyanobacterial crust appeared to be a micro-refugium because it contained a number of non-encysting protozoa feeding on cyanobacterial filaments. The numbers and species richness of protozoa under shrubs were greater than in bare soils, supporting the resource island hypothesis that desert plants create soil heterogeneity by localizing soil fertility under their canopies.

Soil Moisture and Precipitation Patterns

In addition to vegetation affects per se, precipitation patterns in conjunction with vegetation cover have major influences on microbial community dynamics. The soil microbial community structure was examined in three vegetation zones varying in precipitation patterns along the Pine Canyon Watershed, an elevation and vegetation gradient in Big Bend National Park, Chihuahuan Desert by Clark, et al. (2009). Soil samples at each site were obtained in mid-winter (January) and in mid-summer (August) for two years to capture a component of the variability in soil temperature and moisture that can occur seasonally and between years along this watershed. Precipitation patterns and amounts differed substantially between years with a drought characterizing most of the second year. Soils were collected during the drought period and following a large rainfall event and compared to soil samples collected during a relatively average season. Using molecular markers, the fungal amounts (varying between 15% to over 25% of total abundances) differed significantly across seasons and sites and greatly outweighed the

quantity of bacterial (< 10 %) and actinomycete (< 5%) levels for all sites and seasons. The authors concluded that microbial responses to changes in precipitation frequency and amount due to climate change will differ among vegetation zones along this Chihuahuan Desert watershed gradient. The differential susceptibility of the microbial communities to changes in precipitation amounts along the elevation gradient may reflect the interactive effects of the soil moisture window duration following a precipitation event and differences in soil heat loads. In these extreme environments, the amounts and types of C inputs may not be as important in regulating microbial structure among vegetation zones as is the seasonal pattern of soil moisture and the soil heat load profile that characterizes the location.

In a similar study Angel, et al. (2010), surveyed the diversity of soil bacteria and archaea along a steep precipitation gradient ranging from the Negev Desert in the south of Israel (< 100 mm annual rain) to the Mediterranean forests in the north (> 900 mm annual rain) and found that differences in community composition were not statistically significant within sites for bacteria and archaea, but differed profoundly by ecosystem type along the geographic gradient. The authors concluded that the differences could largely be explained by the precipitation gradient combined with the vegetation cover. Studies of desert protozoa in relation to other microbes are more limited.

Desert and Arid Grasslands

Some attention has been given to examining differences in microbial community responses to environmental forcing factors in desert regions and arid grasslands. For example, Clark, et al. (2009) in a biogeographical study using molecular genetic techniques in the Chihuahuan Desert with vegetation spanning desert scrub, oak-pine forest and sotol grasslands, reported that microbial community structure at the low elevation desert scrub and sotol grasslands sites was most strongly related to soil pH with bacterial and actinobacterial levels accounting for site differences along the gradient. Using electrophoresis techniques, DGGE band counts of amplified soil bacterial DNA, microbial communities were found to differ significantly across sites and season with the highest band counts found in the mid-elevation grassland site. Further studies of a desert sotol grassland by Bell, et al. (2008), showed that in addition to soil moisture, the seasonal and yearly variability of soil bacterial activity was most closely associated with levels of soil organic

matter, extractable NH$_4$-N, and soil pH, while variability in fungal activity was related to soil temperatures ranging between 13° C and 26° C. Their findings indicate that changes in soil moisture, coupled with soil temperatures and resource availability, drive the functioning of soil-microbial dynamics in these desert grasslands. Comparative data for bacteria and fungi showed that bacteria were more able to respond to moisture pulses regardless of temperature, while fungi only responded to moisture pulses during cooler seasons, with the exception of substantial increased magnitudes in precipitation occurring during warmer months.

Grasslands

A substantial record of research has been published on the roles of vegetation and moisture, including precipitation events, affecting soil microbial communities in grasslands. Some representative examples are summarized.

Vegetation Aspects

Millard and Singh (2010) reviewed the role of grassland vegetation in microbial diversity at three spatial scales (field level, rhizosophere, and rhizoplane) especially as related to agroecosystems and provided a comprehensive overview of the findings. Among other effects, they report that bacterial diversity is not directly related to plant diversity, although fungal diversity is. In addition, the soil fungal community had an effect upon the composition of the bacterial community. They concluded that at the present time of the study, vegetation influences fungal communities (particularly mycorrhizae) while litter inputs support fungal saprotrophs. However, on a longer time scale, bacterial community structure is influenced more by the quality or composition of soil organic matter, thereby reflecting C inputs to the soil over decades.

On a broader geographic scope, Grayston, et al. (2001) examined the factors regulating soil microbial community organization and function in three temperate upland grassland ecosystems varying in species composition. Variation in C biomass was mainly due to grassland type and site (accounting for 55% of the variance in the data). Microbial C biomass was significantly higher in the unimproved grassland site (237.4 g C m^{-2}) compared to two

agriculturally more improved sites (81.2 g C m^{-2} and 63.8 g C m^{-2}). Soil microbial activity, measured as basal respiration, was also mainly affected by grassland type and site, but in contrast to C biomass, respiration was significantly higher in some of the improved grasslands. Moreover, variation in microbial community structure was due, predominantly, to grassland type and date of sampling. In a further study of the effects of soil fertility, Innes, et al. (2004) reported that soil microbial biomass and activity were found to be significantly greater in more fertile, agriculturally improved soils than in the less productive unimproved meadow soils. Differences in microbial community structure were also evident between the two soils, with fungal abundance being greater in the unimproved soil type. Overall, the authors concluded that effects of plant species on microbial properties differ markedly in soils of differing fertility, making general predictions difficult about how individual plants impact soil properties.

De Deyn, et al. (2011) examined whether changes in grassland vegetation composition resulting from management for plant diversity influences short-term rates of C assimilation and transfer from plants to soil microbes using an in situ ^{13}C-CO$_2$ pulse-labeling approach to measure differential C uptake among different plant species. They also examined the transfer of the plant-derived ^{13}C to key groups of soil microbiota across selected treatments of a long-term plant diversity grassland restoration experiment. Plant taxa differed markedly in the rate of ^{13}C assimilation and concentration. Uptake was greatest and ^{13}C concentration declined fastest in plots planted with *Ranunculus repens*. Assimilation was least and ^{13}C signature remained longest in mosses. Overall, the authors conclude that plant diversity may not directly affect the C assimilation or retention of C by individual plant taxa or groups of soil microbes, but it can impact on the fate of recent C by changing their relative abundances in the plant-soil system. Moreover, across all treatments they found that plant-derived C is rapidly transferred specifically to arbuscular mycorrhizal fungi and decomposer fungi, indicating their consistent key role in the cycling of recent plant derived C. In additional studies of natural vegetation diversity on belowground microbial communities, Dornbush, et al. (2008) compared soil food webs beneath C-3- and C-4-grass plantings by measuring bacterial and fungal biomass and protozoan abundance repeatedly over two years. Bacterial biomass and ciliate abundance were greater beneath C-4 grasses, but there were no differences in fungi, amoebae, or flagellates. The organization of soil food webs varied significantly among sample dates, but differences were unrelated to aboveground plant growth. These findings, in combination with previous work, suggest that preexisting soil properties

moderate the effect of plant inputs on soil food webs. The authors suggest that high levels of soil organic matter provide a stable environment and energy source for soil organisms and thus buffer soil food webs from short-term dynamics of plant communities.

Soil Moisture and Precipitation Patterns

Soil moisture and precipitation effects have been examined in relation to amount of moisture and frequency of precipitation. The role of variability in soil moisture on microbial community C and N resource use was examined by Tiemann and Billings (2011) within a guiding assumption that microbial response to increased soil moisture variability and other forms of stress can induce ecosystem-scale alterations in C and N cycling processes through alterations in microbial function. Four grassland sites in the Great Plains, representing a precipitation gradient of 485 to 1,003 mm y^{-1}, were studied. The frequency and size of water additions and dry down periods were manipulated in a 72-day laboratory incubation in soil samples from the four sites, while keeping water totals constant. Under conditions where water was added in a long interval treatment (added at the beginning of each two-week cycle), 1.4 – 2.0 times more C was mineralized compared to soils undergoing a short interval treatment (four wetting events evenly distributed over each two-week cycle). Further evidence showed increases in microbial N demand and decreases in microbial C use efficiency with increased moisture variability, regardless of the geographic site sampled. This may imply that these grassland systems could experience enhanced heterotrophic CO_2 release and declines in plant-available N with climate change. Moreover, this may have important implications for C budgets in these grasslands when coupled with the declines in net primary productivity reported in other studies as a result of increases in precipitation variability across the region. See additional comments on these effects on respiration in the final section on Carbon Dynamics and the Biogeochemical C-cycle.

McCulley and Burke (2004) obtained additional evidence of the effects of precipitation variability on bacterial community biogeochemical processes in the Great Plains. They sampled at three upland and lowland topographic positions spanning along a 500-mm regional precipitation gradient. Microbial biomass increased across the regional gradient. Different microbial communities were associated with the different grassland community types. The relative abundance of fungi decreased while gram-negative anaerobic

bacteria increased along the increasing moisture gradient from shortgrass steppe to tallgrass prairie. There were no differences in microbial biomass at the landscape-level. The only alteration in microbial community composition between upland and lowland landscape positions was a shift toward more nonspecific bacteria in lowlands. The fact that the trends in microbial biomass and community composition at the landscape-level were less pronounced than regional levels suggests that variability in microbial community composition is larger regionally across the Great Plains than landscape variability associated with topographical features at any particular site. All-totaled, these findings indicate that alterations in microbial communities may play an important role in determining the biogeochemical patterns of grasslands in the Great Plains region.

The effects of stress due to drying and rewetting on microbial communities residing in surface soils was experimentally studied by Fierer, et al. (2003) using two samples of soil collected in Santa Ynez, CA, one from an annual grassland, the other from underneath an oak canopy. Soil water stress in the laboratory was manipulated by exposing the two different soil types to 0, 1, 2, 4, 6, 9, or 15 drying-rewetting cycles over a two-month period. They found that drying-rewetting regimes influenced bacterial community composition in the oak, but not in grass soils. The two soils support inherently different bacterial communities. Only the bacteria residing in the oak soil (less frequently exposed to moisture stress in their natural environment) were significantly affected by drying-rewetting cycles. The community indices of taxonomic diversity and richness, based on molecular genetic evidence, were relatively insensitive to the drying-rewetting frequency. The drying-rewetting induced shifts in bacterial community composition may partly explain the changes in C mineralization rates that are commonly observed following exposure to numerous drying-rewetting cycles.

Grassland and Forest Research

Further field studies of microbial communities along transects between grasslands and forests have been pursued by sampling soil at varying locales along a gradient and by transplanting soil from one locale to another. Some examples are presented. The relationship between plant and soil microbial communities along a successional gradient in a chalk grassland in northwestern France was examined by Chabrerie, et al. (2003). They found that organic matter resources supporting soil microbial communities were

larger in older stages of plant succession (high plant and litter biomass and nitrogen content in plant leaves in forest stages), but biomass and potential level of enzymatic activity of microbial communities were negatively correlated with the presence of shrubs and positively associated with grassland habitats. They also found complex interactions of these microbiological factors with other edaphic and biotic factors, including a positive correlation with soil humidity, soil soluble C and polysaccharides contents, and negative correlation with the presence of lignin in plant biomass and litter. Overall, the results show that microbial communities are independent from the successional gradient of vegetation, at least in terms of functional analyses and genetic structure based on 16S and 18S molecular genetic evidence.

Ingham, et al. (1989) reported that the structure of the belowground detrital food webs was similar in three different semiarid vegetation types along a gradient from shortgrass prairie (*Bouteloua gracilis*), mountain meadow (*Agropyron smithii*), and lodgepole pine (*Pinus contorta* subsp. *latifolia*). The biomass of bacteria was dominant in the meadow and prairie, while that of fungi was dominant in the forest. Biocide reduction in invertebrate predators in the bacterial-dominated grasslands resulted in no apparent effect on bacterial densities, because one group of bacterial consumers (protozoa) increased following the decrease in bacteria-feeding nematodes; but it produced an increase in fungal biomass. Conversely, in the forest, following the biocide-induced reduction in consumers, the total fungal biomass decreased. The meadow appeared to be the most resilient of the three ecosystems, as both nematode and arthropod numbers returned to control levels more rapidly than in the prairie or the forest.

In a study focused on protozoa, Dash and Guru (1980) examined the species composition, distribution and seasonal variation in the population density of Testacea (testate amoebae) associated with tropical grassland, crop field, mountain soil and forest soil in the western part of Orissa, India. Grasslands and forest sites exhibited higher species diversity in comparison to the mountain sites. Furthermore, species found in forest sites and moist grasslands had larger tests than in the mountain sites. Maximum densities (e.g., 1,166 g^{-1} wet soil) were recorded from the grassland and a minimum (620 g^{-1}) was recorded from a mountain site. Soil moisture was positively correlated with active forms (r = 0.89). Dash and Guru also report that secondary production of the testate amoebae was related to soil moisture based on in-situ experiments. In watered plots, secondary production was 1.21 to 2.46 g m^{-2} compared to 0.86 to 1.51 g m^{-2} in unwatered plots. Anderson (2000) also found that active stages of naked amoebae were correlated with soil

moisture content in a northeastern U.S. grassy site. In that longitudinal study with monthly samples over four years, maximum total densities of amoebae during the warmer months reached 3 to 5 x 10^3 g^{-1} soil dry weight. At higher levels of soil moisture (> 25% w/w) the percentage of active amoebae was in the range of 70 – 80%.

In a soil transplant experiment, Waldrop and Firestone (2006) reciprocally transplanted soil cores from under oak forest canopies and adjacent open grasslands in a California oak-grassland ecosystem to determine how microbial communities respond to changes in the soil environment and the potential consequences for the cycling of C. Using bacterial marker lipids and ^{13}C uptake from a C substrate (pyruvate), the microbial community was monitored every three months for up to two years. Soil microbial communities under oak canopies were more sensitive to environmental change than those in adjacent soil from the open grassland. Oak canopy soil communities changed rapidly when transplanted into the open grassland soil environment; but, conversely, grassland soil communities did not change when transplanted into the oak canopy environment. Similarly, microbial biomass and microbial respiration decreased when microbial communities were transplanted from the oak canopy soils to the grassland site, but not when the grassland communities were transplanted to the oak site. The authors proposed that microbial community composition and function is altered when microbes are exposed to new extremes in environmental conditions, that is, environmental conditions outside of their "life history" envelopes.

Tropical and Temperate Forests

Tropical Forests

Given the low nutrients in ancient mineral soils in many tropical forests, and the relative rapid recycling of nutrients in surface litter layers, attention has been given to the role of fertilizing nutrients in tropical forest soil microbial community dynamics. Hafich, et al. (2012) examined the long-term effects of reforestation versus maintained grassland on microbial community structure and nutrient cycling in the heavily deforested Northern Zone of Costa Rica, with particular attention to perspectives on ecosystem health and the C sequestration potential of tropical soils. The soil from the secondary forests had greater levels of phosphate, inorganic nitrogen, organic C, respiratory activity, and evidence of abundance and diversity of basidiomycetes, more abundance of fungal rDNA, and lower abundance but greater diversity of

Rhizobium rDNA than soils from adjacent grasslands of the same age. The data suggested a trend towards the secondary forest soils becoming more fungal-dominant, with greater microbial activity, greater nitrogen mineralization activity and more efficient use of C. The effects of phosphorus addition to soil in three broad leaf and pine tropical forests of southern China were studied by Liu, et al. (2012) with the objective to test the hypothesis that P addition would increase microbial biomass and change the composition of the microbial community; and that the old-growth forests would be more sensitive to P addition due to its higher soil N availability. Addition of P significantly increased the microbial biomass and soil respiration, and altered the microbial community composition in the old-growth forest, suggesting that P availability is one of the limiting factors for microbial growth. In contrast, P addition had no effect on the microbial biomass and the microbial community composition in the pine forests. Likewise, in the mixed forest, the microbial biomass did not significantly respond to P addition, but soil respiration and the ratio of fungi-to-bacteria increased significantly. In a follow-up study, Liu, et al. (2013) added N and P in a N-saturated, old-growth tropical forest in southern China. They found that N and P addition had opposing effects on soil microbial biomass and activity. Additions of P increased soil microbial biomass, while additions of N reduced soil microbial biomass. These effects, however, were transient, disappearing over longer periods. Nitrogen and P addition altered the composition of the microbial community. Nitrogen additions significantly increased relative abundance of fungal PLFAs and P additions significantly increased relative abundance of arbuscular mycorrhizal (AM) fungal PLFAs. The soil light fraction C was diminished by addition of N, but there was no effect on heavy fraction C and total soil C. In contrast, P addition significantly decreased both light fraction C and total soil C. There were no interactions between addition of N and of P on soil microbe composition.

Further studies by Cusack, et al. (2011) on the effects of nitrogen additions on microbial community and soil organic matter (SOM) in two tropical forests, showed that microbial biomass increased in response to N fertilization in both tropical forests and corresponded to declines in pools of low-density SOM. The chemical quality of this soil C pool reflected ecosystem-specific changes in microbial community composition. Specifically, in the lower-elevation forest, gram-negative bacteria biomass increased, and there were significant losses of labile C chemical groups (e.g., O-alkyls). By contrast, in the upper-elevation tropical forest fungal biomass increased with N additions coupled with declines in C groups associated with

increased soil C storage (alkyls). Overall, these findings suggest that ecosystem-specific changes in microbial community composition are likely to have far-reaching effects on soil C storage and cycling. Moreover, microbial communities in N-rich tropical forests can be sensitive to added N, but may exhibit significant variability in how ecosystem structure and function respond to N deposition among different tropical forest types.

In a more comprehensive study of nutrient regimes, Krashevska, et al. (2010) investigated the role of C, nitrogen and phosphorus as limiting factors of microorganisms and microbial grazers (testate amoebae) in a montane tropical rain forest in southern Ecuador. They amended soil in experimental plots with C (as glucose), N (as NH_4NO_3) and P (as NaH_2PO_4) separately and in combination at bimonthly intervals for 20 months. Carbon addition strongly increased ergosterol concentration, and to a lesser extent the amount of linoleic acid as fungal biomarkers, suggesting that saprotrophic fungi are limited by inadequate C. Microbial biomass and fungal ergosterol concentrations reached a maximum in the combined treatment (C, N and P) indicating that both N and P also were limited. In contrast to saprotrophic fungi and microorganisms in total, testate amoebae declined with the addition of C and reached maximum density by the addition of N. The results indicate that saprotrophic fungi in tropical montane rain forests are mainly limited by C; whereas, gram positive and negative bacteria benefit from increased availability of P. Testate amoebae were hindered by increased dominance of saprotrophic fungi in glucose-C treatments, but benefited from increased supply of N.

Several biogeographic studies illuminate some of the differences in microbial community abundance and diversity in varying tropical locales. Bamforth (2006) examined the potential role of protozoa in stimulating bacterial activity and C turnover in a tropical rain forest in the Puerto Rican Luquillo Experimental Forest, within the Caribbean National Forest. He found that Amoebae numbered 69,000 – 170,000, ciliates 1000 – 25,000, and testate amoebae 58,000 – 190,000 g^{-1} dry weight of litter, but were reduced in abundance by one quarter to one half in the underlying soils. About 50% of the individuals of ciliate populations and 33% of testate amoebae populations were small, rapidly growing, r-selected species. This illustrated that functional differences between species determine community composition and potential for bacterial grazing, thus affecting C flow in tropical terrestrial systems. Although tropical forests have high rates of soil C cycling, little information is available on how roots, arbuscular mycorrhizal fungi (AMF), and free-living microorganisms interact and influence organic matter mineralization in these ecosystems. Nottingham, et al. (2013) examined the effects of roots and

arbuscular mycorrhizal fungi (AMF) mycelia on microbial community composition, microbial C utilization, and hydrolytic enzyme activities for large, potted tropical trees with comparison to AMF-only soil. Plant-derived C was incorporated predominantly into bacterial groups in both rhizosphere and AMF-only soils. Gram-positive bacteria incorporated additional soil-derived C in rhizosphere soils, which also contained the highest microbial biomass. For hydrolytic enzymes, glucosidase and N-acetyl -glucosaminidase activities were highest in rhizosphere soils, while phosphomonoesterase activity was highest in AMF-only soil. Further studies on enzyme activities and litter mass loss in a lowland tropical forest showed that leaf litter mass loss was increased by the presence of roots, but not by the presence of AMF mycelia alone. Root-microbial interactions influenced organic matter cycling, including evidence for rhizosphere priming and accelerated leaf litter decomposition in the presence of roots. Although AMF mycelia alone did not stimulate organic matter mineralization, they appeared to be a conduit of C to other soil microorganisms.

Tropical and Temperate Forest Comparative Research

Miyashita, et al. (2013) examined soil bacterial community structure in five tropical forests in Malaysia and one temperate forest in Japan using pyrosequencing analyses of 16S rRNA gene sequence variations. Despite differences in vegetation and aboveground conditions, the composition of bacterial groups was similar across all sampling-sites and forests, with Acidobacteria, Proteobacteria, Verrucomicrobia, Planctomycetes and Bacteroidetes accounting for 90% of all phyla detected. Moreover, at higher taxonomic levels, the same taxa were predominant, although there was significant heterogeneity in relative abundance of specific taxa across sampling-sites within one forest or across different forests. The results suggested that bacterial communities were adapted to specific micro- and macro-environments, but macro-environmental diversity made a larger contribution to total bacterial diversity in forest soil.

In a similar study using fungi, McGuire, et al. (2013) examined ectomycorrhizal fungi in tropical and boreal forests, and analyzed fungal community composition across litter (organic horizons) and underlying soil horizons (0 – 20 cm) using 454 pyrosequencing and clone library sequencing. They found in both forests, there was a significant clustering of fungal communities by site and soil horizons with analogous patterns detected by both sequencing technologies. Free-living saprotrophic fungi dominated the recently-shed leaf litter and ectomycorrhizal fungi dominated the underlying

soil horizons, a finding that is consistent with soil vertical distribution reported in temperate and European boreal forests. Overall, these results apply broadly to ectomycorrhizal-dominated systems, including tropical rain forests. The authors conclude that ectomycorrhizal and free-living saprotrophic fungi have different influences on soil C and nitrogen dynamics, and information on the spatial distribution of these functional groups should improve our understanding of forest nutrient cycling.

Temperate Forests

The role of nutrients in temperate forest soil microbial community structure and function has received attention, largely because N has been identified as a limiting nutrient, although secondarily P can become limiting under conditions of more replete N availability (e.g., Kluber, et al., 2012). Some representative studies of the effects of natural N and anthropogenic N sources on microbial communities are reviewed, followed by more general papers on biogeographical and environmental factors.

Nutrient research. Phillips, et al. (2013) in a broad-based theoretical paper published a new perspective for predicting C-nutrient couplings in temperate forests based largely on associations of trees with either arbuscular mycorrhizal (AM) or ectomycorrhizal (ECM) fungi and noted that forests dominated by AM tree species have nutrient economies that differ in their C-nutrient couplings from those in plots dominated by ECM trees. They demonstrated how the relative abundance of AM and ECM trees can be used to estimate nutrient dynamics across the landscape. Twieg, et al. (2009) examined the influence of soil nutrients on ectomycorrhizal (ECM) communities in a chronosequence of mixed temperate forests, and reported that ECM fungal diversity varied unpredictably with total and mineralizable N or the C to N ratio, but conversely there were significant relationships between ECM fungal diversity and both available and organic P. Overall, they concluded that ECM fungal community structure was more strongly influenced by stand age than specific soil nutrients, but better correlations with soil nutrients may occur at broader spatial scales covering a wider range of site qualities.

A range of papers investigated the effects of either anthropogenic sources of N or experimental addition of N on microbial communities in temperate forest plots (e.g., Compton, et al. 2004; Deforest, et al., 2004; Hasselquist, et al., 2012; Krumins, et al., 2009; Waldrop, et al., 2004; Zak, et al., 2006, 2008). Some major findings are summarized. Experimental fertilization with NO_3-N produced no major effects on microbial community composition, but reduced

microbial biomass (by as much as 18% compared to controls). Effects on decomposition varied with substrate, varying in the amount and kind of enzymes secreted in response to the N-amendments, thus suggesting that accurate prediction of soil C storage requires a better understanding of the physiological response of microbial communities to atmospheric N deposition, especially how different groups of decomposing microbes relate to high versus low concentrations of added N. Although fungal community structure may not vary so markedly in response to nutrient amendment, there is evidence that bacterial communities may be more sensitive, especially in sandy soils. However, there may be sufficient differences across vegetation stands and biogeographic regions, to warrant more detailed investigations. For example, Couteaux, et al. (1998) examined the response of protozoan and microbial communities in six coniferous forest soils after transfer of columns of soil samples to forests with different levels of atmospheric pollution. No effect of enhanced N or S deposition on protozoan numbers and microbial biomass C, basal respiration or caloric quotient was found. However, reciprocal transfer of various soil columns resulted in lower abundance and activity of protozoa and microbes. This reduction was not explained by differences in N and S deposition, but more likely by differences in microclimate and adaptation. In some cases, protozoa correlated with pH, C to N ratio, P and S content and leached mineral N.

Biogeographic and climate studies. O'Hanlon (2012) reviewed current knowledge about the effect of small scale spatial and short term temporal variations in belowground ectomycorrhizal communities in forests focusing on how the niche concept and ECM functional morphology may explain a significant amount of spatial and temporal variations in ECM communities, including a novel sampling strategy to provide representative samples of the actual ECM community in the study area. To more fully explore some of these relations, examples of some focused biogeographic and climate research studies within the scope of the themes of this chapter are reviewed. In a broad-scale study of the relationship between host plant genus-level diversity and ectomycorrhizal fungi along a diversity gradient in a Chinese subtropical forest, Gao, et al. (2013) reported that EM fungal diversity was positively correlated with host plant diversity. Furthermore, this relationship was best predicted by host genus-level diversity, rather than species-level diversity or family-level diversity. Carletti, et al. (2009) examined microbial communities in six spruce (*Picea abies* (L.) Karst.) forests located in the alpine range of Northern Italy, with respect to community profiles and microbial biomass C as related to some edaphic and habitat variables. The geologic parent material

was the most significant factor in defining both soil chemistry and microbial community composition; in contrast the composition of the deeper humus layers (OH, A) was stable and similar within a spruce forest cycle, that is within one generation of a tree stand. Moreover, the north/south-facing, aspect-dependent microclimate features also played a distinct role. Bacterial communities appeared to be shaped first and foremost by the substratum, secondly by mountain slope orientation, and thirdly by forest stage.

Microbial communities in mineral soil within 12 different types of natural forest in Austria were studied by Hackl, et al. (2005). The forests included oak and beech, spruce-fir-beech, floodplain and pine forests. Principal component analysis of phospholipid fatty acid analysis (PLFA) patterns revealed that the microbial communities were compositionally distinct in the floodplain and pine forests (azonal forest types), and were more similar in the oak, beech and spruce-fir-beech forests, which represent the zonal vegetation types of the region.

The pine forest soils were differentiated from other forest soils by PLFA signatures typical of fungi. This may also indicate ectomycorrhizal fungi associated with pine trees, in addition to the presence of PLFA markers common in actinomycetes. These findings suggest that the occurrence of azonal forest types at sites with specific soil conditions possess specific soil microbial communities.

To examine the effects of climate change and global warming on microbial communities and decomposition of soil organic matter (SOM), Vanhala, et al. (2011) transferred surface soil sections and ground vegetation from the northern borders of a boreal forest to corresponding forest sites in the southern borders, and examined the effects on microbial communities after two years in comparison to control soil samples transplanted within the northern boreal zone.

Although ground vegetation and soil microbial community structure and functions were different initially at the two sites, adaptation by the northern soil sections transferred to the warmer south led to changes in vegetation and soil microbial community structure and function. Soils that were transplanted from north to south, lost microbial biomass, even more than soils transplanted in the control groups at the site of origin.

This research provides evidence of likely major changes in high latitude forest microbial communities, especially in relation to C decomposition rates that are correlated with changing microbial biomass and activity associated with global warming.

Antarctic and Arctic Research

With increasing evidence that climate change and global warming are producing major changes in high latitude ecosystems, considerable research has focused on responses of soil microbial communities in the Antarctic and Arctic (e.g., Convey et al., 2012). Some illustrative examples of research are reviewed with the objective of establishing a broader context for a more focused discussion of the role of microbial communities in the biogeochemistry of C dynamics in the last major section of this chapter, where special attention will be given to research on microbial communities and the C biogeochemical cycle in the Arctic tundra. In this section, particular attention is given to microbial community structure and biomass in relation to biogeography and environmental factors within the scope and theme of this chapter.

Antarctic

Biogeographic studies, especially along topographic and climatic gradients, have elucidated some of the environmental forcing functions affecting soil microbial communities and their C decomposition, including bacteria, fungi, protozoa and their nematode predators (e.g., Bamforth, et al., 2005; Foissner, 1996; Petz, 1997; Todorov & Golemansky, 1999; Vincke, et al., 2006; Virginia & Wall, 1999). Chong, et al. (2012b), examining a gradient in Antarctic locales, reported a narrow range of bacterial species compared to the overall sample size, while taxonomic distribution of Antarctic soil bacterial communities were significantly correlated to both the underlying soil parameters and geographic (regional) origins. This general pattern was further illustrated by evidence gathered from a wide variety of geographic locales in Antarctica, showing that in contrast to some recent evidence for a distinct Antarctic microbiome, phylogenetic comparisons indicated that a majority of Antarctic bacterial isolates were highly similar to those retrieved from tropical and temperate regions, suggesting widespread distribution of eurythermal mesophiles in Antarctic environments (Chong, et al., 2013). However, there was evidence that across different Antarctic regions, the dominant bacterial genera exhibit some spatially distinct diversity patterns analogous to those recently proposed for Antarctic terrestrial macroorganisms. Further evidence of spatial variability was reported by van Horn, et al. (2013) who examined factors controlling soil microbial biomass and bacterial diversity and community composition in a cold desert ecosystem, especially the role of geographic scale from snow patches to inter-basin comparisons. Acidobacteria

and Actinobacteria were prevalent at the organic C-rich, mesic, low elevation sites, while Firmicutes and Proteobacteria were dominant at the high elevation, low moisture and low C biomass sites. Microbial community patterns were significantly related with soil water content and edaphic characteristics including soil pH, organic matter, and sulfate concentrations. However, the magnitude, and even the direction of these relationships, varied across basins. The application of mixed effects models revealed evidence of significant contextual effects at local and regional scales.

Yeregeau, et al. (2007a, b) examined bacterial and fungal communities along Antarctic environmental gradients and reported that microbial abundance generally showed a significant positive relationship with vegetation and vegetation-associated soil factors (e.g., water content, organic C, total N). Microbial community structure was mainly related to latitude or location and latitude-dependent factors (e.g., mean temperature, nutrients and pH). Moreover, strong interactions between vegetation cover and location were observed, with the effects of vegetation cover being most pronounced in more extreme sites. Analyses of bacterial diversity along a gradient of increased climate severity over 27 degrees of latitude showed that bacterial diversity was observed to decline with increased latitude, but habitat-specific patterns appeared to also be important. For example, a negative relationship was found between bacterial diversity and latitude for fell-field soils, but no such pattern was observed for vegetated sites. A site previously identified as a biodiversity hotspot within this region (Mars Oasis), proved exceptional within the study transect, with unusually high bacterial diversity. In independent analyses, geographical distance and vegetation cover were found to significantly influence bacterial community composition. Similarly, Bogoev and Gyosheva (1996) studied six relatively harsh climatic sites on Livingston Island and observed that the total quantities of microorganisms, and consequently their biomass, are considerably smaller as compared to their quantity in habitats with more favorable climatic conditions. In general, bacterial biomass quantity was larger than the biomass of microscopic fungi. The total quantity of microorganisms isolated on agar plates varied between $10^5 - 10^6$ cells g^{-1} soil. In other studies of catabolic activity and physiological diversity of bacteria inhabiting the soil of moss, vascular plants, and fell field habitats from Livingston Island, Kenarova, et al. (2013) reported a lack of site-specific distribution of bacterial abundance, in contrast to localized bacterial C catabolic activity and community level physiological profiles. Overall the physiology of soil bacteria was habitat specific, concerning both the rate of catabolic activity and pattern of C source utilization.

Malosso, et al. (2004) investigated organic matter decomposition in Antarctic soil using ^{13}C-labeled plant materials and found no evidence of significant changes in microbial community structure over time during residue decomposition. There was evidence of a marked increase, however, of fungal biomass. Gas chromatography-mass spectrometry confirmed the transformation of the plant residue by showing the incorporation of ^{13}C-plant C into the fungal ergosterol. This incorporation of ^{13}C into the ergosterol increased over the incubation period. Treonis, et al. (2002) studied the rate of C decomposition in the cold desert soil of the McMurdo Dry Valleys of Antarctica, including the effects of changing moisture and temperature on rates of decomposition and the activity and abundance of soil organisms in field studies and laboratory microcosms. They reported that soil respiration, nitrification, and the decomposition of C (cotton strips) were all greater in dry valley soils that were wetted to 10% moisture content, as compared to soils at 0.6%, overall indicating that the decomposition potential for dry valley soils is high when moisture and temperature limitations are removed. In the field, however, decomposition was extremely slow, and biota did not respond to improving environmental conditions. Soil decomposition processes appear to be limited primarily by the extreme desiccation of the dry valleys and are likely restricted to the brief periods following infrequent snowfall, melt, and soil wetting that permit the activity of soil microbes and other biota.

In a more general analysis of environmental parameters likely to influence soil bacterial community composition on the Antarctic Peninsula using a statistical approach, Chong, et al. (2012a) examined soil samples from six environmentally distinct locations near to Rothera Point off the west coast of the Antarctic Peninsula. Using Spearman rank correlation, similar to a correlational approach used by Arenz and Blanchette (2011), they demonstrated that the taxonomic distribution of the soil bacteria among the six study sites was relatively even, especially among the islands within Ryder Bay, although each location possessed a distinct community structure. Significant differences in the environmental conditions and soil chemical parameters gave evidence that differences in location and soil pH were the environmental variables that most probably explained the soil bacterial community patterns, a conclusion consistent with an increasing number of studies from both Antarctic and Arctic locations.

Arctic and Sub-Arctic Tundra

There is a substantial literature base on tundra microbial abundance, C-biomass, and community structure, especially more recently in relation to

climate change and global warming. Some recent comparative reviews across high latitude locales provide a good overview (e.g., Anderson, 2010a; Jahn, et al., 2010; Nemergut, et al., 2005). To remain within the scope of this chapter, only some generalizations from the literature are presented here. Given the vastness of Arctic tundra relative to the somewhat limited biogeographical coverage of current tundra research, some generalizations must be tempered by the realization that future more geographically expansive research may require revisions or complete redescription of our current knowledge.

Tundra microbial communities vary in composition and microbe abundance along geographic and topographic gradients comparable to reports in other ecosystems. For example, Eskelinen, et al. (2009) examined the relationships between habitat variables (plant functional type composition and soil organic matter quality) and decomposer community composition in 10 sites, five from non-acidic and five from acidic heath sites in alpine tundra in northern Europe. Their study demonstrated a strong link between the plant community composition, soil organic matter quality, and microbial community composition. Differences in one compartment were paralleled by changes in others. Variation in the forb-shrub gradient of vegetation paralleled variations in the chemical quality of organic matter and decomposer communities. Soil pH, through its direct and indirect effects on plant and microbial communities, may function as an ultimate environmental driver that gives rise to and amplifies the interactions between above- and belowground systems. Similarly, Zak and Kling (2006) studying an Alaskan tundra gradient differing in topographic position, plant composition, and soil drainage, reported microbial community composition and function were distinct among tundra ecosystems, with tussock tundra containing a significantly greater abundance and activity of soil fungi. Microbial respiration in wet sedge tundra soil was lower than in tussock and birch willow tundra. Similar to the findings of Eskelinen, et al. (2009), Zak and Kling concluded that edaphic factors such as topographic variation in plant litter biochemistry and soil drainage tend to shape the metabolic capability of soil microbial communities, which, in turn, influence the chemical composition of dissolved organic matter across the arctic tundra landscape. In a more localized transect analysis (Anderson, 2013a) along a 10 km transect in northwestern Alaskan tundra, densities (number g^{-1} soil dry weight) of bacteria were 2.7-16 x 10^9, and nanoflagellates 0.7 – 7.9 x 10^7. The corresponding C-biomass expressed as $\mu g\ g^{-1}$ soil dry weight reported for bacteria (358 – 2,114) was much larger than for their predatory nanoflagellates (12 – 37).

Recent interest has focused on climatic issues affecting arctic microbial communities, especially the role of seasonal forcing effects and climate warming, among other factors. Buckeridge, et al. (2013) summarized changes in the composition, structure and nutrient stoichiometry of the soil microbial community in mesic arctic tundra on nine sample dates in six months from winter to fall and found that bacteria and fungi were active and growing in soils between -5° C and 0° C. However, a significant shift occurred after the thaw period, resulting in a distinct community that persisted through the spring, summer and fall sample dates, including increasing amounts of certain bacteria (especially Gram +ves) and a decline in fungal biomass, reflecting changes in soil nutrient status as the seasons progressed. Bjork, et al. (2008) took a similar approach to the analysis of seasonal effects and examined the temporal variation in soil microbial communities in Alpine tundra (northern Sweden). They reported two seasonal shifts in microbial composition. The first shift was associated with snowmelt and mainly related to a decrease in biomarker molecular evidence of fungi but an accompanied increase in biomarkers for Gram-positive- and actinobacteria, resulting in a decrease in the ratio of fungi-to-bacteria The second shift occurred across the growing, season, and was associated with a an increase in biomarkers for Gram-negative bacteria.

Rinnan, et al. (2013), focusing on the relative effects of warming versus soil fertilization, noted that fertilization increased the relative abundance of fungi, but warming caused only a minimal shift in the microbial community composition. The function of the microbial community was also differently affected. Namely, two decades of fertilization have favored fungi relative to bacteria, and increased the turnover of complex organic compounds such as vanillin, while warming has had no such effects. A similar finding was reported by Campbell, et al. (2010) who studied short- and long-term nutrient fertilization in soils on the North Slope of the Brooks Range. They found bacterial diversity was lower in long-term fertilized plots. The Acidobacteria were one of the most abundant phyla in all soils and distinct differences were noted in the distributions of Acidobacteria subgroups between mineral and organic soil layers that were also affected by fertilization. In addition, Alpha- and Gammaproteobacteria were more abundant in long-term fertilized samples compared with control soils. Overall, their study indicated that long-term fertilization resulted in a significant change in microbial community structure and function linked to changes in C and N availability and shifts in aboveground plant communities.

Simard and Mohn (2012) examined the effects of long-term experimental warming on community composition and noted that microbial communities were stable over time but strongly structured by warming, leading to significant reductions in the evenness of bacterial communities, while the evenness of fungal communities increased significantly. These patterns were strongest in the organic horizon, where temperature change was greatest and were associated with a significant increase in the dominance of the Actinobacteria and significant reductions in the Gemmatimonadaceae and the Proteobacteria, but an increase in ectomycorrhizal fungi.

Furthermore, warming may differentially affect the decomposition and turnover of different molecular classes of C compounds (especially new inputs of dissolved C versus microbial C), with microbial C being slower compared to new soluble inputs (Boddy, et al., 2008). Temperature had little effect on the rate of mineralization of low molecular weight soluble C sources used directly as respiratory substrate, whereas the turnover rate of the C immobilized in the microbial biomass prior to mineralization was temperature sensitive. A more detailed analysis of the differential rates of soluble C compounds by arctic microbial communities was presented by Rinnan and Bääth (2009) who traced the rate of incorporation of several low molecular weight organic compounds among bacteria and fungi, indicating that lower molecular weight compounds (e.g., glucose, acetic acid, and amino acids) were rapidly metabolized compared to higher molecular weight soluble compounds (e.g., starch and vanillin). Given the likely increase in soluble C compounds available in arctic soils with the increasingly northward advance of vascular plants, including shrubs and trees, that release soluble organic compounds into the rhizosphere, the fate of C compounds in high latitude microbial communities and their role in the biogeochemical C-cycle becomes of increasing importance, especially in relation to global climate change.

Soil Microbial Ecosystem Services

The foregoing biogeographic sections provide substantial evidence of the importance of microbial communities in soil C dynamics and ecosystem functions across major geographic regions. The microbial community is at the base of soil food webs, and as major consumers of organic C nutrients, they maintain a rapid flow of C and energy through higher levels of the trophic hierarchy. Soil microbes have a major role in remineralization of soil nutrients, and are major decomposers of plant litter and other sources of organic residues

that are mobilized and assimilated into the soil food chain, or become an important part of the soil composition and structure. With increasing evidence of global climate change, and growing interest in maintaining the fertility and productivity of soils in natural and agricultural ecosystems, considerable recent research has focused on the ecosystem services of soil microbial communities.

Some of these services have been nicely summarized in an online book edited by Lal et al. (2013). Among other positive contributions by soil communities, Lal et al. note the following services: 1) supporting productivity (renewal, retention and delivery of nutrients for plants), 2) regulating the environment (regulation of major elemental cycles; buffering, filtering, and moderation of the hydrologic cycle; disposal of wastes and dead organic matter), and provisioning (building material; physical stability and support of plants). Given the specific goals of this chapter, no further exposition is presented on this topic. However, it may be productive to keep these perspectives in mind as we further consider the C biogeochemistry of soil microbial communities.

CARBON DYNAMICS AND THE BIOGEOCHEMICAL C-CYCLE

Three forms of C are particularly significant in microbial C dynamics: 1) organic compounds, soluble (SOM) and particulate (POM), 2) microbial C biomass, and 3) CO_2, largely respiratory in origin. Other forms of C, including inorganic compounds, are also important for soil composition and quality (e.g., mineralogy, bulk and surface chemistry, texture, etc.). Some recent advances in our knowledge of the role of microbial communities in soil C budget, community dynamics, and coupling with the C-cycle are reviewed with a particular emphasis on themes related to climate change and global warming.

Three foci are explored: 1) the sources and fate of organic C in microbial communities, 2) the effects of temperature, soil moisture and precipitation events on respiratory metabolism of C compounds, and 3) coupling of respiratory CO_2 production with the atmosphere, especially in relation to global warming. Specific attention is given to examples of some recent research. The role of mineral nutrients has been addressed, where appropriate, in previous biogeographic sections.

Sources and Fate of Carbon Compounds

A substantial amount of C nutrients available to soil microbial communities comes from plant primary production. These nutrients include soluble C compounds (SOM) from root exudates, or through fall from leaves during precipitation, and particulate C compounds (POM) of larger molecular weight from litter and some below-ground plant residues. In general, there is a rapid flow of C within soil microbial communities (e.g., Fitter, et al., 2005), especially for SOM in the form of low molecular weight sugars, amino acids, and organic acids, among other soluble organic compounds. Experimental studies have examined the kinetics of C uptake, flow and fate in field and laboratory research and some illustrative examples are considered. In general, low molecular weight C compounds such as soluble organic matter (SOM) are metabolized most rapidly, largely by bacteria that compete effectively in their uptake through rapid assimilation. Larger molecular weight compounds, including POM and relatively refractory organic compounds, are metabolized more slowly and some are particularly metabolized by fungi and other decomposing microbiota. Neff and Anser (2001) analyzed the dynamics of dissolved C compounds (DOC) in terrestrial ecosystems and synthesized information on the geochemical and biological factors that control DOC fluxes through soils, including conceptual issues and quantitative evaluations of key process rates. These were incorporated in a general numerical model of DOC dynamics. Their paper is a useful comprehensive review of the large-scale issues in C fluxes based on data from field studies.

Stevenson, et al. (2004) discovered that forest and pasture soil, analyzed at a landscape scale, varied in respiratory responses to added low molecular weight organic compounds. Pasture soils responded especially to carbohydrate and amino-acid substrates. However, compared to forest soil microbial communities, they had a significantly lower relative response to carboxylic acid substrates. Forest soils responded particularly to carboxylic acids as a group, including citric acid, alpha-ketobutyric acid, alpha-ketoglutaric acid, and alpha-ketovaleric acid. The relative rate of metabolic turnover of various low molecular weight C components of DOC in grasslands was investigated by Boddy, et al. (2007) who injected ^{14}C- labeled glucose and amino acids into the soil in the field and in laboratory experiments and traced their rate of assimilation. In the field, the glucose and amino acids were taken up more rapidly in initial phases of assimilation compared to uptake in laboratory experiments. They had short half-lives in soil solution (20 – 40 min.), but persisted in soil microbes for much longer. Only small amounts of the added

[14]C were recovered in plant biomass (< 5% of total added to soil), indicating that plant roots are poor competitors for low molecular weight DOC in comparison to soil microorganisms. In sum, the results suggest that some components of the low molecular weight DOC pool are turned over extremely rapidly (ca. 4,000 times annually). In a follow-up study in the Arctic, Boddy, et al. (2008), using ^{14}C-labeled glucose and amino acids, examined the fate of the soil amended organics in laboratory microcosms.

Table 1. Statistical analyses of microbiota means ± s.e. for densities (number g^{-1} soil dry weight) and carbon content combining data from three experiments using Alaskan tundra soil amended with pulses of glucose solution during a one-week experiment

	Densities	Carbon
Bacteria	$(x\ 10^8)$	$(\mu g/g)$
Day 1	2.1 ± 0.8	27.0 ± 10.6
Days 4 & 8	$5.2 \pm 1.2*$	$67.3 \pm 15.2*$
Flagellates	$(x\ 10^5)$	$(\mu g/g)$
Day 1	4.3 ± 1.5	4.1 ± 1.4
Days 4 & 8	$7.4 \pm 1.7*$	$6.9 \pm 1.6*$
Naked Amoebae	$(x\ 10^3)$	(ng/g)
Day 1	4.2 ± 2.6	127.4 ± 60.2
Days 4 & 8	$10.0 \pm 2.5**$	$446.6 \pm 110.2*$
Testate Amoebae	$(x10^2)$	(ng/g)
Day 8	$11.5 \pm 0.9*$	$18.2 \pm 1.2*$

To be concise, the results for Days 4 & 8 have been combined and averaged. Asterisk indicates the treatment mean is significantly different from that of the control lacking glucose, *significant at $p < 0.05$, ** significant at $p < 0.01$, Slower growing testate amoebae were assayed only on the last day (Day 8).

Both forms of DOC had a short half-life; i.e., 1.07 ± 0.10 h for glucose and 1.63 ± 0.14 h for amino acids. While in a second phase of mineralization, assumed to be C that had entered the microbial biomass, the half-life was much less. Temperature had little effect on the rate of mineralization of [14]C-

labeled nutrients used directly as respiratory substrate. In contrast, the turnover rate of the ^{14}C immobilized in the microbial biomass prior to mineralization was temperature sensitive.

In this environment, the flux of labile, low molecular weight DOC through the soil solution was extremely rapid and relatively insensitive to temperature. However, the turnover of C incorporated into higher molecular weight microbial C pools appeared to show greater temperature sensitivity.

In a similar laboratory experimental study of the fate of a pulsed addition of glucose-C in samples of Alaskan arctic soil maintained at 20° C, Anderson (2012b) traced the total amount of glucose-C that was respired, relative to the amount accumulated in microbial C-biomass, including bacteria, flagellates, naked amoebae, and testate amoebae (Table 1).

Although there was an immediate burst of respiration after a pulsed addition of glucose solution, lasting for less than four hours, it accounted for < 1% of the glucose-C added. The percentage of glucose incorporated into the total microbial fraction (bacteria and protists) after eight days of laboratory culture varied from 5 to 39% based on comparisons with the amount in control cultures, not enriched with glucose solutions. Larger amounts of microbial sequestered glucose-C occurred in summer tundra soil samples compared to those obtained in early spring.

This study suggests that sporadic bursts of plant root exudates of low molecular weight C compounds may produce only a transitory loss by microbial respiratory CO_2 to the atmosphere, and more may be sequestered in microbial particulates and soil compartments. A replication of this laboratory experiment with soil from a temperate site in Northeastern U. S. A. produced similar results, with a burst of respiration immediately following addition of the glucose solution rapidly declining within the first four to five hours (Anderson 2012a). Additional studies of the rate of flow and fate of C compounds with varying composition in soil at a variety of geographic sites have been published by Berger, et al. (2010), Dilly and Zyakun (2008), Martinez-Trinidad, et al. (2010), Meier and Bowman (2008), and O'Dowd, et al. (1997).

Temperature and Soil Moisture Effects

Soil temperature and available moisture (including duration and periodicity of precipitation) are among the most significant factors, beyond C nutrients, affecting C compound metabolism by microbial communities. Some

general findings are reviewed within the scope of this chapter, while also recognizing the variations that occur across different soil environments. Some of the differential effects of temperature have been cited in the previous biogeographic section, but the focus here is more specifically on C dynamics, respiration and evidence based on experiments. From a systems perspective, many of these environmental factors act in complex multivariate ways to influence the fate of soil C compounds. However, this aspect is not addressed here fully due to the particular focus of the chapter.

Temperature

In general, C metabolism and respiration increase with increasing temperature within the tolerance range of soil microbiota as expected based on physiological principles and the Arrhenius equation (Fang & Moncrieff, 2001; Lloyd & Taylor, 1994; Perrin, et al., 2004; Sjøgersten & Wookey, 2002). However, as may be expected, elevating temperatures well above the norm in short-term experiments tends to reduce respiration rates, especially in alpine and higher latitude environments (e.g., Chang, et al., 2012). In general Q_{10} values are in the range of 2.0 for many experimental assays under controlled laboratory conditions, but can vary widely depending on geographic locale, soil composition and extent of water saturation (e.g., Anderson 2010b, 2011; Bekku, et al., 2003; Chang, et al., 2012; Liu, 2013; Winkler, et al., 1996). With increasing concern about the effects of global warming on soil microbial communities, there has been an impetus to study the adaptive capacity of soil organisms in various geographic regimes to increasing temperatures; that is, whether the increases in respiration with moderate increases in temperature are sustained, or if an adaptive response produces an eventual decline leading to a lower steady state. In general there is some evidence of long-term adaptation, but results have varied depending on experimental design, geographic locale, soil composition and nutrient status. Some examples are reviewed.

Bradford, et al. (2008, 2010) reported evidence of mass-specific respiration rates adjusting downward over extended periods of experimental soil warming for 77 days, including field-based and laboratory studies. However, Hartley, et al. (2008) criticized the mass-based respiration analyses and suggested that there is insufficient evidence of adaptation as well as no evolutionary rationale for its occurrence. In response, Bradford, et al. (2010) counter argued that evolutionary trade-offs between maximum catalytic rates and the stability of the binding structure of enzymes must be taken into consideration in interpreting the results. Moreover, they concluded that the downward changes in respiration may be due to population or species shifts

during the experimental incubations, in addition to other possible alternate mechanistic explanations. Clearly, more research is needed on the way soil microbial communities adjust to longer-term changes in temperature, especially for different climatic regimes. This is all the more important toward making accurate predictions about temperature-induced changes in atmospheric CO_2 concentrations resulting from soil respiration.

Soil Moisture and Precipitation Patterns

On a global scale soil respiration tends to increase linearly with increasing annual precipitation (Raich and Schlesinger, 1992), but within a given locale, the relationship is more complex, reaching a plateau at higher levels of moisture and eventually declining at elevated amounts due in part to water logging of the soil and reduced aerobic respiration (Bowden, et al., 1998; Xu, et al., 2004). At higher soil moisture saturations, however, methane production by archaebacteria can be increased, as is characteristic of some wetlands. It is a major concern due to the more serious effects of methane relative to CO_2 for global warming. The linear increase in respiration with moderate soil moisture content has been documented at major geographic locations, including Arctic tundra (e.g., Oberbauer, et al., 1992; Shaver, et al., 2006; Anderson 2008, 2010a,b), especially under conditions of experimental warming at field-based sites (e.g., Welker, et al., 1999) or during natural thermal cycles in environmental settings (Grogan and Chapin, 1999). Similar data have been presented for grasslands and meadows (Risch and Frank, 2006; McCulley, et al., 2007; Suh, et al., 2009), and temperate and tropical forests (e.g., McCulley, et al., 2007; Adachi, et al., 2006).

Given possible changes in global precipitation patterns due to climate change, particular attention has been given to the effects of increased occurrences of periodic precipitation events on C metabolism and respiratory CO_2 production. There is substantial evidence that pulsed precipitation events produce transitory peaks in soil respiration, known as the "Birch effect" (e.g., Birch 1958; Griffiths & Birch, 1961), largely based on laboratory studies. A useful review of some recent and historical literature has been published by Cook and Orchard (2008). The moisture-pulsed peak of CO_2 release becomes more significant for arid and semi-arid regions where periodic rain events stimulate rapid increases in soil microbial metabolism and may be sufficiently large to affect annual increments in atmospheric CO_2 concentrations, thus exacerbating global warming. The initial respiration peak, largely caused by bacterial respiration, is followed within a day or two by a sharp decline in respiration, attributed partially to bacterial grazing by protozoa (heterotrophic

flagellates and amoebae) that increase markedly in response to the more favorable growth conditions (e.g., Anderson, 2011). However, a complete explanation for the eventual decline remains to be determined. Experimental research has provided additional more detailed evidence of the temporal events, particularly within the context of microbial communities. For example, in a laboratory-based experiment using temperate forest soil (Anderson, 2011), the initial peak in respiratory release of CO_2 within 24 hours after re-wetting of the soil was approximately 11 nmol min^{-1} g^{-1} soil dry weight, but declined to about 6 nmol min^{-1} g^{-1} by 72 hours later. The decline in respiratory C loss was accompanied by increased sequestration of C in the eukaryotic microbes as the bacterial C passed up the food chain, with flagellate C increasing from 5.7 to 6.8 µg g^{-1} soil and naked amoeba C content increasing five-fold (60 to 300 ng g^{-1} soil). More detailed data on the relationship between respiration and soil moisture are available in a review paper, including some new data Anderson (2011).

Coupling of Above-Ground and Below-Ground C Dynamics

A major role of atmospheric CO_2 in global warming is well established, and future scenarios of its effects on a global scale depend critically on the balance of CO_2 fixation by vegetation and the output from major sources, including respiration of belowground microbial communities, heterotrophic metabolism of plant roots and the soil-dwelling invertebrate communities. Consequently, considerable recent attention has been given to the coupling of aboveground primary production and belowground respiratory production of CO_2. While it is clear that microbial communities contribute a substantial amount of soil respiratory CO_2 efflux, it is essential to clearly identify how much soil respiratory CO_2 comes from belowground plant sources such as roots and how much from microbial contributions. A variety of methods have been used to eliminate plant root contributions in field-based studies as a way of determining the remaining contributions by the microbial community. These include girdling of trees (Olsson, et al., 2005) to stop translocation of nutrients to the roots, thus causing their death; trenching to cut roots distal to the tree base resulting in less metabolism and eventual death (Schaefer, et al., 2009), and application of stable isotopic labeling of CO_2 with ^{13}C to trace the concentration of respiratory C from the plant roots, thus separating the sources and sinks of C in vegetative sites where autotrophic sources of soluble organic matter are of importance (e.g., Andrews, et al., 2000; Burke, et al., 2003; Leake, et al., 2006; Paterson, et al., 2009). However, use of ^{13}C tracers may incur technical limitations in partitioning of autotrophic and heterotrophic

sources of soil respiration due to diffusive fractionation produced by non-steady state gas transport effects (Moyes, et al., 2010). Some modifications of field-based techniques and improved analyses have been recommended to overcome some of these limitations (e.g., Albanito, et al., 2012). Moreover, given the rapid uptake kinetics of some organics by microorganisms and the relatively small amount of bio-available organics in most soils, Hobbie and Hobbie (2012) have offered some critical reflections on the use of radioisotopes in microbial metabolic studies.

In general, based on the best current evidence, as much as 40 to 60% of soil CO_2 efflux in some forests can be traced to plant root respiration and must be subtracted from the total source to better account for non-plant contributions (e.g., Olsson, et al., 2005). In some forest field-based studies, the non-plant heterotrophic contribution, presumably largely microbial, can exceed 60% of total respiration (e.g., Lalonde & Prescott, 2007); and variation in total soil respiration may be mainly influenced by heterotrophic respiration (77%) throughout the year, while the influence of root respiration (46%) can be strongest in summer (e.g., Tomotsune, et al., 2013).

The effects of increasingly elevated atmospheric CO_2 on plant productivity, increased root exudates, and consequent enhanced microbial metabolic CO_2, have become a major focus of recent research (e.g., Anderson, 2013b; Ceulemans, et al., 1999; Oneill, 1994). Some of the complexities theoretically and environmentally of the influence of atmospheric CO_2 on vegetation responses and belowground effects have been substantially reviewed by Metcalfe, et al. (2011), including some priorities for future research. In general, there is experimental evidence that increased atmospheric CO_2 can yield higher densities of heterotrophic protists, at least in the rhizosphere of some herbaceous plants based on laboratory controlled experiments (e.g., Anderson and Griffin, 2001). However, there is mixed evidence that such enhanced aboveground productivity will have an enduring influence on increased belowground respiratory CO_2 efflux (e.g., Bader & Koerner, 2010), although the responses may vary considerably among different soil habitats and vegetation regimes (e.g., Metcalfe, et al., 2011). Given the likely positive feedback on the greenhouse effect and global warming due to soil respiratory efflux into the atmosphere, there is a strong impetus to do research that may resolve more fully the C dynamics associated with aboveground and belowground CO_2 exchange. Additional information on climate change in relation to aboveground and belowground coupling, and its relationship to C dynamics can be found in recent reviews (e.g., Anderson, 2010a, Raich & Schlesinger, 1992; Schröter, et al., 2004).

CONCLUSION

A broad range of biogeographic studies have been reported examining the contribution of climatic, edaphic, vegetation, and microbial community composition on the abundance, diversity and functioning of the soil dwelling microbes. While some generalizations emerge from the studies reviewed here, it is also very evident that too few studies involve a synoptic analysis of all of the major microbial groups in soils, i.e. a composite and comparative analysis of heterotrophic bacteria, cyanobacteria, fungi, and major protozoan groups (flagellates, naked amoebae and testate amoebae). Based on the overall review of published data, the composition and physiological activity of soil-dwelling microbes is strongly influenced by climate, especially the amount and frequency of precipitation as well as temperature, within a reasonable survival range, along biogeographic gradients. Studies of geographic and topographic gradients point particularly to climate variables as a major factor. Some evidence indicates that forest and grassland microbial communities, however, may differ in response to precipitation pattern frequencies. In addition, the kind of vegetation (clearly also influenced by climate) is a further contributory factor in microbial community organization, biomass, and respiration. Associations with a plant rhizosphere, or beneath a plant canopy, can be different from that in the surrounding soil, although there is good evidence that there is recruitment from the surrounding soil populations.

Fungi in some locales contribute most to microbial biomass. There is fairly good evidence, for example, that fungi account for a major part of microbial biomass in moss-rich Arctic tundra, and it may also be the case in some lower latitude locations varying in kind of vegetation. Some fungi, for example, are abundant in relation to roots of pine stands as well as among different trees in tropical forests, for example. Fungi associated with plant roots in some locales may be the first to assimilate plant root sources of organic compounds (especially if they are arbuscular fungi) and transfer some of the nutrients to bacteria and protozoa in the vicinity of the roots. However, bacteria are a major source of nutrition for protozoan food webs, and their rapid uptake of soluble organic nutrients provides a rich and rapid source of nutrition for microbes higher up the microbial food chain. In general, low molecular weight C compounds such as soluble organic matter (SOM) are metabolized most rapidly, largely by bacteria that compete effectively in their uptake through rapid assimilation. Larger molecular weight compounds, including POM and relatively refractory organic compounds, are metabolized more slowly and some are particularly metabolized by fungi and other

decomposing microbiota. Laboratory studies indicate that soluble organic nutrients, as for example exuded by roots, are rapidly assimilated in microbial communities, resulting initially in a burst of substrate-induced respiration. This, however, is brief in the range of a few hours, and in some cases accounts for less than one percent of the total C added. The remainder is sequestered in the microbial biomass or bound to soil particles. Because some of this biomass C is contained in refractory biotic substances such as cyst walls, organic tests, chitinous coverings, and in higher invertebrate exoskeletons, egg cases, etc., a good proportion may be largely sequestered in the soil and not lost relatively soon to the atmosphere as respiratory CO_2.

Some evidence indicates that increasing levels of atmospheric CO_2 may promote plant primary production and increased exudation of root organic nutrients, thus enhancing the abundance of rhizosphere-dwelling microbes. It is not certain, however, whether this effect is sustained over longer time periods, or only temporary. Nonetheless, laboratory and field studies clearly indicate that increased soil organic nutrients will produce increased release of microbial respiratory CO_2 to the atmosphere leading to a positive feedback effect on global warming and climate change. Likewise, increasing temperatures, within the tolerance range of the soil microbes, produces increasing respiration rates with a Q_{10} of approximately 2.0 on a global scale, and also based on some laboratory analyses. However, Q_{10} values can vary within a broad range around 2.0 across different biogeographic locales. Whether increasing temperatures will cause a sustained increase in microbial respiratory CO_2 efflux, or whether the microbes will adjust downward through adaptation, appears to be a topic of continued research. Nonetheless, substantial research confirms that soil microbial communities are a major contributor to C flow in the biogeochemical C-cycle.

It is difficult to make generalizations about the relative effects of environmental variables on the different microbial taxa in soils at different geographic locations. This is due in part to the limited number of studies that address them in composite. Also there is increasing evidence that although there are clearly predictable effects of climatic and geographic variables on microbial communities, including niche and edaphic factors, there are also stochastic aspects that must be taken into consideration. Some of the predictable variation relative to random factors may be clarified by more carefully designed studies on soil from varied geographic locales combining simultaneous, coordinated laboratory and field experiments under carefully defined and controlled conditions.

ACKNOWLEDGMENTS

Some of the research published by the author, and reviewed here, was supported by grants from the United States National Science Foundation. Prof. Hugh Ducklow, Department of Earth and Environmental Sciences at Columbia University, provided valuable advice as a reviewer of this manuscript. This is Lamont-Doherty Earth Observatory Contribution Number 7750.

REFERENCES

Adachi, M., Bekku, Y. S., Rashidah, W., Okuda, T. & Koizumi, H. (2006). Differences in soil respiration between different tropical ecosystems. *Applied Soil Ecology, 34*, 258-265.

Adl, S. M. (2003). *The Ecology of Soil Decomposition.* Wallingford, UK: CAB International.

Albanito, F., McAllister, J. L., Cescatti, A., Smith, P. & Robinson, D. (2012). Dual-chamber measurements of delta C-13 of soil respired CO_2 partitioned using a field-based three end-member model. *Soil Biology & Biochemistry, 47*, 106-115.

Anderson, O. R. (2000). Abundance of terrestrial gymnamoebae at a northeastern U. S. site: A four-year study, including the El Niño winter of 1997-1998. *Journal of Eukaryotic Microbiology, 47*, 148-155.

Anderson, O. R. (2008). The Role of amoeboid protists and the microbial community in moss-rich terrestrial ecosystems: Biogeochemical implications for the carbon budget and carbon cycle, especially at higher latitudes. *Journal of Eukaryotic Microbiology, 55*, 145-150.

Anderson, O. R. (2010a). The reciprocal relationships between high latitude climate changes and the ecology of terrestrial microbiota: Emerging theories, models, and empirical evidence, especially related to global warming. In: B. Gutierrez & C. Pena (Eds.), *Tundras: Vegetation, Wildlife and Climate Trends* (47-79). New York: Nova Science Publishers.

Anderson, O. R. (2010b). An Analysis of Respiratory Activity, $Q_{10,}$ and Microbial Community Composition of Soils from High and Low Tussock Sites at Toolik, Alaska. *Journal of Eukaryotic Microbiology, 57*, 218-219.

Anderson, O. R. (2011). Soil Respiration, Global Warming and the Role of Microbial Communities. *PROTIST, 162*, 679-690.

Anderson, O. R. (2012a). The role of bacterial-based protist communities in aquatic and soil ecosystems and the carbon biogeochemical cycle, with emphasis on naked amoebae. *Acta Protozoologica, 51*, 209-221.

Anderson, O. R. (2012b). The Fate of Organic Sources of Carbon in Moss-rich Tundra Soil Microbial Communities: A Laboratory Experimental Study. *Journal of Eukaryotic Microbiology, 59*, 564-570.

Anderson, O. R. (2013a). Bacterial and heterotrophic nanoflagellate densities and C-biomass estimates along an Alaskan tundra transect with prediction of respiratory CO_2 efflux. *Journal of Eukaryotic Microbiology*, doi:10.1111/jeu.12081.

Anderson, O. R. (2013b). Soil respiration, climate change, and the role of microbial communities. In: F. J. De Bruijn, (Ed.) *Molecular microbial ecology of the Rhizosphere* (1055-1062). Hoboken, N. J.: John Wiley and Sons, Inc.

Anderson, O. R. & Griffin, K. (2001). Abundances of protozoa in soil of laboratory-grown wheat plants cultivated under low and high atmospheric CO_2 concentrations. *Protistology, 2*, 76-84.

Andrew, D. R., Fitak, R. R., Munguia-Vega, A., Racolta, A., Martinson, V. G. & Dontsova, K. (2012). Abiotic factors shape microbial diversity in Sonoran Desert soils. *Applied & Environmental Microbiology, 78*, 7527-7537.

Andrews, J. A., Matamala, R., Westover, K. M. & Schlesinger, W. H. (2000). Temperature effects on the diversity of soil heterotrophs and the δ^{13}C of soil respired CO_2. *Soil Biology & Biochemistry, 32*, 699-706.

Angel, R., Soares, M. I. M., Ungar, E. D. & Gillor, O. (2010). Biogeography of soil archaea and bacteria along a steep precipitation gradient. *ISME Journal, 4*, 553-563.

Arenz, B. E. & Blanchette, R. A. (2011). Distribution and abundance of soil fungi in Antarctica at sites on the Peninsula, Ross Sea Region and McMurdo Dry Valleys. *Soil Biology & Biochemistry, 43*, 308-315.

Bader, M. K.-F. & Koerner, C. (2010). No overall stimulation of soil respiration under mature deciduous forest trees after 7 years of CO_2 enrichment. *Global Change Biology, 16*, 2830-2843.

Bamforth, S. S. (1984). Microbial distributions in Arizona USA deserts and woodlands. *Soil Biology & Biochemistry, 16*, 133-138.

Bamforth, S. S. (1985). Soil protozoa of two Utah USA cool deserts. *Pedobiologia, 28*, 423-426.

Bamforth, S. S. (1995). Interpreting soil ciliate biodiversity. *Plant & Soil, 170*, 159-164.

Bamforth, S. S. (2006). Protozoa from aboveground and ground soils of a tropical rain forest in Puerto Rico. *Pedobiologia, 50,* 515-525.

Bamforth, S. S., Wall, D. H. & Virginia, R. A. (2005). Distribution and diversity of soil protozoa in the McMurdo dry valleys of Antarctica. *Polar Biology, 28,* 756-762.

Barness, G., Rodriguez Zaragoza, S., Shmueli, I. & Steinberger, Y. (2009). Vertical distribution of a soil microbial community as affected by plant ecophysiological adaptation in a desert system. *Microbial Ecology, 57,* 36-49.

Bekku, Y. S., Nakatsubo, T., Kume, A., Adachi, M. & Koizumi, H. (2003). Effect of warming on the temperature dependence of soil respiration rate in arctic, temperate, and tropical soils. *Applied Soil Ecology, 22,* 205-210.

Bell, C., McIntyre, N., Cox, S., Tissue, D. & Zak, J. (2008). Soil microbial responses to temporal variations of moisture and temperature in a Chihuahuan Desert grassland. *Microbial Ecology, 56,* 153-167.

Berger, T. W., Inselsbacher, E. & Zechmeister-Boltenstern, S. (2010). Carbon dioxide emissions of soils under pure and mixed stands of beech and spruce, affected by decomposing foliage litter mixtures. *Soil Biology & Biochemistry, 42,* 986-997.

Birch, H. F. (1958). The effect of soil drying on humus decomposition and nitrogen. *Plant & Soil, 10,* 9–31.

Bjork, R. G., Bjorkman, M. P., Andersson, M. X. & Klemedtsson, L. (2008). Temporal variation in soil microbial communities in Alpine tundra. *Soil Biology & Biochemistry, 40,* 266-268.

Boddy, E., Roberts, P., Hill, P. W., Farrar, J. & Jones, D. L. (2007). Fast turnover of low molecular weight components of the dissolved organic carbon pool of temperate grassland field soils. *Soil Biology & Biochemistry, 39,* 827-835.

Boddy, E., Roberts, P., Hill, P. W., Farrar, J. & Jones, D. L. (2008). Turnover of low molecular weight dissolved organic C (DOC) and microbial C exhibit different temperature sensitivities in Arctic tundra soils. *Soil Biology & Biochemistry, 40,* 1557-1566.

Bogoev, V. & Gyosheva, M. (1996). Selected data characterizing microbiocoenoses in soil samples from the Antarctic Region. *Bulgarian Antarctic Research Life Sciences, 1,* 13-18.

Bowden, R. D., Newkirk, K. M. & Rullo, G. M. (1998). Carbon dioxide and methane fluxes by a forest soil under laboratory controlled moisture and temperature conditions. *Soil Biology & Biochemistry, 30,* 1591-1587.

Bradford, M. A., Davies, C. A., Frey, S. D., Maddox, T. R., Melillo, J. M., Mohan, J. E., Reynolds, J. F., Treseder, K. K. & Wallenstein, M. D. (2008). Thermal adaptation of soil microbial respiration to elevated temperature. *Ecology Letters, 11*, 1316-1327.

Bradford, M. A., Watts, B. W. & Davies, C. A. (2010). Thermal adaptation of heterotrophic soil respiration in laboratory microcosms. *Global Change Biology, 16*, 1576-1588.

Bridgham, S. D., Cadillo-Quiroz, H., Keller, J. K. & Zhuang, Q. (2013). Methane emissions from wetlands: biogeochemical, microbial, and modeling perspectives from local to global scales. *Global Change Biology, 19*, 1325-1346.

Buckeridge, K. M., Banerjee, S., Siciliano, S. D. & Grogan, P. (2013). The seasonal pattern of soil microbial community structure in mesic low arctic tundra. *Soil Biology & Biochemistry, 65*, 338-347.

Burke, R. A., Molina, M., Cox, J. E., Osher, L. J. & Piccolo, M. C. (2003). Stable carbon isotope ratio and composition of microbial fatty acids in tropical soils. *Journal of Environmental quality, 32*, 198-206.

Campbell, B. J., Polson, S. W., Mack, M. C. & Schuur, E. A. G. (2010). The effect of nutrient deposition on bacterial communities in Arctic tundra soil. *Environmental Microbiology, 12,* 1842-1854.

Carletti, P., Vendramin, E., Pizzeghello, D., Concheri, G., Zanella, A., Nardi, S. Squartini, A. (2009). Soil humic compounds and microbial communities in six spruce forests as function of parent material, slope aspect and stand age. *Plant & Soil, 315*, 47-65.

Caruso, T., Chan, Y., Lacap, D. C., Lau, M. C. Y., McKay, C. P. & Pointing, S. B. (2011). Stochastic and deterministic processes interact in the assembly of desert microbial communities on a global scale. *ISME Journal, 5,* 1406-1413.

Ceulemans, R., Janssens, I. A. & Jach, M. E. (1999). Effects of CO_2 enrichment on trees and forests: lessons to be learned in view of future ecosystems studies. *Annals of Botany, 84*, 577-590.

Chabrerie, O., Laval, K., Puget, P., Desaire, S. & Alard, D. (2003). Relationship between plant and soil microbial communities along a successional gradient in a chalk grassland in north-western France. *Applied Soil Ecology, 24*, 43-56.

Chan, Y., Lacap, D. C., Lau, M. C. Y., Ha, K. Y., Warren-Rhodes, K. A., Cockell, C. S., Cowan, D. A., McKay, C. P. & Pointing, S. B. (2012). Hypolithic microbial communities: between a rock and a hard place. *Environmental Microbiology, 14*, 2272-2282.

Chang, X., Wang, S., Luo, C., Zhang, Z., Duan, J., Zhu, X., Lin, Q. & Xu, B. (2012). Responses of soil microbial respiration to thermal stress in alpine steppe on the Tibetan plateau. *European Journal of Soil Science, 63*, 325-331.

Chong, C. W., Goh, Y. S., Convey, P., Pearce, D. & Tan, I. K. P. (2013). Spatial patterns in Antarctica: what can we learn from Antarctic bacterial isolates? *Extremophiles, 17*, 733-745.

Chong, C. W., Pearce, D. A., Convey, P. & Tan, I. K. P. (2012a). The identification of environmental parameters which could influence soil bacterial community composition on the Antarctic Peninsula – a statistical approach. *Antarctic Science, 24*, 249-258.

Chong, C. W., Pearce, D. A., Convey, P., Yew, W. C. & Tan, I. K. P. (2012b). Patterns in the distribution of soil bacteria 16S rRNA gene sequences from different regions of Antarctica. *Geoderma, 181*, 45-55.

Clarholm, M. (2002). Bacteria and protozoa as integral components of the forest ecosystem: their role in creating a naturally varied soil fertility. *Antonie van Leeuwenhoek, 81*, 309-318.

Clark, J. S., Campbell, J. H., Grizzle, H., Acosta-Martinez, V. & Zak, J. C. (2009). Soil microbial community response to drought and precipitation variability in the Chihuahuan Desert. *Microbial Ecology, 57*, 248-260.

Compton, J. E., Watrud, L. S., Porteous, L. A. & DeGrood, S. (2004). Response of soil microbial biomass and community composition to chronic nitrogen additions at Harvard forest. *Forest Ecology & Management, 196*, 143-158.

Convey, P., Aitken, S., di Prisco, G., Gill, M. J., Coulson, S. J., Barry, T., Jonsdottir, I. S., Dang, P. T., Hik, D., Kulkarni, T. & Lewis, G. (2012). The impacts of climate change on circumpolar biodiversity. *Biodiversity, 13*, 134-143.

Cook, F. J. & Orchard, V. A. (2008). Relationships between soil respiration and soil moisture. *Soil Biology & Biochemistry, 40*, 1013-1018.

Couteaux, M.-M., Raubuch, M. & Berg, M. (1998). Response of protozoan and microbial communities in various coniferous forest soils after transfer to forests with different levels of atmospheric pollution. *Biology & Fertility of Soils, 27, 179-188.*

Cusack, D. F., Silver, W. L., Torn, M. S., Burton, S. D. & Firestone, M. K. (2011). Changes in microbial community characteristics and soil organic matter with nitrogen additions in two tropical forests. *Ecology, 92,* 621-632.

Dash, M. C. & Guru, B. C. (1980). Distribution and seasonal variation in numbers of testacea protozoa in some Indian soils. *Pedobiologia, 20*, 325-342.

De Bruijn, F. J. (Ed.) (2013). *Molecular microbial ecology of the Rhizosphere*. Sections 13 and 14 (pp. 1047-1146). Hoboken, N. J.: John Wiley and Sons, Inc.

De Deyn, G. B., Quirk, H., Oakley, S., Ostle, N. & Bardgett, R. D. (2011). Rapid transfer of photosynthetic carbon through the plant-soil system in differently managed species-rich grasslands. *Biogeosciences, 8,* 1131-1139.

Deforest, J. L., Zak, D. R., Pregitzer, K. S. & Burton, A. J. (2004). Atmospheric nitrate deposition, microbial community composition, and enzyme activity in northern hardwood forests. *Soil Science Society of America Journal, 68*, 132-138.

Dilly, O. & Zyakun, A. (2008). Priming effect and respiratory quotient in a forest soil amended with glucose. *Geomicrobiology Journal, 25*, 425-431.

Dornbush, M., Cambardella, C., Ingham, E. & Raich, J. (2008). A comparison of soil food webs beneath C-3 and C-4 dominated grasslands. *Biology & Fertility of Soils. 45*, 73-81.

Eskelinen, A., Stark, S. & Mannisto, M. (2009). Links between plant community composition, soil organic matter quality and microbial communities in contrasting tundra habitats. *Oecologia, 161*, 113-123.

Fang C. & Moncrieff, J. B. (2001). The dependence of soil CO_2 efflux on temperature. *Soil Biology & Biochemistry, 33*, 155–165.

Fierer, N., Schimel, J. P. & Holden, P. A. (2003). Influence of drying-rewetting frequency on soil bacterial community structure. *Microbial Ecology, 45*, 63-71.

Fierer, N., Leff, J. W., Adams, B. J., Nielsen, U. N., Bates, S. T., Lauber, C. L., Owens, S., Gilbert, J. A., Wall, D. H. & Caporaso, J. G. (2012). Cross-biome metagenomic analyses of soil microbial communities and their functional attributes. *Proceedings of the National Academy of Sciences of the United States of America, 109*, 21390-21395.

Fitter, A. H., Gilligan, C. A., Hollingworth ,K., Kleczkowski, A., Twyman, R. M., Pitchford J. W. & the Members of the NERC Soil Biodiversity Programme. (2005). Biodiversity and ecosystem function in soil. *Functional Ecology, 19*, 369—377.

Foissner, W. (1996). Faunistics, taxonomy and ecology of moss and soil ciliates (Protozoa, Ciliophora) from Antarctica, with description of new

species, including *Pleuroplitoides smithi* gen n, sp n. *Acta Protozoologica, 35*, 95–123.

Gao, C., Shi, N.-N., Liu, Y.-X., Peay, K. G., Zheng, Y. Ding, Q., Mi, X.-C., Ma, K.-P., Wubet, T., Buscot, F. & Guo, L.-D. (2013). Host plant genus-level diversity is the best predictor of ectomycorrhizal fungal diversity in a Chinese subtropical forest. *Molecular Ecology, 22*, 3403-3414.

Gonzalez-Franco, A. C., Robles-Hernandez, L., Nunez-Barrios, A., Strap, J. L. & Crawford, D. L. (2009). Molecular and cultural analysis of seasonal actinomycetes in soils from *Artemisia tridentata* habitat. *Phyton, 78*, 83-90.

Grayston, S. J., Griffith, G. S., Mawdsley, J. L., Campbell, C. D. & Bardgett, R. D. (2001). Accounting for variability in soil microbial communities of temperate upland grassland ecosystems. *Soil Biology & Biochemistry, 33*, 533-551.

Griffiths, E & Birch, H. F. (1961). Microbiological changes in freshly moistened soil. *Nature, 189*, 424.

Grogan, P. & Chapin III, F. S. (1999). Arctic soil respiration: effects of climate and vegetation depend on season. *Ecosystems, 2*, 451-459.

Hafich, K., Perkins, E. J., Hauge, J. B., Barry, D. & Eaton, W. D. (2012). Implications of land management on soil microbial communities and nutrient cycle dynamics in the lowland tropical forest of northern Costa Rica. *Tropical Ecology, 53*, 215-224.

Hackl, E., Pfeffer, M., Donat, C., Bachmann, G. & Zechmeister-Boltenstern, S. (2005). Composition of the microbial communities in the mineral soil under different types of natural forest. *Soil Biology & Biochemistry, 37*, 661-671.

Hartley, I. P., Hopkins, D.W., Garnett, M. H., Sommerkorn M. & Wookey, P.A. (2008). Soil microbial respiration in arctic soil does not acclimate to temperature. *Ecology Letters, 11*, 1092–1100.

Hasselquist, N. J., Metcalfe, D. B. & Hogberg, P. (2012). Contrasting effects of low and high nitrogen additions on soil CO_2 flux components and ectomycorrhizal fungal sporocarp production in a boreal forest. *Global Change Biology, 18*, 3596-3605.

Hobbie, J. E. & Hobbie, E. A. (2012). Amino acid cycling in plankton and soil microbes studied with radioisotopes: measured amino acids in soil do not reflect bioavailability. *Biogeochemistry, 107*, 339-360.

Ingham, E. R., Coleman, D. C, & Moore, J. C. (1989). An analysis of food-web structure and function in a shortgrass prairie a mountain meadow and a lodgepole pine forest. *Biology & Fertility of Soils, 8*, 29-37.

Innes, L., Hobbs, P. J. & Bardgett, R. D. (2004). The impacts of individual plant species on rhizosphere microbial communities in soils of different fertility. *Biology & Fertility of Soils, 40*, 7-13.

Jahn, M., Sachs, T., Mansfeldt, T. & Overesch, M. (2010). Global climate change and its impacts on the terrestrial Arctic carbon cycle with regards to ecosystems components and the greenhouse-gas balance. *Journal of Plant Nutrition & Soil Science, 173*, 627-643.

Jin Z., Lei J., Li, S., Xu, X. & Chang, Q. (2010). Variation in rhizosphere microbes of three shelter shrubs in drift desert hinterland in Xinjiang, China. *Chinese Journal of Applied & Environmental Biology,16*, 759-764.

Kenarova, A., Encheva, M., Chipeva, V., Chipev, N., Hristova, P. & Moncheva, P. (2013). Physiological diversity of bacterial communities from different soil locations in Livingston Island, South Shetland archipelago, Antarctica. *Polar Biology, 36*, 223-233.

Kluber, L. A., Carrino-Kyker, S. R., Coyle, K. P., DeForest, J. L., Hewins, C. R., Shaw, A. N., Smemo, K. A. & Burke, D. J. (2012). Mycorrhizal response to experimental pH and P manipulation in acidic hardwood forests. *PLoS ONE, 7*, e48946.

Krashevska, V., Maraun, M., Ruess, L. & Scheu, S. (2010). Carbon and nutrient limitation of soil microorganisms and microbial grazers in a tropical montane rain forest. *Oikos, 119*, 1020-1028.

Krumins, J. A., Dighton, J., Gray, D., Franklin, R. B., Morin, P. J. & Roberts, M. S. (2009). Soil microbial community response to nitrogen enrichment in two scrub oak forests. *Forest Ecology & Management, 258*, 1383-1390.

Lal, R., Lorenz, K., Hüttl, R. F., Schneider, B. U. & von Braun, J. (2013). *Ecosystem Services and Carbon Sequestration in the Biosphere.* New York: Springer Verlag. Retrieved 10 Nov. 2013 from: http://link.springer.com/book/10.1007/978-94-007-6455-2.

Lalonde, R. G. & Prescott, C. E. (2007). Partitioning heterotrophic and rhizospheric soil respiration in a mature Douglas-fir (*Pseudotsuga menziesii*) forest. *Canadian Journal of Forest Research, 37*, 1287-1297.

Leake, J. R., Ostle, N. J., Rangel-Castro, J. L. & Johnson, D. (2006). Carbon fluxes from plants through soil organisms determined by $^{13}CO_2$ pulse-labelling in an upland grassland. *Applied Soil Ecology, 33*, 152-175.

Liu, H. (2013). Thermal response of soil microbial respiration is positively associated with labile carbon content and soil microbial activity. *Geoderma, 193*, 275-281.

Liu, L., Gundersen, P., Zhang, T. & Mo, J. (2012). Effects of phosphorus addition on soil microbial biomass and community composition in three forest types in tropical China. *Soil Biology & Biochemistry, 44*, 31-38.

Liu, L., Zhang, T., Gilliam, F. S., Gundersen, P., Zhang, W., Chen, H. & Mo, J. (2013). Interactive effects of nitrogen and phosphorus on soil microbial communities in a tropical forest. *PLoS ONE, 8*, e61188.

Lloyd, J. & Taylor, J. A. (1994). On the temperature dependence of soil respiration. *Functional Ecology, 8*, 315–323.

Makhalanyane, T. P., Valverde, A., Lacap, D. C., Pointing, S. B., Tuffin, M. I. & Cowan, D. A. (2013). Evidence of species recruitment and development of hot desert hypolithic communities. *Environmental Microbiology Reports, 5*, 219-224.

Malosso, E., English, L., Hopkins, D. W. & O'Donnell, A. G. (2004). Use of ^{13}C-labelled plant materials and ergosterol, PLFA and NLFA analyses to investigate organic matter decomposition in Antarctic soil. *Soil Biology & Biochemistry, 36*, 165-175.

Martinez-Trinidad, T., Watson, W. T., Arnold, M. A. & Lombardini, L. (2010). Microbial activity of a clay soil amended with glucose and starch under live oaks. *Arboriculture & Urban Forest, 36*, 66-72.

McCulley, R. L. & Burke, I. C. (2004). Microbial community composition across the Great Plains: Landscape versus regional variability. *Soil Science Society of America Journal, 68*, 106-115.

McCulley, R. L., Boutton, T. W. & Archer, S. R. (2007). Soil respiration in a subtropical savanna parkland: response to water additions. *Soil Science Society of America Journal, 71*, 820-828.

McGuire, K. L., Allison, S. D., Fierer, N. & Treseder, K. K. (2013). Ectomycorrhizal-dominated boreal and tropical forests have distinct fungal communities, but analogous spatial patterns across soil horizons. *PLoS ONE, 8*, e68278.

Meier, C. L. & Bowman, W. D. (2008). Links between plant litter chemistry, species diversity, and below-ground ecosystem function. *Proceedings of the National Academy of Sciences-Biology, U. S. A., 105*, 19780-19785.

Metcalfe, D. B., Fisher, R. A. &Wardle, D. A. (2011). Plant communities as drivers of soil respiration: pathways, mechanisms, and significance for global change. *Biogeosciences, 8*, 2011. 2047-2061.

Millard, P. & Singh, B. K. (2010). Does grassland vegetation drive soil microbial diversity? *Nutrient Cycling in Agroecosystems, 88*, 147-158.

Miyashita, N. T., Iwanaga, H., Charles, S., Diway, B., Sabang, J. & Chong, L. (2013). Soil bacterial community structure in five tropical forests in

Malaysia and one temperate forest in Japan revealed by pyrosequencing analysis of 16S rRNA gene sequence variation. *Genes & Genetic Systems, 88*, 93-103.

Moyes, A. B., Gaines, S. J., Siegwolf, R. T. W. & Bowling, D. R. (2010). Diffusive fractionation complicates partitioning of autotrophic and heterotrophic sources of soil respiration. *Plant Cell & Environment, 33*, 1804-1819.

Neff, J. C. & Asner, G. P. (2001). Dissolved organic carbon in terrestrial ecosystems: Synthesis and a model. *Ecosystems, 4*, 29-48.

Nemergut, D. R., Costello, E. K., Meyer, A. F., Pescador, M. Y., Weintraub, M. N. & Schmidt, S. K. (2005). Structure and function of alpine and arctic soil microbial communities. *Research in Microbiology, 156*, 775-784.

Norovsuren, Zh., Zenova, G. M. & Mosina, L. V. (2007). Actinomycetes in the rhizosphere of semidesert soils of Mongolia. *Eurasian Soil Science, 40*, 415-418.

Nottingham, A. T., Turner, B. L., Winter, K., Chamberlain, P. M., Stott, A. & Tanner, E. V. J. (2013). Root and arbuscular mycorrhizal mycelial interactions with soil microorganisms in lowland tropical forest. *FEMS Microbiology Ecology*, 85, 37-50.

Oberbauer, S. F., Gillespie, C. T., Cheng, W., Gebauer, R., Sala Serra, A. & Tenhunen, J. D. (1992). Environmental effects on carbon dioxide efflux from riparian tundra in the northern foothills of the Brooks Range, Alaska, USA. *Oecologia, 92*, 568-577.

O'Dowd, R. W., Parsons, R. & Hopkins, D. W. (1997). Soil Respiration induced by the D- and L-isomers of a range of amino acids. *Soil Biology & Biochemistry, 29*, 1665-1671.

O'Hanlon, R. (2012). Below-ground ectomycorrhizal communities: the effect of small scale spatial and short term temporal variation. *Symbiosis, 57*, 57-71.

Olsson, P., Linder, S., Giesler, R. & Högberg, P. (2005). Fertilization of boreal forest reduces both autotrophic and heterotrophic soil respiration. *Global Change Biology, 11*, 1745-1753.

O'Neill, E. G. (1994). Response of soil biota to elevated atmospheric carbon dioxide. *Plant & Soil, 165*, 55-65.

Paterson, E., Midwood, A. J. & Millard, P. (2009). Through the eye of the needle: a review of isotope approaches to quantify microbial processes mediating soil carbon balance. *New Phytologist, 184*, 19-33.

Perrin, D., Laitat, É., Yernaux, M. & Aubinet M. (2004). Modelling of the response of forest soil respiration fluxes to the main climatic variables. *Biotechnologie Agronomie Societe et Environment, 8*, 15-25.

Petz, W. (1997). Ecology of the active soil microfauna (Protozoa, Metazoa) of Wilkes Land, East Antarctica. *Polar Biology, 18*, 33-44.

Phillips, R. P., Brzostek, E. & Midgley, M. G. (2013). The mycorrhizal-associated nutrient economy: a new framework for predicting carbon-nutrient couplings in temperate forests. *New Phytologist, 199*, 41-51.

Raich, J. W. & Schlesinger, W. H. (1992). The global carbon dioxide flux in soil respiration and its relationship to vegetation and climate. *Tellus, Series B - Chemical and Physical Meteorology, 44B*, 81-99.

Rinnan, R. & Bääth, E. (2009). Differential utilization of carbon substrates by bacteria and fungi in tundra soil. *Applied & Environmental Microbiology, 75*, 3611-3620.

Rinnan, R., Michelsen, A. & Bääth, E. (2013). Fungi benefit from two decades of increased nutrient availability in tundra heath soil. *PLoS ONE, 8*, e56532.

Risch, A. C. & Frank, D. A. (2006). Carbon dioxide fluxes in a spatially and temporally heterogeneous temperate grassland. *Oecologia, 147*, 291-302.

Robinson, B. S., Bamforth, S. S. & Dobson, P. J. (2002). Density and diversity of protozoa in some arid Australian soils. *Journal of Eukaryotic Microbiology, 49*, 449-453.

Rodriguez-Zaragoza, S., Mayzlish, E. & Steinberger, Y. (2005). Vertical distribution of the free-living amoeba population in soil under desert shrubs in the Negev Desert, Israel. *Applied & Environmental Microbiology, 71*, 2053-2060.

Saul-Tcherkas, V., Unc, A. & Steinberger, Y. (2013). Soil microbial diversity in the vicinity of desert shrubs. *Microbial Ecology, 65*, 689-699.

Schaeffer, D., Feng, W. & Zou, X. (2009). Plant carbon inputs and environmental factors strongly affect soil respiration in a subtropical forest of southwestern China. *Soil Biology & Biochemistry, 41*, 1000-1007.

Schröter, D., Brussaard, L., De Deyn, G. Proveda, K., Brown, V. K., Berg, M. P., Wardle, D. A., Moore, J. & Wall, D. H. (2004). Trophic interactions in a changing world: modeling aboveground –belowground interactions. *Basic and applied Ecology, 5*, 515-528.

Shaver, G. R., Giblin, A. E., Nadelhoffer, K. J., Thieler, K. K., Downs, M. R,, Laundre, J. A. & Rastetter, E. B. (2006). Carbon turnover in Alaskan

tundra soils: effects of organic matter quality, temperature, moisture and fertilizer. *Journal of Ecology*, *94*, 740-753.

Simard, S. W. & Mohn, W. W. (2012). Long-term warming alters the composition of Arctic soil microbial communities. *FEMS Microbiology Ecology, 82*, 303-315.

Sjögersten, S. & Wookey, P. A. (2002). Climatic and resource quality controls on soil respiration across a forest-tundra ecotone in Swedish Lapland. *Soil Biology & Biochemistry, 34*, 1633-1646.

Stevenson, B. A., Sparling, G. P., Schipper, L. A., Degens, B. P. & Duncan, L. C. (2004). Pasture and forest soil microbial communities show distinct patterns in their catabolic respiration responses at a landscape scale. *Soil Biology & Biochemistry, 36*, 49-55.

Suh, S., Lee, E. & Lee, J. (2009). Temperature and moisture sensitivities of CO_2 efflux from lowland and alpine meadow soils. *Journal of Plant Ecology, 2*, 225-231.

Swift, M. J. (1996). Soil biodiversity and the fertility of tropical soils. Biol. International 33: 24-28.

Swift M. J. & Anderson J. M. (1993). Biodiversity and ecosystem function in agricultural systems. In: E. D. Schultze & H. Mooncy (Eds.), *Biodiversity and Ecosystem Function* (pp. 15-42). Berlin: Springer-Verlag.

Tiemann, L. K. & Billings, Sharon A. (2011). Changes in variability of soil moisture alter microbial community C and N resource use. *Soil Biology & Biochemistry, 43*, 1837-1847.

Todorov, M. & Golemansky, V. (1999). Biotopic distribution of testate amoebae (Rhizopoda: Testacea) in continental habitats of the Livingston Island (the Antarctic). *Bulgarian Antarctic Research Life Sciences, 2*, 48-56.

Tomotsune, M., Yoshitake, S., Watanabe, S. & Koizumi, H. (2013). Separation of root and heterotrophic respiration within soil respiration by trenching, root biomass regression, and root excising methods in a cool-temperate deciduous forest. *Ecological Research, 28*, 259-269.

Treonis, A. M., Wall, D. H. & Virginia, R. A. (2002). Field and microcosm studies of decomposition and soil biota in a cold desert soil. *Ecosystems, 5*, 159-170.

Trumbore S. E. (1993). Comparison of carbon dynamics in tropical and temperate soils using radiocarbon measurements. *Global Biogeochemical Cycles, 7*, 275–290.

Twieg, B. D., Durall, D. M., Simard, S. W. & Jones, M. D. (2009). Influence of soil nutrients on ectomycorrhizal communities in a chronosequence of mixed temperate forests. *Mycorrhiza, 19*, 305-316.

van Horn, D. J., Van Horn, M. L., Barrett, J. E., Gooseff, M. N., Altrichter, A. E., Geyer, K. M., Zeglin, L. H. & Takacs-Vesbach, C. D. (2013). Factors controlling soil microbial biomass and bacterial diversity and community composition in a cold desert ecosystem: role of geographic scale. *PLoS ONE, 8*, e66103.

Vanhala, P., Karhu, K., Tuomi, M., Bjorklof, K., Fritze, H., Hyvarinen, H. & Liski, J. (2011). Transplantation of organic surface horizons of boreal soils into warmer regions alters microbiology but not the temperature sensitivity of decomposition. *Global Change Biology, 17*, 538-550.

Vincke, S., Van de Vijver, B., Nijs, I. & Beyens, L. (2006). Changes in the testacean community structure along small soil profiles. *Acta Protozoologica, 45*, 395-406.

Virginia, R. A. & Wall, D. H. (1999). How soils structure communities in the Antarctic Dry Valleys. *Bioscience, 49*, 973–983.

Waldrop, M. P. & Firestone, M. K. (2006). Response of microbial community composition and function to soil climate change. *Microbial Ecology, 52*, 716-724.

Waldrop, M. P., Zak, D. R. & Sinsabaugh, R. L. (2004). Microbial community response to nitrogen deposition in northern forest ecosystems. *Soil Biology & Biochemistry, 36*, 1443-1451.

Welker, J. M., Brown, K. B. & Fahnestock, J. T. (1999). CO_2 flux in arctic and alpine dry tundra: comparative field responses under ambient and experimentally warmed conditions. *Arctic Antarctic & Alpine Research, 31*, 272-277.

Wierzchos, J., de los Rios, A. & Ascaso, C. (2012). Microorganisms in desert rocks: the edge of life on Earth. *International Microbiology, 15*, 172-182.

Winkler, J. P., Cherry, R. S. & Schlesinger, W. H. (1996). The Q-10 of microbial respiration in a temperate forest soil. *Soil Biology & Biochemistry, 28*, 1067-1072.

Xu, L., Baldocchi, D. D. & Tang, J. (2004). How soil moisture, rain pulses and growth alter the response of ecosystems respiration to temperature. *Global Biogochemical Cycles, 18*, GB4002, doi:10.1029/2004GB002281.

Yergeau, E., Bokhorst, S., Huiskes, A. H. L., Boschker, H. T. S., Aerts, R. & Kowalchuk, G. A. (2007a). Size and structure of bacterial, fungal and nematode communities along an Antarctic environment gradient. *FEMS Microbiology and Ecology, 59*, 436–451.

Yergeau, E., Newsham, K. K., Pearce, D. A. & Kowalchuk, G. A. (2007b). Patterns of bacterial diversity across a range of Antarctic terrestrial habitats. *Environmental Microbiology, 9*, 2670–2682.

Zak, D. R. & Kling, G. W. (2006). Microbial community composition and function across an arctic tundra landscape. *Ecology, 87*, 1659-1670.

Zak, D. R., Holmes, W. E., Burton, A. J., Pregitzer, K. S. & Talhelm, A, F. (2008). Simulated atmospheric NO_3- deposition increases soil organic matter by slowing decomposition. *Ecological Applications, 18*, 2016-2027.

Zak, D. R., Holmes, W. E., Tomlinson, M. J., Pregitzer, K. S. & Burton, A. J. (2006). Microbial cycling of C and N in northern hardwood forests receiving chronic atmospheric NO_3-depostion. *Ecosystems, 9*, 242-253.

Zhang, W., Zhang, G., Liu, G., Dong, Z., Chen, T., Zhang, M., Dyson, P. J. & An, L. (2012). Bacterial diversity and distribution in the southeast edge of the Tengger Desert and their correlation with soil enzyme activities. *Journal of Environmental Sciences, 24*, 2004-2011.

In: Soil Carbon
Editor: Aquila Margit

ISBN: 978-1-63117-438-4
© 2014 Nova Science Publishers, Inc.

Chapter 2

LAND USE EFFECTS ON SOIL CARBON AND NITROGEN STOCKS AND FLUXES IN THE PAMPAS: IMPACT ON PRODUCTIVITY

R. Alvarez[1,2,], J. L. De Paepe[1], H. S. Steinbach[1], G. Berhongaray[1], M. M. Mendoza[1], A. A. Bono[3], N. F. Romano[3], R. J. C. Cantet[1,2] and C. R. Alvarez[1]*

[1]Facultad de Agronomía-Universidad de Buenos Aires, Argentina
[2]CONICET, Argentina
[3]EEA Anguil, INTA, Argentina

ABSTRACT

The Pampas in Argentina is a vast plain of around 60 Mha that is considered as one of the most important grain producing regions in the World. Cultivation began by the last quarter of the 19th Century occupying 50% of the surface at present. Low input agriculture, in combination with livestock production, was performed till 1970, afterwards soybean (*Glicyne max* (L). Merr.) crop was introduced. This crop replaced pastures and at present it accounts for 60% of the grain crop seeded surface. During the 2007-2008 growing season 382 sites, widespread over the region, were sampled up to 1 m depth and carbonate

---

* Corresponding author: ralvarez@agro.uba.ar. Av. San Martín 4453 (1417) Buenos Aires, Argentina.

carbon, organic carbon, and total nitrogen stocks were determined in these soil samples. Land use had no effect on inorganic carbon content, which accounted for one third of total soil carbon. Aggregation of soil map information showed that the carbonate carbon stock of the region was estimated to be 1.9 Gt. Organic carbon was affected by land use. Cropped soils lost 16% of their organic carbon stock up to 50 cm depth. The overall organic carbon stock of the Pampas was estimated to be 4.2 Gt. This estimation was achieved combining an artificial neural network model for carbon estimation that used as inputs climate, textural, and land use information, and satellite image classification that referred to land use per pampean county. The net flux of $C-CO_2$ from soils to the atmosphere as the consequence of cultivation was estimated to be 0.33 Gt. Cropping effect on total nitrogen stock was similar to the effect on organic carbon. The average C/N ratio of the soils was 8.9, with only small differences between land use types and depth layers. When comparing present organic carbon and nitrogen stocks with those estimated from soil maps corresponding to surveys performed ca. 40 years ago, we observed that areas with high carbon and nitrogen contents tended to loose carbon and nitrogen and areas with low levels tended to maintain the levels or increase them. The maintenance or increase of carbon could be explained by increases in rainfall due to climate change and the increase of crop residue inputs of higher yielding crops, despite the soybean adoption in rotations. Additionally, the adoption of no-till as the most common tillage system favored carbon gain of many soils. The Pampean agroecosystems have a positive nitrogen balance with an annual gain of ca. 1 Mt N yr^{-1}. The main nitrogen input to soils is through biological nitrogen fixation, both by leguminous pasture and soybean, which represents ca. 5 times the fertilizer input. In sites with annual crop rotations the nitrogen balance is near neutral. Productivity of Pampean soils is regulated by available water holding capacity and organic carbon content. Changes in carbon levels produced by the agricultural use have little impact on productivity. The effects of different management practices on organic carbon and some labile fractions (microbial biomass, in vitro mineralized carbon, particulate carbon, light fraction carbon) were assessed in numerous field experiments, performed mainly in the humid portion of the Pampas. The labile fractions of organic matter are not good indicators of future changes in the carbon content of Pampean soils. These fractions change more pronounced than organic carbon only when carbon changes are greater than 25-50% but vary in equal measure or even with an opposite trend below this threshold. Neither these fractions, nor the potentially mineralizable nitrogen or the organic matter/clay+silt ratio are good predictors of the nitrate level at planting or crop yields. Crop yield prediction could not be improved by using the mentioned labile fractions instead of organic carbon in models. Conversely, biological fractionations, determined by in vitro nitrogen mineralization, allowed a

better soil nitrogen mineralization prediction under field conditions than using only carbon and nitrogen stocks. An artificial neural network model was fitted that predicted field nitrogen mineralization using as inputs, land use information, soil texture, temperature, water content, total nitrogen stock, and the in vitro mineralization test results, with very good performance (R^2= 0.76). The impact of cultivation on soil nitrogen supplying capacity was assessed for by applying this network. In average, cultivation produced an 80% decrease in the nitrogen mineralization capacity of soils. Despite soil nitrogen was only moderately affected by cropping, the biological fractionation method, coupled with artificial intelligence modeling, showed great significant soil degradation in the Pampas. Present levels of nitrogen mineralization capacity of soils are not enough to sustain high to medium wheat (*Triticum aestivum* L.) and corn (*Zea mayz* L.) yield and fertilization has become a required practice. Our results show that pampean soils loose only a small fraction of their organic matter content, but this fraction had a deep impact on soil fertility. Biological fractionation, not physical fractionation of organic matter, allowed to asses cropping effects on soil quality, and resulted in the development of suitable methods for agronomic purposes in the Pampas.

DESCRIPTION OF THE PAMPEAN REGION
AND THE PROBLEM ADDRESSED

The Pampean region in Argentina is a 60 Mha plain located between 57ºW and 68ºW and 28ºS and 40ºS (Figure 1). Around 90% of the national grain production is carried out in this region. Because of its extension and yield potential it has been considered as one of the most important grain crop producing areas of the World (Satorre & Slafer, 1999). Grasslands, in which graminaceous species predominate, are the natural vegetation with some minor areas occupied by natural forest that cover ca. 7% of the total surface (INDEC, 2002b). Foreign trees, introduced ca. 150 years ago, are used as wind barriers occupying 0.2% of the area. The climate is warm-temperate, with a mean temperature ranging from 14ºC in the South to 20ºC in the North, and humid, with annual rainfall ranging from 500 mm in the West to 1200 mm in the East.

The relief is flat or slightly rolling with Mollisols as predominant soils, formed on loess-like materials (Alvarez & Lavado, 1998). Because of the eolian origin of sediments from Southwest to Northeast and the climatic rainfall gradient from West-East, the depth and texture of soils varies from shallow and sandy in the West to deep and clayed in the East (Alvarez &

Lavado, 1998; Berhongaray et al., 2013). The predominant clay mineral in the region is illite (Alvarez & Lavado, 1998). Within the upper 1 m of the soil profile a petrocalcic horizon appears in many soils along the West and the South ends of the region (Teruggi, 1957). Soil organic carbon is strongly associated to rainfall (Berhongaray et al., 2013). Conversely, soil pH is not climate regulated, with a range from 6 to 9, and no signals of acidification by cultivation have been detected (Berhongaray et al., 2013). High pH values are usually associated to hydromorphic lands (Hall et al., 1992).

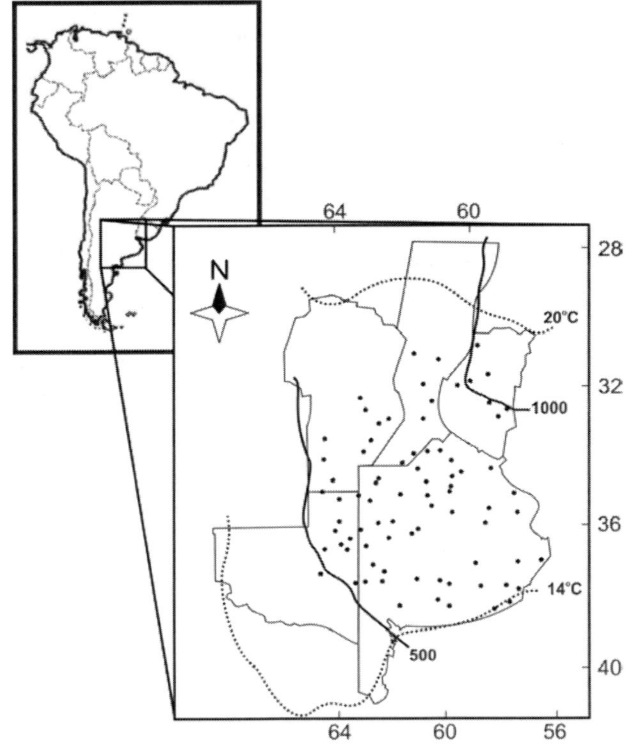

Figure 1. Map of the Pampean Region indicating sampled sites during the 2007-2008 growing season with mean isotherms and isohyets (mm) (redrawn from Berhongaray et al., 2013).

In 1870 agriculture was introduced in the region and the cultivated area exponentially increased (Figure 2). Agricultural expansion was faster in the humid portion and started later and increased at a slower rate in the semiarid area (Alvarez & Steinbach, 2012). In the humid and semiarid portions of the

Pampas, rainfed crops are cultivated, where annual rainfall is above 600 mm, on well-drained soils while hydromorphic soils are used for grazing (Hall et al., 1992; Alvarez & Lavado, 1998). Currently ca. 50% of the Pampas is under annual crop production, with soybean (*Glicyne max* (L). Merr.), wheat (*Triticum aestivum* L.), and maize (*Zea mayz* L.) as major crops (MinAgri, 2013). In recent decades agricultural land use intensified and soybean was adopted as the main component of crop rotations (Viglizzo et al., 2001, 2011); at present this crop occupies 60% of the area under cultivation. Agricultural expansion in recent decades has been attributed to rainfall increases and technological improvements that also lead to 2 to 3-fold yield increases, and this observed increase is crop dependent (Magrin et al., 2005, Viglizzo et al., 1997). Because of the natural high fertility of soils and in combination with mixed cropping-livestock production, a low-external input regional agriculture developed (Viglizzo et al., 2001, 2011). A widespread adoption of no-till and fertilizer use occurred since 1990.

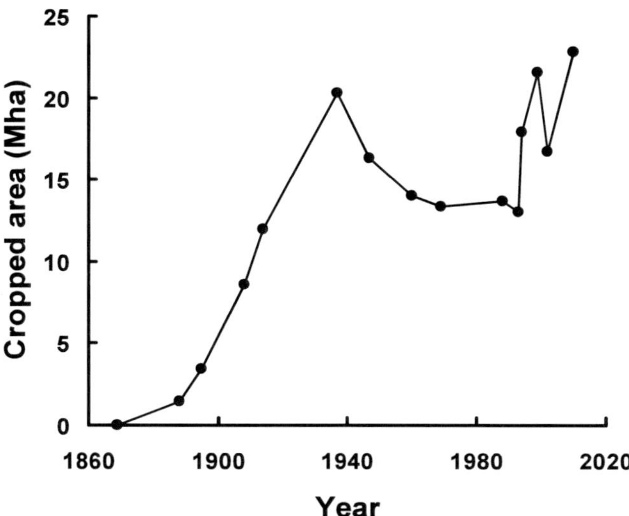

Figure 2. Evolution of cultivated surface in the Pampean Region (Provinces of Buenos Aires, Córdoba, Entre Ríos, La Pampa y Santa Fe). Elaborated with data from Anonymous (1983, 1898, 1909, 1939, 1947, 1964) e INDEC (1969, 1988, 1988, 1988, 1988, 1988, 1994, 1995a, b, c; 2002a).

Extensive rainfed agriculture generates a reduction in organic matter content (Davidson & Ackerman, 1993; Houghton et al., 1983) and soil nutrients (McLauchlan, 2006) that affects productivity adversely (Follett &

Stewart, 1985; Pan et al., 2009). It is also important to establish the effect of land use on carbon sequestered in soils because of the possibility of transforming them into C-CO$_2$ sinks (Mishra et al, 2009; Meersmans et al., 2008.). Additionally, in semi-arid and arid areas, cropping can modify the level of soil carbonates. Irrigation with carbonated water can produce increases (Wu et al., 2009); while decreases can be observed as a result of tillage systems that expose previously buried soil to the atmosphere (Moreno et al., 2006), and acidification due to fertilization (Wu et al., 2009).

In the Pampas, the regional expansion of soybean production can lead to soil degradation because of the increased agricultural use in combination with a low residue crop input (Viglizzo et al., 2001; 2010). Conversely, soil aggradation can be expected when the additional input of nitrogen to Pampean agroecosystems due to biological nitrogen fixation is considered (Alvarez et al., 2012a). Possible adverse effects of this soybean production expansion are not evident at the moment on grain crop yields, which in fact have markedly increased over the last 40 years (Alvarez, 2011; MinAgri, 2013). However, it is possible that the degradation was offset by technological improvements, such as the widespread adoption of the fertilization practice (Alvarez et al, 2012a.).

Research at site scale have reported declines of soil organic carbon stock in the Pampas by cultivation in the upper profile layer, but at regional scale, only two studies have been made using modeling techniques (Alvarez, 2001; Caride et al., 2012). These researches estimated organic carbon decreases in the 0-20 cm depth due to cropping in some pampean sub-regions and a net flow of C-CO$_2$ to the atmosphere of 200 Mt was estimated. The effect of agriculture on carbon stocks in depth and total soil nitrogen stocks and how it influences soil productivity has been little studied. Our goal is to summarize results of recent studies and other not published information to address this problem.

SOIL CARBON STOCK CHANGES

During the 2007-2008 growing season a regional soil survey was performed by sampling 82 farms widespread over the Pampas, which were considered to be representative of the different pampean sub-regions (Figure 1). At each farm, five common land uses were sampled: forested areas, never cultivated soils under graminaceus vegetation, cultivated soils under the pasture phase of a mixed rotation, cultivated sites under the agricultural phase

of a mixed rotation or under continuous agriculture, and hydromorphic soils devoted to grazing with natural vegetation (Berhongaray et al., 2013). Samples were taken up to 100 cm depth or until the upper limit of a petrocalcic horizon in layers of 25 cm. Bulk density, organic carbon, carbonate carbon, total nitrogen, texture, pH in water, and electrical conductivity were determined (Berhongaray et al., 2013). Nitrogen mineralization potential of samples was determined in an incubation test (Alvarez et al., 2012b). For all the sampled sites (n = 382) mean annual temperature and precipitation were estimated.

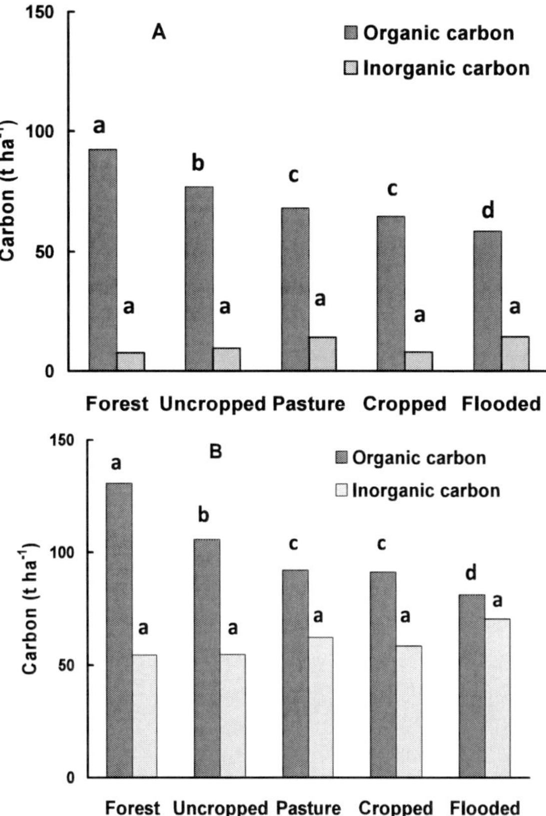

Figure 3. Soil organic and inorganic carbon contents as a function of soil use for two different soil layers: A) 0-50 cm and B) 0-100 cm. Different letters indicate significant differences (P= 0.01) between land uses (redrawn from Berhongaray et al., 2013).

Organic carbon was affected by land use (Figure 3). Never cultivated soils, which functioned as the control treatment, had an average content of 101 t C ha^{-1} up to 1 m deep. On average, carbon content of cultivated sites was significantly lower (P= 0.01), 87 t C ha^{-1}, with no significant differences between pasture and agricultural phases of rotations. Forested areas had more organic carbon than the controls and hydromorphic soils had the lowest contents. Around 50% of the stock of organic carbon was sequestered in the 0-25 cm layer, regardless of land use. When testing significance by soil layer, differences between never cultivated soils and cultivated soils were significant only up to 50 cm depth. Organic carbon stock of soils under cultivation was 16 % lower than in controls up to 1 m depth. Differences between never cultivated sites and cultivated sites ranged between -22 and +64 %. Cultivated soils had on average 9% less organic carbon than controls for the 50-100 cm soil layer (P = ns). Land use did not affect the carbonate carbon stock, which averaged 50 t C ha^{-1} up to 1 m depth (Figure 3). On average, 35% of soil carbon was present in the form of carbonates. In soils of the semiarid portion of the Pampas this percentage rose up to 80%. Carbonate carbon was stratified inversely than organic carbon with the greatest accumulation in the 75-100 cm soil layer (50% of the carbon sequestered to 1 m depth).

The spatial distribution of organic carbon was modeled using artificial neural networks that used land use, depth, climate and textural data as inputs (R^2= 0.64) (Berhongaray et al., 2013). The model showed that carbon increased with rainfall and finer textured soils and decreased at higher temperature or in depth. With other conditions being equal, land use regulates the level of carbon in the following order: forest > controls never cultivated > cultivated soils under pasture phase = cultivated soils under crop phase > flooding lands. Satellite image classification and data from the National Agricultural Census 2002 were used to estimate land use type per pampean county. Combining the generated neural network model with this land use information, organic carbon stock for the whole Pampean Region was estimated (Berhongaray et al., 2013). For an area surveyed of 48.2 Mha, which includes most of the cultivated soils of the region, the organic carbon stock was estimated of 4.22 Gt C (uncertainty = 0.14) during 2007-2008. With information from soil surveys performed by the National Institute of Agricultural Technology, mostly made in the 1960-1980 period, an estimation of 3.96 Gt C (uncertainty = 0.22) for the same area could be performed (Berhongaray et al., 2013). The stocks of the 0-25 cm layer were 2.04 and 1.93 Gt respectively. The uncertainty analysis indicated no differences in carbon stocks between the two sampling periods because the confidence intervals of

95 % of past and present estimates overlap. At county scale, when the stock of organic carbon in 1960-1980 was greater than 95 t ha^{-1} up to 1 m depth, soils lost carbon in 2007-2008 and below this threshold increases were more frequent (Figure 4). From the integration of past soil survey data, a regional carbonate carbon stock was estimated of 1.87 Gt C up to 1 m depth (Berhongaray et al., 2013).

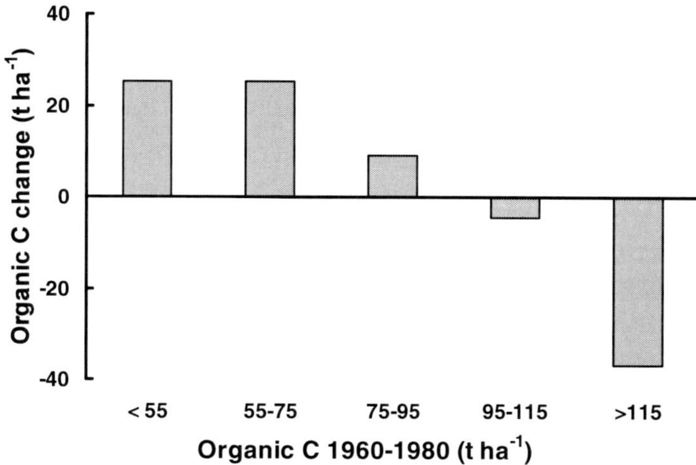

Figure 4. Average variation in the organic carbon stock of pampean soils up to 1 m depth produced between 1960-1980 and 2007-2008 as a function of organic carbon content during 1960-1980 (redrawn from Berhongaray et al., 2013).

The average estimated soil organic carbon decrease of 16 % for the Pampas is low relative to that observed in other cultivated areas of the World where reductions of 30-50% were reported in the upper 20-30 cm of the soil profile (Guo & Gifford, 2002). This decrease is attributed to the reduction of carbon inputs to soils under cultivation (Lauenroth et al., 2000), mainly due to the fallow period and crop harvest. In the Pampas, carbon inputs of annual crops account for 30 to 70% of those under natural grasslands (Alvarez & Steinbach, 2012), leading to decreases of soil carbon levels compared to uncultivated soils. In areas where permanent agriculture is practiced, the input difference between natural and cultivated systems is greater than under situations where mixed systems are maintained. For example, in the humid environment of the so called Rolling Pampa, where permanent agriculture was installed since the last 30-40 years in most places, input of natural vegetation was estimated at 6.3 t C ha^{-1} yr^{-1}. Compared to this value, the average input

under agriculture during the 1970-2004 period was estimated at 2.8 t C ha^{-1} yr^{-1} (44 % of natural input) (Alvarez & Steinbach, 2012). By contrast, in the Semiarid Pampa, where mixed systems are still common, the carbon input of the natural vegetation is 3.9 t C ha^{-1} yr^{-1} and the mixed system input during the 1970-2004 period was estimated as 2.6 t C ha^{-1} yr^{-1} (67% of natural input). Soil erosion is another factor reducing organic carbon. Erosion losses of 8% of organic carbon sequestered up to 1 m depth were estimated in the Rolling Pampa (Alvarez et al., 1995).

The agricultural or pasture phase of rotations does not affect significantly the soil organic carbon stock in the Pampas. Long-term experiments conducted in the region have shown that when the pasture phase lasts a few years (2-4), it does not affect markedly the organic carbon stock (Casanovas et al., 1995; Galantini, 2005). Assuming that the area occupied by hydromorphic and forested soils remained unchanged in the last 140 years, during which the agricultural use expanded, and that grain crops replaced natural grasslands, a net loss of carbon from soils or C-CO$_2$ flux to the atmosphere of 326 Mt was estimated for the surveyed area. This flow is equivalent to a regional consumption of fossil fuels during 9 years (CIA World Factbook, 2008). Considering that the regional soil carbon stock at present is similar to the stock during the 1960-1980 period, it appears that the change of the soil organic carbon level have occurred prior to 1960. Agricultural intensification and the adoption of soybean as a main component of many rotations have not impacted negatively the carbon stock sequestered at the regional level. Areas with great carbon levels 30-50 years ago tended to lose carbon, while areas with low levels maintained or increased their carbon content. Commonly, it is easy to maintain or increase carbon in low carbon soils but this is not the case in soils with great carbon contents (Alvarez, 2005). Cultivation did not affect the inorganic carbon stock sequestered in pampean soils, which can be attributed to the short agricultural history and the small rates of applied fertilizer which did not affect soil pH (Berhongaray et al., 2013).

TOTAL NITROGEN STOCK CHANGES

Land use impact on total soil nitrogen was similar than the effect on soil organic carbon. Nitrogen content was lower in cultivated sites than under controls sites up to 50 cm depth, without significant differences between soils under the pasture or the agricultural phase of rotations (Figure 5). The ratio organic C/total N was not affected by land use or depth, averaging 8.9 (Figure

5). Forest soils had more nitrogen than controls and flooded lands had the lowest contents. Only in flooded soils the C/N ratio was higher than in well-drained soils below 25 cm depth (P= 0.01).

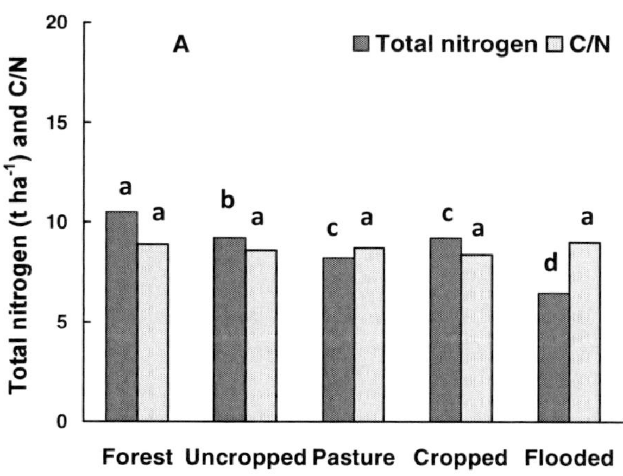

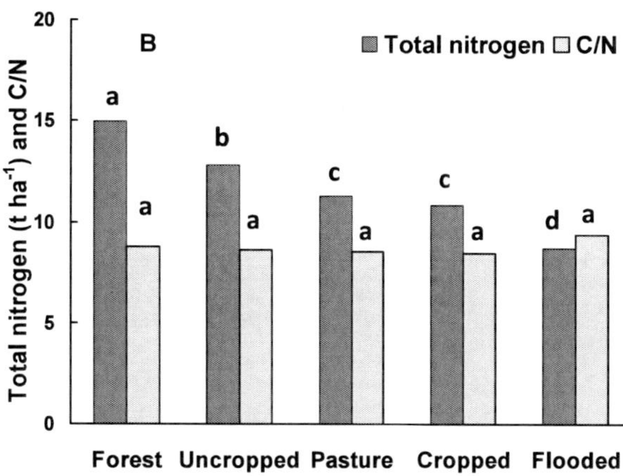

Figure 5. Total nitrogen stock and C/N ratio as a function of land use for two soil layers: A) 0-50 cm and B) 0-100 cm. Different letters indicate significant differences (P = 0.01) between land uses (redrawn from Mendoza et al. 2012 b with unpublished data incorporated).

The spatial distribution of total nitrogen could be modeled using artificial neural networks with good performance ($R^2 = 0.81$) and as a function of land use, depth, climate and texture with data from the 2007-2008 survey (Mendoza

et al., 2012 a; b). The model showed that nitrogen increased with greater rainfall and in fine textured sites and decreased in areas of higher temperature and with depth. Regarding the impact of land use, soil nitrogen level varied in the following order: forest soils > uncultivated controls > cultivated soils under the pasture phase of rotation = cultivated soils under the annual crops > flooded lands.

Again, neuronal network modeling information was combined with assigned areas of each land use per pampean county by satellite imaginary information. This resulted in a regional total nitrogen stock estimation for the 0-25 cm layer for 2007-2008 of 0.250 Gt (uncertainty = 0.014), which was compared with that calculated by soil map integration corresponding to the 1960-1980 period; 0.231 Gt (uncertainty = 0.023) (De Paepe & Alvarez, 2010). As observed for the regional carbon stock, there were no significant differences between the two sampling periods, since the uncertainties of the estimates overlapped. Conversely, contrasting results at county scale were observed, and when the nitrogen stock during 1960-1980 was greater than ca. 12 t ha^{-1} soils lost nitrogen in 2007-2008 and below that threshold increases were more frequent (Figure 6).

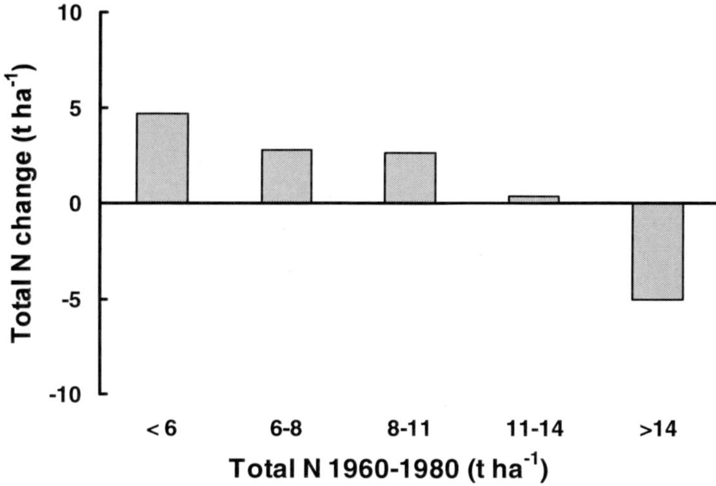

Figure 6. Average changes of total nitrogen stock in the 0-25 cm soil layer of pampean soils (nitrogen stock in 2007-2008 minus stock in 1960-1980) in relation to nitrogen content during 1960-1980 period. Calculated with data from Mendoza et al. (2012 b) and following the methodology described in Berhongaray et al. (2013) for organic carbon.

CARBON FLUX CHANGES

Integrating results from 112 field experiments, in which production of stubble and roots of major pampean crops was assessed a neural network model ($R^2 = 0.85$) was developed for predicting carbon inputs to the soil for wheat, corn and soybean (Alvarez et al. 2011). Another neural network was fitted ($R^2 = 0.86$) using information of soil heterotrophic respiration measurements in 7 long-term experiments of several years, to estimate the C-CO_2 emission from soils to the atmosphere originated in the organic pools (Alvarez et al. 2011). The combination of both tools allowed carbon balance estimation of pampean soils (Alvarez et al., 2011) using as input past crop yields and crop rotation composition that was estimated using seeded area (MinAgri, 2013).

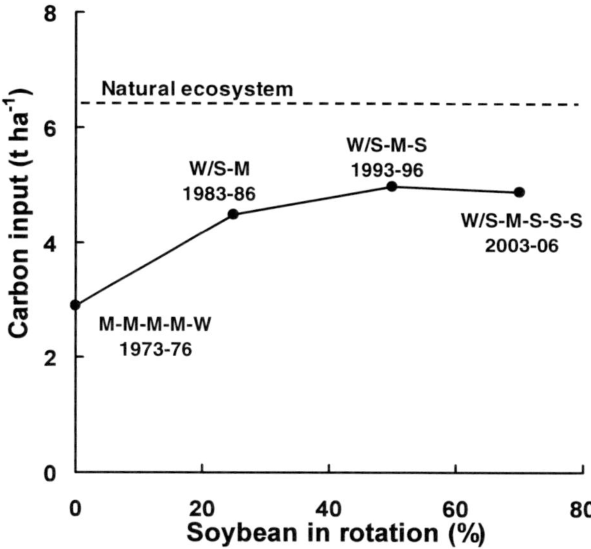

Figure 7. Evolution of carbon input to soil from residues of the common agricultural rotations historically used in the Humid Pampa as a function of the percentage of soybean in crop rotation. Time periods and rotations for which carbon inputs were calculated are indicated. C = corn, W= wheat, S = soybeans, W/S = double cropping wheat/soybean. For comparative reasons the carbon input of the natural ecosystem has been marked (redrawn with data from Alvarez et al., 2011).

Despite the adoption of soybean as main component of crop rotations, increases of carbon inputs to soils were produced over the last 30 years (Figure 7) and a less negative carbon balance was estimated at present compared to the balance of 30-50 years ago (Figure 8). This process was the result of greater carbon inputs associated with yield increases of wheat, corn and soybean. These greater carbon inputs balanced losses, and resulted in increases of organic carbon in soils that were originally poor in soil carbon but this was not observed in carbon-rich soils. Additionally, rainfall increases registered in the Pampas (Magrin et al., 2005), mainly in the semiarid portion of the region during the last 40 years (Figure 9), may have led to an increase in productivity in sandy soils of low initial carbon level. Another factor that produces increased levels of organic carbon is the widespread adoption of no-till as the predominant regional tillage method (Figure 10). A meta-analysis that integrated results from numerous experiments of short, medium, and long-term periods under this management in the Pampas showed that carbon increases between 3% and 15% % in the topsoil were found, and the largest increases corresponded to soils from the semiarid portion of the region (Steinbach & Alvarez, 2006).

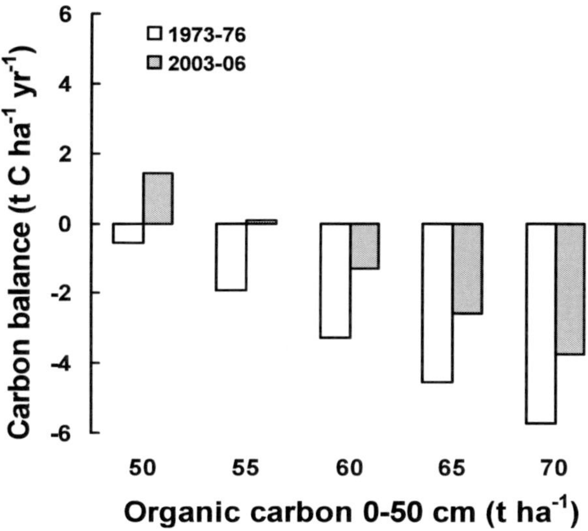

Figure 8. Soil carbon balance of the Humid Pampa for two time periods as a function of organic carbon level (redrawn from Alvarez et al., 2011).

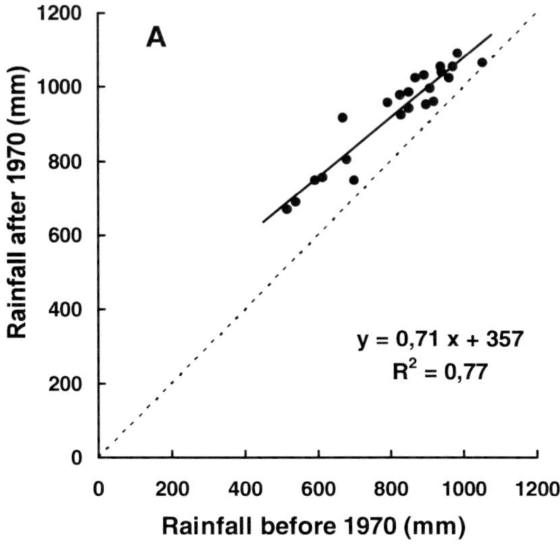

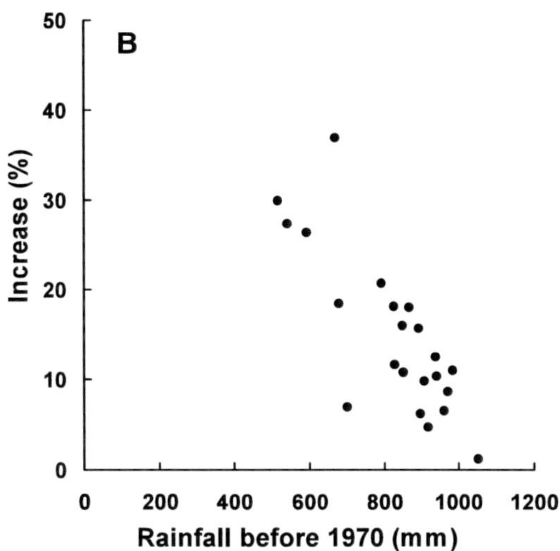

Figure 9. Changes in annual rainfall levels for various locations in the Pampas corresponding to two periods: the early twentieth century-1970 and 1970-2010. A) total amount, B) relative change. Records form the following observatories were used: 9 de Julio, Anguil, Azul, B. Blanca, Balcarce, Bolivar, Bordenave, Buenos Aires, Ceres, Dolores, Gral Acha, Gral Pico, La Plata, Laprida, Las Flores, Mar del Plata, Miaramar, Pehuajo parchment Pigue, Rafaela, Rosario, San Pedro, Santa Rosa , Tandil, Lauquen Dam, Victorica, Zavalla.

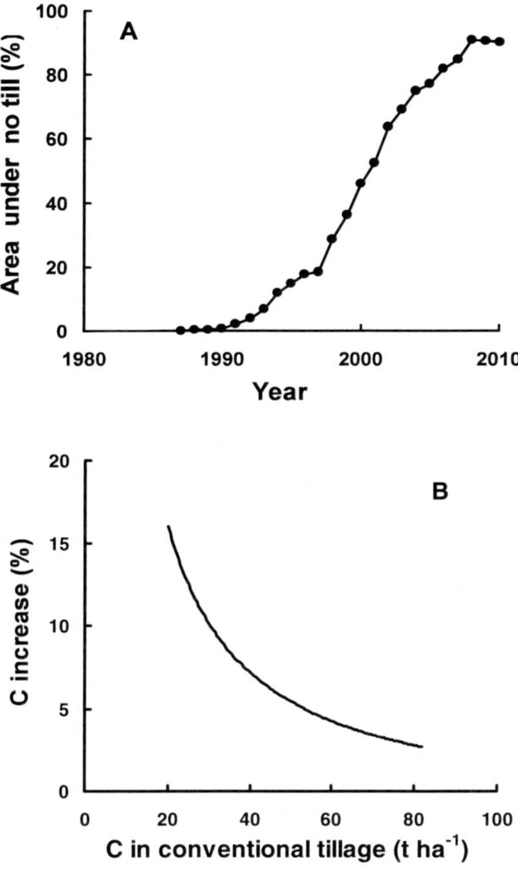

Figure 10. A): evolution of the area under no-till in the Pampas (made with data from MinAgri, 2013 and AAPRESID, 2013). B): effect of changing tillage system from systems with soil tillage to no-till on carbon stock of the soil in the tilled layer (redrawn from Steinbach & Alvarez, 2006).

NITROGEN FLUX CHANGES

The mass balance methodology allows to understand the functioning of the nitrogen cycle at regional (Galloway et al., 2004) and ecosystem scale (Spiertz, 2010). It can predict future trends in fluxes (Howarth et al., 2002), its possible impact on the environment (Eickhout et al., 2006), and point management options for increasing nitrogen use efficiency (Spiertz, 2010).

The soil nitrogen balance is estimated including all inputs and outputs of nitrogen to and from the system (Oenema et al, 2003). A simplified version, known as surface balance, includes all these inputs but the losses from the system are ignored as these are very difficult to estimate (OECD, 2001). The latter methodology, aimed to develop agricultural policies, allows to detect nitrogen surplus, which should be avoided in order to improve nutrient use efficiency, as this surplus may be retained in the soil, emitted to the atmosphere or leached into the groundwater (OECD, 2001; Panten et al, 2009). In the Pampas the surface nitrogen balance has been calculated at regional scale in order to account for changes in land use occurred in the last 50 years and its impact on nitrogen fluxes (Alvarez et al., 2014).

During the past four decades cropped area under grain crops doubled in the Pampas, mainly because of soybean expansion, at the expense of the area devoted to seeded pastures and annual forages; while the area under natural pasture grasses or forest remained unaltered (Figure 11 A). Agricultural advance was associated with a yield increase of 200-300%, according to the crop, that leads to a 4-fold grain production increase (Figure 11 B). During the same period fertilization was adopted as a common practice, mainly for cereal crops (Alvarez et al. 2012a). All these factors led to great changes in nitrogen fluxes in pampean agroecosystems.

The nitrogen input estimation by fertilization and grain output by harvest products is simple, consumption of nitrogen fertilizers and nitrogen concentration in grain are taken into account (Alvarez et al., 2014). Nitrogen input by rainfall has been carefully measured in a site located in the humid portion of the Pampas and it has been assumed that elsewhere the input is proportional to the rainfall measured value (Alvarez, 2001). For biological nitrogen fixation by leguminous pastures and soybean local models have been developed based on experiments with ^{15}N. Alfalfa (*Medicago sativa* L.), is the main forage legume in the region, it fixes on average 19 kg N t^{-1} DM produced and productivity can be estimated using site level weather variables easily obtainable. Regional fixation estimations can be performed accounting for percentages of pure and mixed alfalfa pastures (Alvarez et al., 2014). Soybean crops produced in the Pampas fix on average 52 kg N t^{-1} DM grain produced, this average allows regional nitrogen fixation estimation (Alvarez et al., 2014).

Despite the displacement of pastures by agriculture, nitrogen inputs to the pampean agroecosystems have increased during the last decades mainly due to soybean nitrogen fixation and, to a lesser extent, to the fertilizer use (Figure 12A).

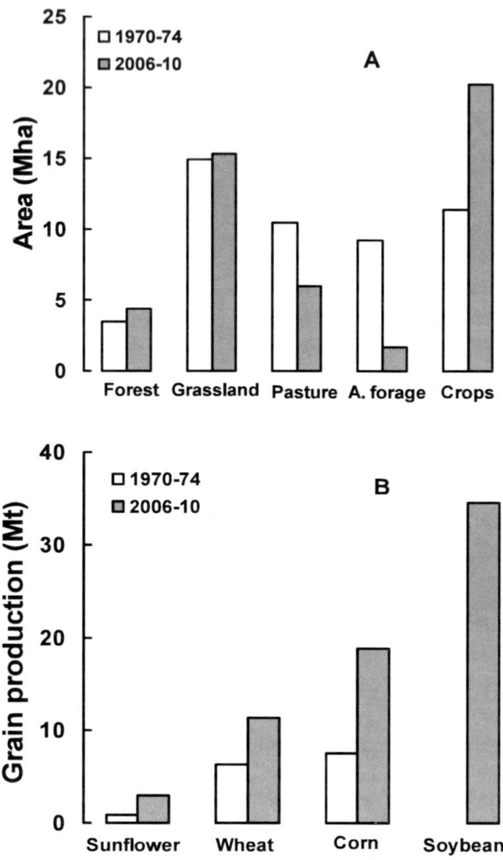

Figure 11. A. Evolution is the area devoted to different uses in the Pampas (A. forage= annual forage) (Redrawn from Alvarez et al., 2014). B. Evolution of major grain crops production in the Pampas (made with data from MinAgri, 2013).

Currently, 42% of the nitrogen inputs result from soybean fixation, 28% by legume pastures, 17% by rainfall, and 13% by fertilizers. Nitrogen outputs by harvested grain have also increased up to 8 times, 76% of this output is accounted for by soybean grains (Figure 12B). The balance between inputs and outputs has been positive in the past and remains positive at present, tending to be less positive today than 50 years ago (Figure 13).

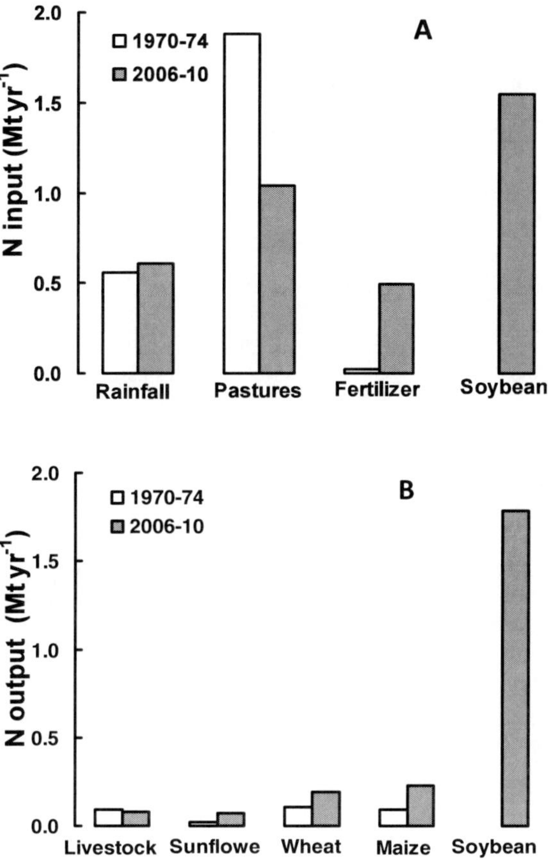

Figure 12. A. Nitrogen inputs estimation to the Pampean agroecosystems by rainfall, seeded pastures, nitrogen fertilizer, and nitrogen fixation by soybean (redrawn from Alvarez et al., 2014). B. Nitrogen output estimation from Pampean agroecosystems by harvested grain of the main produced crops and livestock products (redrawn from Alvarez et al., 2014).

The average gain for the period 1960-2010 was 1.3 Mt N yr⁻¹; equivalent to 0.065 Gt N. This difference is about three times the difference between total nitrogen stock in the soils (0-25 cm) between the samplings during 1960-1980 and 2007-2008. Although this difference is not significant, assuming that the change in the stock of nitrogen 0-25 cm represents 50% of the total change of the profile, the Pampas soils may have won ca. 0.040 Gt N during the agricultural expansion and ca 0.020 Gt N would have been released into the environment. In agroecosystems under annual crops, nitrogen balance has been historically close to neutral (Figure 13). Calculating the regional balance per crop, the difference between inputs and outputs is +12 kg N ha^{-1} yr^{-1} for wheat and varies between -6 and -10 kg N ha^{-1} yr^{-1} for corn, sunflower, and soybean.

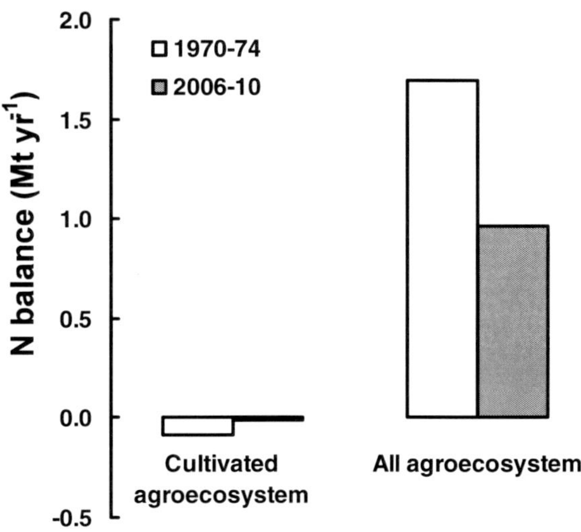

Figure 13. Pampean nitrogen balance accounting for only cultivated agroecosystems and all agroecosystems (redrawn from Alvarez et al., 2014).

EFFECT OF CARBON CHANGES ON SOIL PRODUCTIVITY

The productivity of soils is commonly assessed for by indices including variables with slow temporal change. In the Pampas a model has been developed to estimate soil productivity for wheat using an artificial neural network (De Paepe & Alvarez, 2013). This model uses as inputs soil water holding capacity and organic carbon level, which are stable variables not impacted by management in the short term and it achieved good performance (R^2= 0.75). Both soil variables positively interact defining productivity; the greater the water holding capacity and organic carbon content the greater soil productivity (Figure 14).

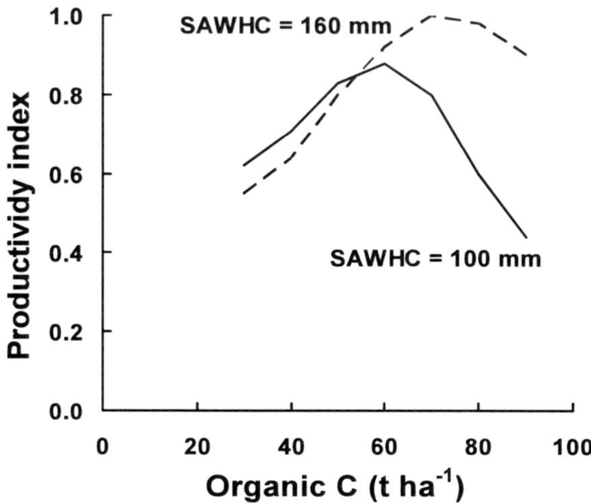

Figure 14. Effect of the interaction between the soil available water holding capacity (0-100 cm) (SWHC) and organic carbon content (0-50 cm), on productivity for wheat in the Pampean Region (redrawn from De Paepe & Alvarez, 2013).

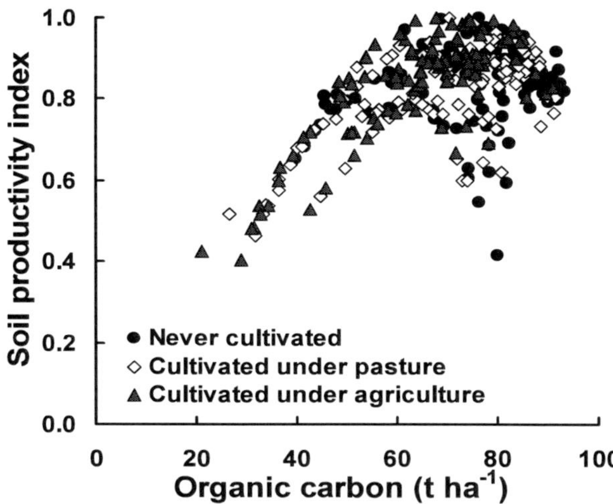

Figure 15. Estimated productivity of never cultivated sites and cultivated sites under the agricultural or pasture phase of crop rotations related to their organic carbon content during the sampling performed in the 2007-2008 period (redrawn from Alvarez et al., 2012b).

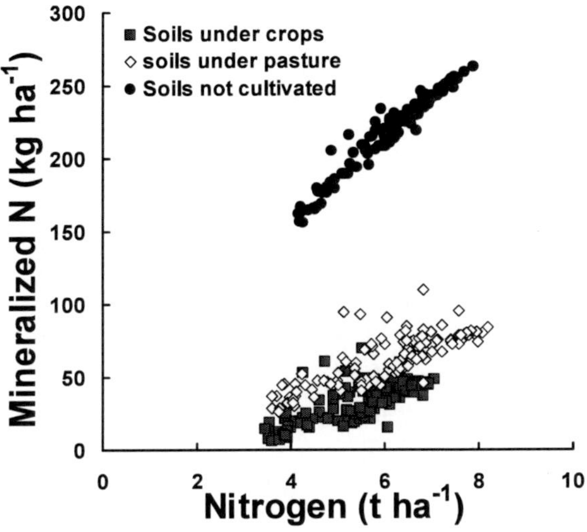

Figure 16. Relationship between soil nitrogen mineralization capacity during the wheat growing cycle estimated with a locally developed model and total nitrogen stock in the 0-25 cm layer under different land uses for sites sampled during the 2007-08 growing cycle (redrawn from Alvarez et al., 2012b).

Using this model, the productivity of the sampled sites in 2007-2008 was estimated (Figure 15). There were no significant differences between estimated productivity rates for uncultivated and cultivated soils, which differed on average only 2%. Cultivation did not result in significant carbon losses at regional scale and this is reflected in the low impact on soil productivity in the Pampas.

It is possible to predict nitrogen mineralization under field conditions during the growing cycle of wheat in the Pampas using a neural network model developed for the soils of the region (Alvarez & Steinbach, 2011) with good performance ($R^2 = 0.78$). Using this model the mineralization capacity of soils sampled in 2007-2008 was assessed (Figure 16). Agricultural use strongly impacted the ability of soils to mineralize nitrogen. In cultivated soils under pasture phase this capacity was 29% of the control soils never cultivated, and in cultivated soils under the agricultural phase this percentage was reduced to 16%. The differences in the ability to mineralize nitrogen between soils in pasture or agriculture rotation phase were significant, doubling the former to the latter. These differences are partly due to the effects of land use on total nitrogen stock but mainly on its resistance to mineralization. Although the total soil nitrogen stock is not greatly affected by

agriculture, labile fractions were strongly reduced, leading to nitrogen fertility losses (Heumann et al., 2003). While nitrogen decreased 20% for agricultural use in the 0-25 cm soil layer, mineralization was reduced about five times. In the West pampean portion mineralization is much lower than in the East because of the lower nitrogen content of soils. The model showed that mineralization was low in many pampean soils indicating limitations to wheat production and the nitrogen fertilization requirement (Alvarez et al., 2012b). This has been the main impact of agriculture on the Pampas soils. Cultivation did not markedly affect the total carbon and nitrogen content of soils but it did reduce drastically their ability to mineralize nitrogen and contribute to crop nutrition.

ASSOCIATION BETWEEN SOIL ORGANIC CARBON CHANGES AND ITS LABILE FRACTIONS

It has long been recognized that labile fractions of organic matter may be useful as indicators of management effects on soil organic pools and possible predictors of future changes in the total content of organic carbon and nitrogen (Cambardella & Elliott, 1992; Haynes, 2000; Janzen et al., 1992). In the Pampas, the effects of management practices on total soil organic carbon and some labile fractions has been evaluated with information of many experiments. The compilation of published results from the region can generate suitable data sets to test whether some labile fractions, like microbial biomass, mineralizable carbon fraction, particulate carbon or carbon in the light fraction, can serve to predict soil organic carbon changes produced by land use. Data from 69 field experiments were integrated. In each experiment the treatment with the lowest soil carbon received a value of 100 and all other treatments in the experiment were calculated relative to this control treatment. The relative change (%) related to the control treatment was calculated. Labile fraction data were processed in the same way, assigning a value of 100 to the same control treatment chosen for organic carbon and calculating the relative change of the other treatments relative to this value. By this procedures 112 paired comparisons of carbon vs. microbial biomass, 116 paired comparison of carbon vs. in vitro carbon mineralized, 162 paired comparison of carbon vs. particulate carbon (soil fraction > 53 μm), and 52 paired comparison of carbon vs. soil light fraction carbon (density < 2 g ml^{-1}) were obtained.

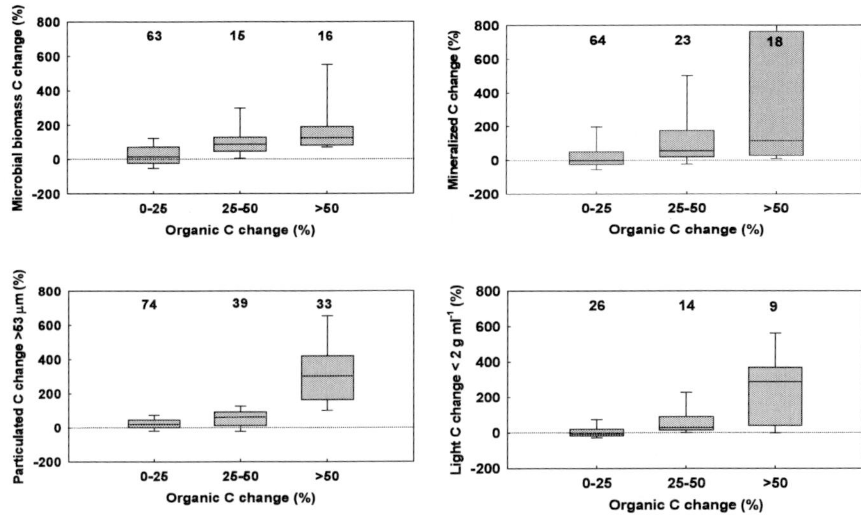

Figure 17. Box plot (percentiles 5, 25, 50, 75 and 95%) of the comparison of the relative variation of four labile fractions against the variation of soil organic carbon in pampean experiments. The value above the box-plots indicates the number of comparisons integrated. Data of microbial biomass were taken from: Echeverria et al. (1992, 1993), Alvarez et al. (1993), Alvarez et al. (1995a), Alvarez et al. (1995b), Alvarez et al. (1995c), Benintende et al. (1995), Costantini et al. (1995), Chagas et al. (1995), Costantini et al. (1996), Alvarez et al. (1998ª), Alvarez et al. (1998b), Abril et al. (2005), Ferreras et al. (2009), Benintende et al. (2012). Data of carbon mineralization were taken from: Alvarez et al. (1993), Chagas et al. (1995), Costantini et al. (1995), Alvarez et al. (1995a), Alvarez et al. (1995b), Alvarez et al. (1995c), Alvarez et al. (1998a), Alvarez et al. (1998b), Palma et al. (2000), Gómez et al. (2001), Abril et al. (2005), Costantini et al. (2006), Ferreras et al. (2009), Alvarez et al. (2011). Data of particulate carbon were taken from: Casanovas et al. (1995), Galantini et al. (2002), Fabrizzi et al. (2003), Eiza et al. (2005), Diovisalvi et al. (2008), Dominguez et al. (2009), Ferreras et al. (2009), Cozzoli et al. (2010), Alvarez et al. (2011), Ciampitti et al. (2011), Divito et al. (2011), Fernandez at al. (2011), Wyngaard et al. (2012). Data of soil light carbon were taken from: Alvarez et al. (1993), Alvarez et al. (1995c), Casanovas et al. (1995), Alvarez et al. (1998a), Alvarez et al. (1998b), Cosentino et al. (2007).

For the four labile fractions analyzed a similar pattern of change in relation to the change rate of organic carbon was observed (Figure 17). When the carbon change between treatments in an experiment was lower than 25%, the labile fractions varied in average with the same magnitude than carbon but in some particular cases they did not vary or even change in contrary direction. Between 20 and 40% of cases showed, labile fractions changes with a

complete opposite behavior to organic carbon change; they decreased when organic carbon increased. When the organic carbon change ranged from 25 and 50% labile fractions, in average, followed these changes. In cases in which organic carbon showed great variations, over 50%, labile fractions changed much more. On average for all pairs of comparisons, the biomass carbon, particulate carbon, and the light fraction carbon changed 4 to 5-fold more than organic carbon, while mineralizable carbon changes were 24-fold greater than carbon changes. Only when carbon changes are very large, greater than 25-50%, labile fractions copied the trend observed for carbon, but in these cases it is easy to detect changes by direct evaluation of organic carbon. These results indicated that the labile fractions analyzed are not suitable tools for predicting organic carbon changes in the pampas.

RELATIONSHIPS BETWEEN SOIL ORGANIC MATTER FRACTIONS, MINERALIZATION INDICES AND PRODUCTIVITY

Nitrogen mineralization tests and some indexes have been proposed in the Pampas as methods to predict soil nitrogen supplying capacity to crops and productivity. In the Semiarid Pampa it has been reported that the level of nitrate-N at planting and barley yield are correlated with the ratio organic matter/clay+silt (Quiroga et al., 2006). In the humid portion of the region it has also been reported that this ratio is useful in predicting nitrogen mineralization during the growing season of wheat (Gonzalez Montaner et al., 1997). Labile fractions, such as the carbon light fraction or the potentially mineralizable nitrogen could explain the yield of wheat in some humid areas (Alvarez et al., 2002).

To test the effect of the organic matter/clay+silt ratio on several crops commonly cultivated in the Semiarid Pampa results from 257 fertilization experiments with wheat, maize, and sunflower were collected (Figure 18). The level of nitrate-N at planting and crop yields achieved by the unfertilized control treatments did not correlate with the mentioned ratio. The attempt to generate multiple regression models that include this variable and other properties of soil and climate, could not detect any possible significant relationships of the ratio and the dependent variables. Apparently, this ratio is not a useful tool for regional nitrate-N or yield prediction in the semiarid portion of the Pampas.

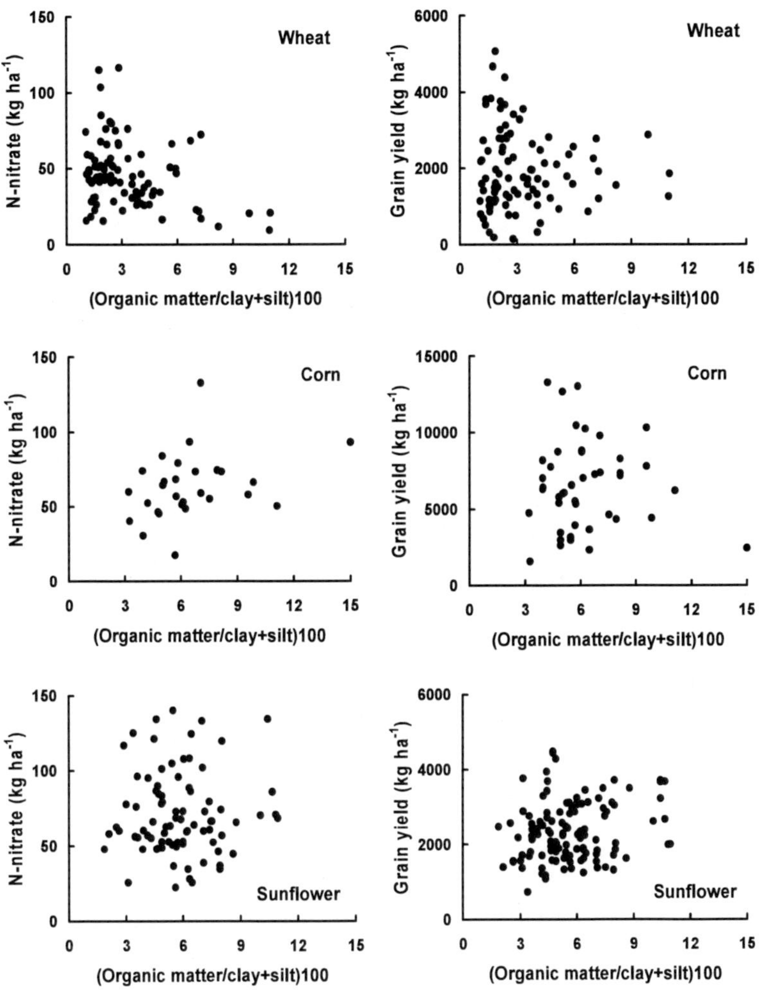

Figure 18. Relationships between of nitrate-N content (0-60 cm depth) or wheat, maize, and sunflower yield in the Semiarid Pampas with the ratio organic matter/clay+silt (0-20 cm depth). Information was obtained from 92 experiments with wheat, 46 with corn, and 119 with sunflower. Plotted results correspond only to unfertilized control treatments. The characteristics of the experimental networks have been described for wheat in Bono et al. (2011) and Romano et al. (2014), for corn in Bono & Alvarez (2012), and for sunflower in Bono & Alvarez (2007).

Integrating information from 69 fertilization experiments conducted in the Humid Pampa for wheat it was established that there are significant relationships between soil organic carbon with the carbon light fraction or

potentially mineralizable nitrogen, although the fits were not high (Figure 19). These labile fractions also failed to improve the prediction of nitrate-N at planting or crop yield compared to the organic carbon prediction (Figure 20).

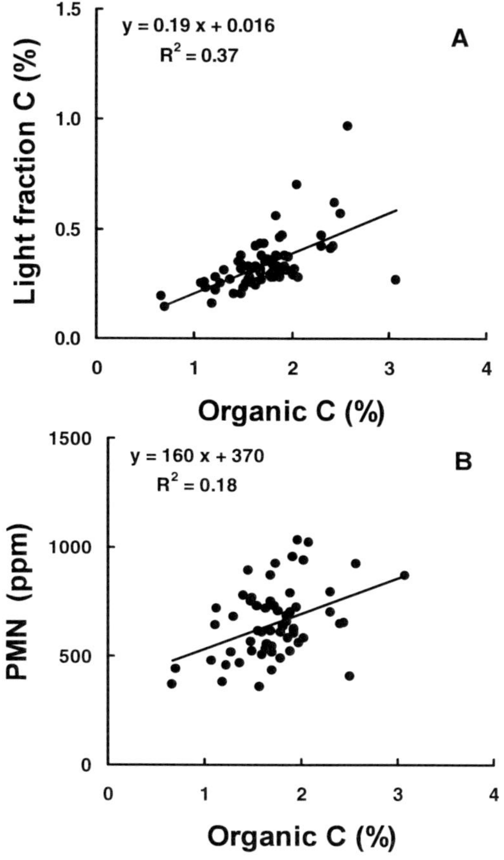

Figure 19. A) relationship between organic carbon in the soil light fraction (density <2 g ml^{-1}) or B) potentially mineralizable nitrogen (PMN) with the organic carbon in the 0-20 cm layer in 69 experiments performed in the Humid Pampa. Only the unfertilized control treatments were plotted (made with data from Barberis et al., 1983 a; b).

Another data set generated from 93 fertilization experiments with wheat and maize in the humid portion of the region showed a weak relationship of organic carbon with light carbon and no significant relationship was found between the results of a short aerobic mineralization test and organic carbon (Figure 21).

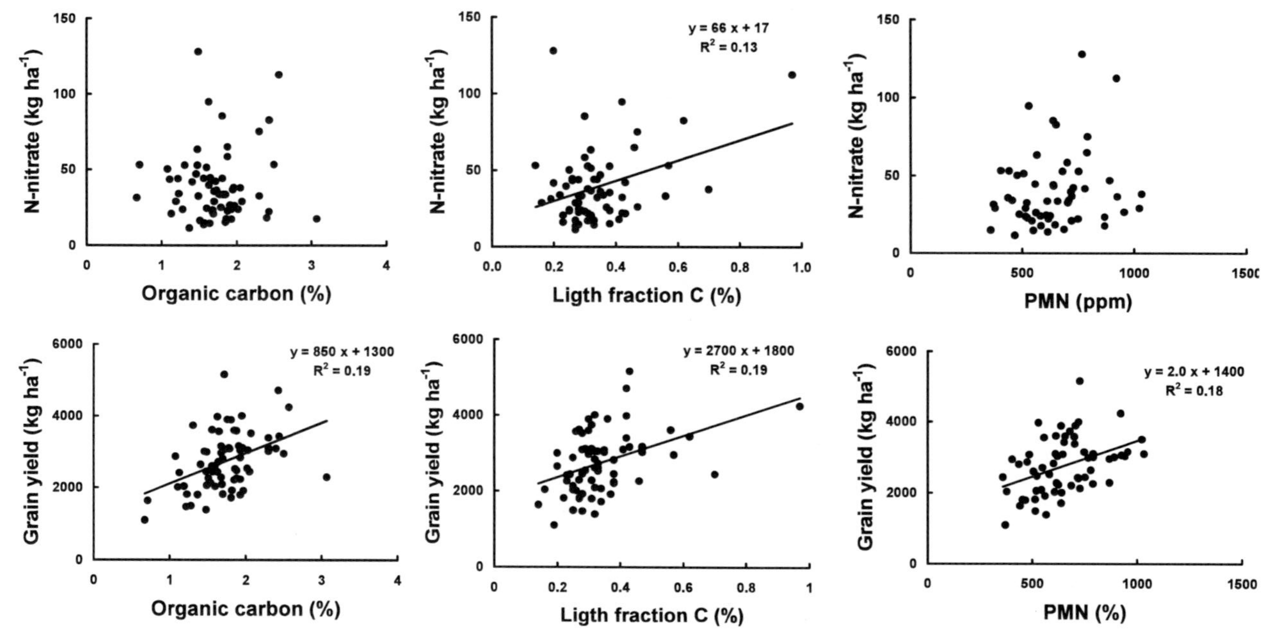

Figure 20. Relationship between nitrate level (0-60 cm) at wheat sowing time or crop yield with organic carbon, carbon in the soil light fraction (density <2 g ml^{-1}) and potentially mineralizable nitrogen (PMN) (all in the 0-20 cm layer) in 69 experiments performed in the Humid Pampa. Only the unfertilized control treatments were plotted (made with data from Barberis et al., 1983 a; b).

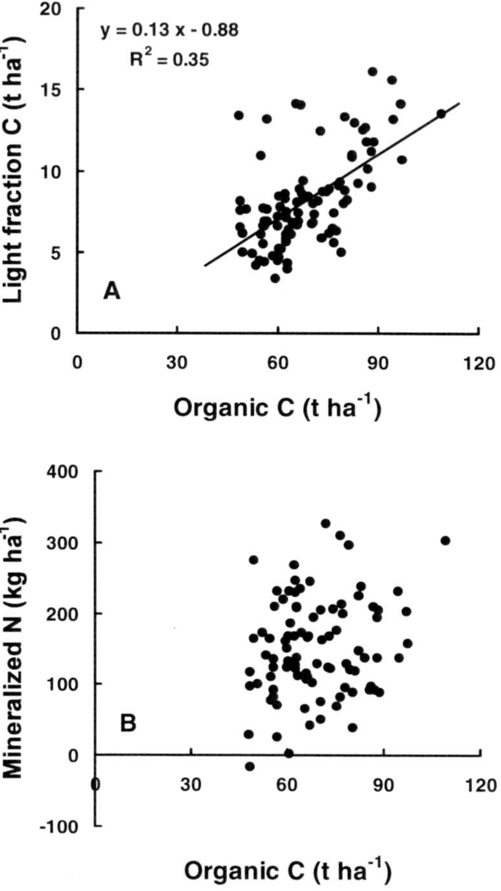

Figure 21. A) relationship between carbon in the soil light fraction (density <2 g ml⁻¹) and B) mineralized nitrogen in a 17-day aerobic test with soil organic carbon content. Data from 93 experiments conducted in the Humid Pampa with wheat and maize described in Alvarez & Steinbach (2011b).

However, the combination of the ratio total N/clay+silt with other soil and climatic variables, used as inputs to an artificial neural network model, allowed obtaining a very good fit for estimating nitrogen field mineralization (Figure 22). The effect of the incubation test on field mineralization, and of some other inputs, is much greater than the impact of total N/clay+silt ratio.

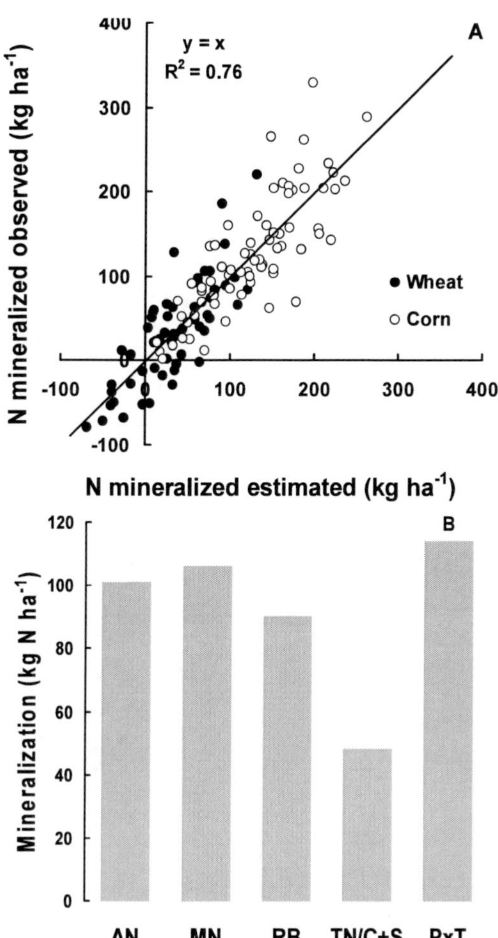

Figure 22. A) performance of an artificial neural network model developed to estimate the apparent nitrogen mineralization at field conditions during the growing cycles of wheat and maize in the Humid Pampa. B) effect of neural network inputs on the apparent mineralization estimated (AN: available mineral N at planting (0-60 cm), NM: N mineralized in aerobic soil test (0-30 cm layer), RB: previous crop residues biomass (0-30 cm), TN/C+S: total N/clay+silt (0-30 cm); PxT. precipitation x temperature along the growing cycle. Both subfigures made with data from Alvarez & Steinbach (2011b).

It has been also reported that a short anaerobic nitrogen mineralization test improved predictability of the nitrogen supplying capacity of soils in another portion of the Humid Pampa (Sainz Rozas et al., 2008). Of the various indices

and labile fractions studied in the Pampas to predict soil productivity or associated factors, the in vitro tests of mineralization have been the most useful methods so far.

CONCLUSION

Agricultural land use had little effect on the total carbon and nitrogen stocks in the Pampean Region. Despite the adoption of soybean as the main crop in recent decades, soils carbon balances are less negative than in the past decades, and this results from the compensation effect of greater carbon inputs accounted for by greater yields. The nitrogen input to agroecosystems by soybean biological fixation offsets the reduction caused by eliminating pastures with legumes. Regional nitrogen balance remains positive in the Pampas. Soil organic carbon regulates productivity and attempts to improve the prediction performance of future changes produced by management using labile fractions changes have not been successful. Only mineralization tests are useful for improving mineralization prediction at field conditions.

ACKNOWLEDGMENTS

This work was granted by UBACYT 20020100617, FONCyT PICT 37164, y CONICET PIP 112-200801-02608.

REFERENCES

AAPRESID. 2013. http://www.aapresid.org.ar/.

Abril, A.; P. Salas; E. Lovera; S. Kopp & N. Casado-Murillo. 2005. Efecto acumulativo de la siembra directa sobre algunas características del suelo en la Región Semiárida Central de la Argentina. *Ciencia del Suelo* 23: 179-188.

Alvarez, R. 2001. Estimation of carbon losses by cultivation from soils of the Argentine Pampa using the Century model. *Soil Use Management* 17, 62-66.

Alvarez, R. 2005. Balance de carbono en suelos de la Pampa Ondulada: efecto de la rotación de cultivos y la fertilización nitrogenada. Actas Simposio

Fertilidad 2005. Nutrición, Producción y Ambiente. Argentina, pág. 61-70.

Alvarez, R; C.R. Alvarez & H.S. Steinbach. 2002. Association between soil organic matter and wheat yield in the Humid Pampa of Argentina. *Comm. Soil Sci. Plant Anal.* 33: 749-757.

Alvarez, C,R,; H.S. Steinbach & R. Alvarez. 2012a. Capítulo 3. El rol de los fertilizantes en la agricultura. In: Alvarez, R., Prystupa, P., Rodríguez, M., Alvarez C.R. 2012. (Eds.). Fertilización de cultivos y pasturas. Diagnóstico y recomendación en la Región Pampeana. Editorial Facultad de Agronomía-UBA, pág. 51-64.

Alvarez, C.R.; A.O. Costantini; A. Bono; M.A. Taboada; F.H. Gutierrez Boem; P.L. Fernandez & P. Prytupa. 2011. Distribution and vertical stratification of carbon and nitrogen in soil under different managements in the pampean region of Argentina. Revista Brasileira de *Ciencia do Solo* 35: 1985-1994.

Alvarez, C.R.; R. Alvarez; M.S. Grigera & R.S. Lavado. 1998a. Associations between organic matter fractions and the active soil microbial biomass. *Soil Biology and Biochemistry* 30: 767-773.

Alvarez; R.; P.E. Daniel; O.J. Santanatoglia & R. Garcia. 1993. Mineralización de carbono en un suelo agrícola: relación entre la disponibilidad del substrato y la biomasa microbiana. *Agrochimica* XXXVII: 655-62.

Alvarez, R.; R.A. Diaz; N. Barbero; O.J. Santanatoglia & L. Blotta. 1995a. Soil organic matter, microbial biomass and CO_2-C production from three tillage systems. *Soil and Tillage Research* 33: 17-28.

Alvarez, R. & Lavado R.S. 1998. Climate, organic matter and clay content relationships in the Pampa and Chaco soils, Argentina. *Geoderma* 83: 127-141.

Alvarez, R.; M.E. Russo; P. Prytupa; J.D. Scheiner & L. Blotta. 1998b. Soil carbon pools under conventional and no tillage systems in the Argentine Rolling Pampa. *Agronomy Journal* 90: 138-143.

Alvarez R, Santanatoglia OJ, Daniel PE, Garcia R. 1995c. Respiration and specific activity of soil microbial, biomass under conventional and reduced tillage. Pesquisa Agropecuaria Brasilia 30: 701-709.

Alvarez, R.; O.J. Santanatoglia & R. Garcia. 1995b. Soil respiration and carbon inputs from crops in a wheat-soyabean rotation under different tillage systems. *Soil Use and Management* 11: 45-50.

Alvarez, R.; H.S. Steinbach & A. Bono. 2011. An artificial neural network approach for predicting soil carbon budget in agroecosystems. *Soil Sci. Soc. Am. J.* 75: 965-975.

Alvarez, R. & H.S. Steinbach. 2011. Modeling apparent nitrogen mineralization under field conditions using regressions an artificial neural networks. *Agron. J.* 103:1159-1168.

Alvarez, R.; H.S. Steinbach & J. De Paepe. 2014. A regional audit of nitrogen fluxes in pampean agroecosystems. *Agric. Ecosys. Environ.* 184: 1-8.

Alvarez, R. & H.S. Steinbach. 2012. Capitulo 3. Efecto del uso agrícola sobre el nivel de materia orgánica. En: Fertilidad de suelos. Caracterización y manejo en la Región Pampeana. Editorial Facultad de Agronomía-UBA, Pág. 201-230.

Alvarez, R.; G. Berhongaray; J. De Paepe: M.R. Mendoza: H.S. Steinbach: C. Caride & R. Cantet 2012b. Productividad, fertilidad y secuestro de carbono en suelos pampeanos: efecto del uso agrícola. Anales de la Academia Nacional de Agronomía y Veterinaria, en prensa, 29 pág.

Anonymous. 1983. Censo general de la Provincia de Buenos Aires. Demográfico Agrícola. Industrial, Comercial. Verificado el 9 de octubre de 1881. Bajo la administración del Dr. Dardo Rocha. Buenos Aires Imprenta de El Diario, 116 San Martín, 118.

Anonymous. 1898. Segundo Censo de la República Argentina, 10 de mayo de 1895. Tomo II. Censos Complementarios. Taller Tipográfico de la Penitenciaria Nacional. 122-132.

Anonymous. 1909. Censo Agropecuario Nacional. La Ganadería y la Agricultura en 1908. Agricultura. Tomo II. Buenos Aires. Taller de publicaciones de la oficina meteorológica Argentina.

Anonymous. 1939. Censo Nacional Agropecuario año 1937. Comisión nacional del Censo Agropecuario Ministerio de Agricultura. República Argentina. Buenos Aires.

Anonymous. 1947. IV Censo General de la Nación. Tomo II. Censo Agropecuario. Presidencia de la nación. Ministerio de Asuntos técnicos. Dirección nacional del Servicio Estadístico. Buenos Aires.

Anonymous. 1964. Censo Nacional Agropecuario. Tomo I. Poder Ejecutivo Nacional. Secretaria de Estado de Hacienda. Dirección Nacional de Estadística y Censo. Buenos Aires.

Barberis, L.; A. Nervi; H. del Campo; S. Urricarriet; S. Sierra; P. Daniel; M. Vazquez & D. Zourarakis. 1983a. Análisis de la respuesta de trigo a la fertilización nitrogenada en la Pampa Ondulada y su predicción. *Ciencia del Suelo* 1: 51-64.

Barberis, L.; A. Nervi; A. Sfeir; P. Daniel; S. Urricarriet; M. Vazquez & D. Zourarakis. 1983b. Análisis de la respuesta de trigo a la fertilización nitrogenada en la Pampa Arenosa y su predicción. *Rev. Facultad de Agronomía-UBA* 4: 325-334.

Benintende, S.; M. Benintende; D. David; M. Sterren; & M. Saluzzio. 2012. Caracterización de indicadores biológicos y bioquímicos en Alfisoles, Molisoles y Vertisoles de Entre Ríos. *Ciencia del Suelo* 30: 23-29.

Benintende, M.; O. Borgetto & S. Benintende. 1995. Mineralización de nitrógeno y contenido de biomasa microbiana en diferentes sistemas de laboreo. *Ciencia del Suelo* 13: 98-100.

Berhongaray, G.; R. Alvarez; J. De Paepe; C. Caride & R. Cantet. 2013. Land use effects on soil carbon in the Argentine Pampas. *Geoderma* 192: 97-110.

Bono, A. & R. Alvarez. Recomendaciones de fertilización para girasol en las Regiones Semiárida y Subhúmeda Pampeanas. Informaciones Agronómicas 35: 1-5 2007.

Bono, A. & R. Alvarez. 2012. Fertilización de maíz en la Región Semiárida y Subhúmeda Pampeana. XXIII Congreso Argentino de la Ciencia del Suelo, Actas 4 pág.

Bono, A.; J. De Paepe & R. Alvarez. 2011. In-season wheat yield prediction in the Semiarid Pampa of Argentina using artificial neural networks. Progress in Food Science and Technology. Vol. 1. Pág. 133-149. Ed. A.J. Greco, Nova Science Publishers, Inc.

Cambardella, C.A. & E.T. Elliott. 1992. Particulate soil organic matter changes across a grassland cultivation sequence. *Soil Sci. Soc. Am. J.* 56: *777-783.*

Caride, C.; G. Piñeiro & J.M. Paruelo. 2012. How does agricultural management modify ecosystem services in the Argentine Pampas? The effects on soil C dynamics. *Agric. Ecosys. and Environ.* 154: 23-33.

Casanovas, E.M.; G.A. Suddert & H.E. Echeverria. 1995. Materia orgánica del suelo bajo rotaciones de cultivos. I Contenido total y de distintas fracciones. *Ciencia del Suelo* 13: 16-20.

CIA World Factbook. 2008. (available in: www.cia.gov).

Chagas, C.I.; O.J. Santanatoglia; M.G. Castiglioni & H.J. Marelli. 1995. Tillage and cropping effects on selected properties of an Argiudoll in Argentina. *Communications in Soil Science and Plant Analysis* 26: 643-655.

Ciampitti, I.A.; F.O. Garcia; L. Picone & G. Rubio. 2011. Soil carbon and phosphorus pools in field crop rotations in Pampean Soils of Argentina. *Soil Science Society of America Journal* 75: 616-625.

Cosentino, D.J.; M.E. Conti & L. Giuffre. 2007. Forty years of soil degradation in Vertic Argiudolls in Entre Ríos province, Argentina. *Ciencia del Suelo* 25: 133-138.

Costantini, A.; D. Cosentino D & A. Segat. 1996. Influence of tillage systems on biological properties of a Typic Argiudoll soil under continous maize in central Argentina. *Soil and Tillage Research* 38: 265-271.

Costantini, A.; H. De Poli; C. Galarza; R. Pereyra Rossiello & R. Romaniuk. 2006. Total and mineralizable soil carbon as affected by tillage in the Argentinean Pampas. *Soil and Tillage Research* 88: 274-278.

Costantini, A.; A. Segat & D. Cosentino. 1995. The effect of different soil management procedures on carbon cycle components in an Entic Hapludoll. Communications in Soil Science and Plant Analysis 26: 2761-2767.

Cozzoli, M.V.; N. Fioriti; G.A. Studdert; G.F. Dominguez & M.J. Eiza. 2010. Nitrógeno liberado por incubación anaeróbica en macro y micro agregados bajo distintos sistemas de cultivo. *Ciencia del Suelo* 28: 155-167.

De Paepe, J.L. & R. Alvarez. 2010. Uso de datos de mapas de suelos para determinar el stock de carbono orgánico y otras variables edáficas a nivel regional. XXII Congreso Argentino de la Ciencia del Suelo, Actas en CD, 4 pág.

De Paepe, J.L. & R. Alvarez. 2013. Developments of a soil productivity index using an artificial neural network approach. *Agron. J.* 105: 1803-1813.

Davidson, E.A. & I.L. Ackerman. 1993. Changes on soil carbon inventories following cultivation of previously untilled soils. *Biogeochemistry* 20: 161-193.

Di Ciocco, C.; E. Penón; C. Coviella; S. López; M. Díaz-Zorita; F. Momo & R. Alvarez. 2012. Nitrogen fixation by soybean in the Pampas: relationship between yield and soil nitrogen balance. *Agrochimica* 40: 305-313.

Divito, G.A.; H.R. Sainz Rozas; H.E. Echeverria; G.A. Studdert & N. Wyngaard N. 2011. Long term nitrogen fertilization: Soil property changes in an Argentinean Pampas. *Soil and Tillage Research* 114: 117-126.

Diovisalvi, N.V.; G.A. Studdert; G.F. Dominguez & M.J. Eiza. 2008. Fracciones de carbono y nitrógeno anaerobico bajo agricultura continua con dos sistemas de labranza. *Ciencia del Suelo* 26: 1-11.

Dominguez, G.F.; N.D. Diovisalvi; G.A. Studdert & M.G. Monterubbianedi. 2009. Soil organic C and N fractions under continuous cropping with contrasting tillage systems on mollisols of the southeastern Pampas. *Soil and Tillage Research* 102: 93-100.

Echeverria, H.; R. Bergonzi & J. Ferrari. 1992-93. Carbono y nitrógeno en la biomasa microbial de suelos del Sudeste Bonaerense. *Ciencia del Suelo* 10-11: 36-41.

Eickhout, B.; A.F. Bouwman & H. van Zeijts. 2006. The role of nitrogen in world food production and environmental sustainability. *Agric. Ecosys. Environ.* 116, 4-14.

Eiza, M.J.; N. Fioriti; G.A. Studdert & H.E. Echeverria. 2005. Fracciones de carbono orgánico en la capa arable: efecto de los sistemas de cultivo y de la fertilización nitrogenada. *Ciencia del suelo* 23: 59-67.

Fabrizzi, K.P.; A. Morón & F.O. Garcia. 2003. Soil carbon and nitrogen organic fractions in degradated vs non degradated Molisols in Argentina. *Soil Science Society of America Journal* 67: 1831-1841.

Fernandez, P.L.;, C.R. Alvarez & M.A. Taboada. 2011. Assesment of topsoil properties in integrated crop-livestock and continous cropping systems under zero tillage. *Soil Research* 49: 143-151.

Ferreras, L.; S. Toresani; B. Bonel; E. Fernandez; S. Bacigaluppo, V. Faggioli & C. Beltrán. 2009. Parámetros químicos y biológicos como indicadores de calidad del suelo en diferentes manejos. *Ciencia del Suelo* 27: 103-114.

Follett, R.F. & B.A. Stewart BA. 1985. (Eds.). Soil erosion and crop productivity. Amer. Soc. Agron. INC. Madison, USA, 533 pág.

Galantini J. 2005. Calidad y dinámica de las fracciones orgánicas en sistemas naturales y cultivados. Proceedings Jornadas Materia Orgánica y Sustancias Húmicas. Argentina. Pp. 6.

Galantini, J.A.; R.A. Rosell; G. Brunetti & N. Senesi. 2002. Dinámica y calidad de las fracciones orgánicas de un Haplustol durante la rotación trigo-leguminosas. *Ciencia del suelo* 20: 17-26.

Galloway, J.N.; F.J. Dentener; D.G. Capone; E.W. Boyer; R.W. Howarth; S.P. Seitzinger; G.P. Asner; C.C. Cleveland; P.A. Green; E.A. Holland; D.M. Karl; A.F. Michaels; J.H. Portner; A.R. Townsend & C.J. Vörösmarty. 2004. Nitrogen cycles: past, present, and future. *Biogeochemistry* 70, 153-226.

Gomez, E.; L. Ferreras; S. Toresani; A. Ausilio & V. Bisaro. 2001. Changes in some soil properties in a Vertic Argiudoll under short-term conservation tillage. *Soil and Tillage Research* 61: 179-186.

González Montaner, J.H.; G.A. Maddonni & M.R. DiNapoli. 1997. Modelling grain yield and grain yield response to nitrogen in spring wheat crops in the Argentinean Southern Pampa. *Field Crops Res.* 51: 241-252.

Guo, L,B, & M, Gifford. 2002. Soil carbon stocks and land use change: a meta analysis. *Glob. Change Biol.* 8: 345-360.

Hall, A.J.; Rebella, C.; Guersa, C. & Culot, J. 1992. Field-crop system of the Pampas. In: Pearson CJ, editor. Field crop ecosystems. Elsevier, Amsterdam.

Heumann, S.; J. Bóttcher & G. Springob. 2003. Pedotransfer functions for the pool size of slowly mineralizable N in sandy arable soils. *J. Plant. Nutr. Soil Sci.* 166: 308-318

Haynes, R.J. 2000. Labile organic matter as an indicator of organic matter quality in arable and pastoral soils in New Zealand. *Soil Biology and Biochemistry* 32: 211-219.

Houghton, R,A,; J.E. Hobbie; J.M. Melillo; B. Moore;, B.J. Peterson; G.R. Shaver & G.M. Woodwell. 1983. Changes in the carbon content of terrestrial biota and soils between 1860 and 1980: a net flux release of CO_2 to the atmosphere. *Ecol. Monograph.* 53: 235-262.

Howarth, R.W.; E.W. Boyer; W.J. Pabich & J.N. Galloway. 2002. Nitrogen use in the United States from 1961-2000 and potential future trends. *Ambio*, 31, 88-96.

INDEC. 1969. Censo Nacional Agropecuario 1969. Datos del Relevamiento agrícola.

INDEC. 1988. Censo Nacional Agropecuario 1988. Censo Nacional Agropecuario 1988. Resultados Generales. Provincia de Buenos Aires 5.

INDEC. 1988. Censo Nacional Agropecuario 1988. Censo Nacional Agropecuario 1988. Resultados Generales. Provincia de La Pampa 8.

INDEC. 1988. Censo Nacional Agropecuario 1988. Censo Nacional Agropecuario 1988. Resultados Generales. Provincia de Córdoba 6.

INDEC. 1988. Censo Nacional Agropecuario 1988. Censo Nacional Agropecuario 1988. Resultados Generales. Provincia de Santa Fe 4.

INDEC. 1988. Censo Nacional Agropecuario 1988. Censo Nacional Agropecuario 1988. Resultados Generales. Provincia de Entre Ríos 7.

INDEC. 1995a. Encuesta Nacional Agropecuaria 93. Región Pampeana por zonas agroestadística. Republica Argentina. Ministerio de Economía y obras y servicios Públicos. Secretaría de Programación Económica INDEC.

INDEC. 1995b. Encuesta Nacional Agropecuaria 1994. Resultados generales. Vol. 1. Republica Argentina. Ministerio de Economía y obras y servicios Públicos. Secretaria de Programación Económica INDEC.

INDEC. 1995c. Encuesta Nacional Agropecuaria 1994. Resultados generales. Vol. 2. Republica Argentina. Ministerio de Economía y obras y servicios Públicos. Secretaria de Programación Económica INDEC.

INDEC. 1994. Encuesta Nacional Agropecuaria 1994. Resultados generales. Vol. 2. Republica Argentina. Ministerio de Economía y obras y servicios Públicos. Secretaria de Programación Económica INDEC.

INDEC. 2002a. Censo Nacional Agropecuario 2002.

INDEC. 2002b. *Censo nacional* de población, hogares y viviendas. (available at http://www.indec.gov.ar/agropecuario/cna.asp (10/12/1912).

Janzen, H.H.; C.A. Campbell; S.A. Brandt; G.P. Lafond & L. Townley-Smith. 1992. Light-Fraction Organic Matter in Soils from Long-Term Crop Rotations., *Soil Sci. Soc. Am. J.* 56: *1799-1806.*

Lauenroth, W,K.; I.C. Burke & J.M. Paruelo. 2000. Patterns of production and precipitation-use efficency of winter wheat and native grasslands in the central great plains of the united states. *Ecosystems* 3: 334-351.

Magrin, G.O., M.I. Travasso & G.R. Rodríguez. 2005. Changes in climate and crop production during the 20TH Century in Argentina. *Climatic Change* 72: 229-249.

McLauchlan, K. 2006. The nature and longevity of agricultural impacts on soil carbon and nutrients: a review. *Ecosystems* 9: 1364-1382.

Mendoza, M.R.; G. Berhongaray & R. Alvarez. 2012 a. Organic nitrogen stock in pampean soil profiles. XXIII Congreso Argentino de la Ciencia del Suelo. Actas 4 pág.

Mendoza, M.R.; G. Berhongaray & R. Alvarez. 2012 b. Stratification of the organic nitrogen stock in pampean soil profiles as a function of vegetation type and land use. Actas 19th ISTRO Conference, 1 pág., Uruguay.

Meersmans, J.; B. Van Wesemael; F. De Ridder & M. Van Molle. 2009. Modelling the three-dimensional spatial distribution of soil organic carbon (SOC) at regional scale (Flanders, Belgium). *Geoderma* 152: 43-52.

MinAgri. 2013. Series y estadísticas agrícolas. (disponible en: www.minagri.gob.ar)

Mishra, U.; R. Lal; B. Slater; F. Calhoun; D. Liu & M. Van Meirvenne. 2009. Predicting soil organic carbon stock using profile depth distribution functions and ordinary kriging. *Soil Sci. Soc. Am. J.* 73: 614-621.

Moreno, F.; J.M. Murillo; F. Pelegrín & I.F. Girón. 2006. Long-term impact of conservation tillage on stratification ratio of soil organic carbon and loss of total and active CaCO₃. *Soil Till. Res.* 85: 86-93.

OECD. 2001. OECD national soil surface nitrogen balances. Available at: www.oecd.org/arg/env/indicators.htm.

Oenema, O.; H. Kross & W. de Vries. 2003. Approaches and uncertainties in nutrient budgets: implications for nutrient management and environmental policies. *Eur. J. Agron.* 20, 3-16.

Palma, R.M.; N.M. Arrigo; M.I. Saubidet & M.I. Conti. 2000. Chemical and biochemical properties as potential indicators of disturbances. *Biology and Fertility of Soils* 32: 381-384.

Pan, G.; P. Smith & W. Pan. 2009. The role of soil organic matter in maintaining the productivity and yield stability of cereals in China. *Agric. Ecosys. Environm.* 129: 344-348.

Panten, K.; J. Rogasik; F. Godlinski; U. Funder; J. Greef & E. Schung. 2009. Gross soil surface nutrient balances: the OECD approach implemented under German conditions. *Agric. Forest Res.* 59, 19-28.

Quiroga, A.; D. Funaro; E. Noellemeyer & N. Peinemann. 2006. Barley yield response to oil organci matter and texture in the Pampas of Argentina. *Soil Till. Res.* 90: 63-68.

Romano, N. F; R. Alvarez & A. Bono. 2014. Mineralización de nitrógeno en suelos de la Región Semiárida Pampeana. Actas XXIII Congreso Argentino de la Ciencia del Suelo, 6 pág.

Sainz Rozas, H.; P.A. Calviño, H.E. Ehceverría; P.A. Barbieri & M. Redolatti. 2008. Contribution of anaerobically mineralizad nitrógeno to the reliable of planning or presidedress soil nitrogen test in maize. *Agron. J.* 100: 1020-1025.

Satorre, E.H. & G.A. Slafer. 1999. Wheat Production systems of the Pampas. In Wheat. Ecolocy and physiology of yield determination. E.M. Satorre and G.A. Slafer Eds. The Haworth Press, Inc. New York. Pp.333-348.

Spiertz, J.H.J. 2010. Nitrogen, sustainable agriculture and food security: a review. *Agron. Sustain. Dev.* 30, 43-55.

Steinbach, H. & R. Alvarez. 2006. Changes in soil organic carbon contents and nitrous oxide emissions alter introduction of no till in pampean agroecosystems. *J. Environm. Quality* 35: 3-13.

Teruggi, M.E. 1957. The nature and origin of argentine loess. *J. Sed. Petrol.* 27, 322-332.

Viglizzo, E.F.; F.C. Frank; L.V. Carreño; E.G. Jobbágy; H. Pereyra; J. Clatt; D. Pincén & F. Ricard F. 2011. Ecological and environmental footprint of

50 years of agricultural expansion in Argentina. *Global Change Biol.* 17: 959-973.

Viglizzo, E,F.; F. Létora; A.J. Pordomingo; J.N. Bernardos; Z.E. Roberto & H. Del Valle. 2001. Ecological lessons and applications from one Century of low external-input farming in the pampas of Argentina. *Agric. Ecosys. Environ.* 83: 64-81.

Viglizzo, E.F.; Z.E. Roberto; Z.E. Lértora; E. López Gay & J. Bernardos. 1997. Climate and land-use change in field-crop ecosystems of Argentina. *Agric. Ecosys. Environm.* 66: 61-70.

Wu, H.; Z. Guo; Q. Gao & C. Peng. 2009. Distribution of soil inorganic carbon storage and its changes due to agricultural land use activity in China. *Agric. Ecosyst. Environ.* 129: 413-421.

Wyngaard, N.; H.E. Echeverria; H.R. Sainz Rozas & G.A. Divito. 2012. Fertilization and tillage effect on soil properties and maize yield in a Southern Pampas Argiudoll. *Soil and Tillage Research* 119: 22-30.

In: Soil Carbon
Editor: Aquila Margit

ISBN: 978-1-63117-438-4
© 2014 Nova Science Publishers, Inc.

Chapter 3

STATE OF SOIL CARBON IN HUNGARIAN SITES: LOSS, POOL AND MANAGEMENT

C. Centeri[1,], B. Szabó[1], G. Jakab[2,†], J. Kovács[2], B. Madarász[2,‡], J. Szabó[2], A. Tóth[2], G. Gelencsér[3,§], Z. Szalai[2,††] and M. Vona[4,‖]*

[1]Szent Istvan University, Faculty of Environmental and Agricultural Sciences, Institute of Environmental and Landscape Management, Department of Nature Conservation and Landscape Ecology, Hungary
[2]Hungarian Academy of Sciences, Research Centre for Astronomy and Earth Sciences, Geographical Institute, Budapest, Hungary

---

[*] Cs. Centeri: Szent Istvan University, Faculty of Environmental and Agricultural Sciences, Institute of Environmental and Landscape Management, Department of Nature Conservation and Landscape Ecology, 2100 Gödöllő, Páter K. u. 1., Hungary. E-mail: Centeri.Csaba@kti.szie.hu.
[†] G. Jakab: Hungarian Academy of Sciences, Research Centre for Astronomy and Earth Sciences, Geographical Institute, 1112 Budapest, Budaörsi út 45., Hungary. E-mail: jakab.gergely @csfk.mta.hu.
[‡] B. Madarász: Hungarian Academy of Sciences, Research Centre for Astronomy and Earth Sciences, Geographical Institute, 1112 Budapest, Budaörsi út 45., Hungary. E-mail: madaraszb@sparc.core.hu.
[§] G. Gelencsér: Vox Vallis Development Association, 7285 Törökkoppány, Kossuth L. u. 66., Hungary.
[††] Z. Szalai: Hungarian Academy of Sciences, Research Centre for Astronomy and Earth Sciences, Geographical Institute, 1112 Budapest, Budaörsi út 45., Hungary. E-mail: szalaiz@mtafki.hu.
[‖] M. Vona: Abiotika Bt., H-1122 Budapest, Városmajor u. 10., Hungary. E-mail: vonamarton@ gmail.com.

[3] Vox Vallis Development Association, Kossuth L. u., Hungary
[4] Abiotika Bt., Budapest, Városmajor u. 10., Hungary

ABSTRACT

In the 19th century, in the Austrian-Hungarian Monarchy, Hungary has always been referred to as the food chamber. The reason for this is the huge areas having high agricultural potentials. Nowadays, 2/3 of the country's present area is suitable for agricultural production and 50 % has good potential for arable farming. The present study intend to present the state of soil carbon on some arable lands where it is lost due to unfavorable management and to present its state where it is protected in undisturbed, natural or semi-natural spots. Soil carbon was measured on various sloping areas of Hungary. Shallow drillings, up to 3.6 meters have been made where soil organic matter has been measured in order to present the amount of buried – and thus lost for agricultural production etc. – material. The above methodologies have been applied to areas with different agricultural management types. This allow us to show the differences among the effects of land use on soil carbon loss and to formulate suggestions for soil carbon protecting and still, economically viable land use forms which might be called optimal. There are experimental areas where different land use forms and land management have been examined, providing information on soil carbon, too.

INTRODUCTION

Since soils are the third largest global carbon pool, their carbon-related processes have primary importance. Soil organic carbon (SOC) research has high priority in recent years as it can store most of the human produced, non-favorable extra carbon in the atmosphere (Merino et al., 2004). Numerous authors estimate the total amount of global soil carbon between 700 and 2,946 × 1015 g with several intermediate estimates (Post et al., 1982).

As the loss of SOC is related to human activity (Nagy et al., 2012), investigation of land use change and the most appropriate approximation of former soils, biota, climate etc. widened the number of related research (Barczi et al., 2006). The newest of these is the soil-phytolith examination (Pető, 2013) that can add valuable input for the investigation the past of SOC on a certain area.

The "other side" of carbon management is when there is no mentionable human disturbance concerning soil use. These are the nature conservation areas (in general) from biosphere reserves to national parks, from nature reserves to natural parks (Fonseca et al., 2012; Fonseca and de Figueiredo, 2012) etc.

There are areas with human exploitation but considered as some of the most natural areas (e.g. meadows (Malatinszky et al., 2013), forests etc.) where soil is far less disturbed than those on arable fields but more disturbed than nature conservation areas.

Forested areas are one of the best examples of these types and, as timber and other wood production has high importance here, effects of species on soil production is severely investigated (Fonseca and de Figueiredo, 2010; Merino et al., 2008).

SOIL CULTIVATION AND SOIL ORGANIC CARBON

Cultivation of soil, one of the main natural resources of the Earth, became more and more mechanic and intensive in the last 50–70 years. Perceptible negative effects of this process on the environment became evident by now. Soil cultivation, as a physical intervention into soil's natural state, has chemical and biological consequences as well.

While it may act positively concerning productivity, its injurious effects are obvious: reduction of soil organic matter (SOM), enhancing erosion risk, soil compaction, destroying soil structure, mixing up infertile soil layers, pollution of surface and subsurface waters, reducing biodiversity etc.

This degradation of the land resources has caused crop yield and factor productivities to decline and has forced farmers and stakeholders to search for a new, ecologically sustainable and in the same time profitable technology.

As alternatives of conventional agricultural technologies, new, environmental friendly conservation farming methods appeared in the 70's and 80's, with the aim of reducing environmental pressure, soil erosion and costs of cultivation, maintaining soil moisture and increasing SOM. Recognizing its benefits, Conservation Agriculture (CA) conquers more and more ground by this time not only in North and South America, but also in Asia, Africa and with some delay in Europe as well.

In past years number of scientific papers reported about the advantages of CA (Andarski et al. 1985; Franzluebbers 2002; Lal et al. 1999; Birkás et al, 2004; Holland 2004; Li et al. 2011; Kassam et al. 2012 etc.). Despite its

increased propagation due to its positive environmental and economic effects, conservation agriculture is still not current in Hungary.

In Hungary multiple examinations have been carried out for already ten years to compare conventional and conservation agricultural methods from different point of view.

Erosion measurements, ecological investigations (weeds, birds, earthworms, microorganisms etc.) were studied and economic analysis was performed on a study area (Kertész et al. 2007; Bádonyi et al. 2008; Kertész et al. 2010; Madarász et al. 2011).

Which are the main differences between widely applied conventional and conservation agriculture? The first one essentially aims increasing soil water holding capacity and reducing weeds by eliminating soil packing, applying ploughing to loose (mix and turn over) topsoil. In consequence of the applied instrument structure, soil, as habitat, is considered as subsidiary. The system is characterized by the dominancy of mechanical interventions, which in most cases significantly change and disturb natural soil life, thus forcing it to regeneration (retarded functioning) for some time. CA technologies strive to produce in such a way that conserves habitats naturally formed by soil flora and fauna. Conservation farming minimize soil's mechanical dusting and preserves good aggregate structure. It keeps soil alive and effective, by not forcing it to retarded functioning after each elemental operation. In the cases of shallow- or no-tillage, significant part of plant residues remain on the top, which protect soil surface and increase SOM content.

SOIL WATER EROSION VERSUS SOIL ORGANIC CARBON LOSS AND POOL

Recent data suggest that soil erosion is one of the most effective factors in soil organic carbon (SOC) dynamics (Szatmári and Barta, 2013, Skianis et al., 2008, Szilassi et al., 2006). Soil erosion can cause noteworthy redistribution in soil nutrients (e.g. potassium, phosphorus) (Farsang et al., 2012) and SOC even in grass covered hillslopes (Nie et al., 2013). The most widely known SOC loss from soils is triggered by the turn of natural vegetation to arable land. The volume of carbon mineralization is up to 30% (Gregorich et al., 1998). After the first few years the general SOC decrease becomes a spatially patterned, relief and erosion controlled phenomenon and on the lower parts SOC content enlargement can occur. The detached SOC can be deposited on

the lower parts of the landscape, or it can enter the streams and it can be mineralized (Olson et al. 2012). Although SOC content increase in subsoil (below 1m) can be an effective tool for carbon sequestration (Lorenz and Lal, 2005) it generally has less favorable reasons like soil erosion. There are still ambiguities on the role of buried SOC.

Some authors present data on carbon sequestration in the subsoil (Nadeu et al., 2013) while others report results on carbon mineralization (Jacinthe et al., 2004).

Continuous intensive tillage increases soil erosion derived nutrient dynamics even on declivous slopes (Abbaszadeh Afshar et al., 2010) however; the redistribution is dominated by water erosion while the contribution of tillage erosion is subordinate. SOC is strongly related to the finest particles which are especially vulnerable to soil erosion delivery, accordingly the clay and organic component is generally overrepresented in soil loss (Farsang et al., 2012, Wang et al., 2010).

SOC is manifested as organic matter (OM) which has a quite heterogeneous structure with varying carbon content and stability. In general the molecular weight versus stability and mobility are in inverse ratio, however, environmental conditions such as acidity and cation concentration strongly determine the recent values (Tombácz et al., 1996).

Accordingly some components of organic matter are more vulnerable to delivery by soil erosion than others, and some components are easily mineralized during the deposition process while others could remain stable for a long time.

There are quite a few results presented concerning selective organic matter erosion although the process is fundamental from the atmospheric carbon mitigation point of view (Lal, 2005).

Most of the studies concentrate on SOC but on calciferous parent material the role of inorganic carbon could be important as well. The solution, migration and leaching of $CaCO_3$ can cause several times higher carbon redistribution in the soil than SOC.

The aims of this paper are:

- to analyze the effect of soil tillage on soil carbon,
- to compare the results of carbon redistribution measurements on arable field with special emphasis on organic matter forms,
- to compare the soil organic matter content of the upper and the lower third of the slopes to prove that water erosion relocate soil organic matter,

    –   to prove the previous aim, analyze shallow drillings to prove that there is huge amount of sedimentation at the bottom of the slope.

MATERIALS AND METHODS

Effects of Soil Tillage on Soil Carbon, Zala County, Hungary

Our 2 hectares research area is situated in Szentgyörgyvár, SW Hungary, The slope angle is between 9 and 10%. Four experimental plots of 24x50 m size were established for soil erosion investigations: two conventional tillage plots (CV) and two conservation tillage plots (CS – non-inversion shallow tillage) in 2003. Recent investigations were done in 2005 and 2010. In 2004 (–2005) winter wheat and in 2010 maize were planted on the experimental plots. The only difference between the two types of plots was the tillage technique. The same amount of fertilizer was applied, and the tillage operations following the contour lines. On the CV plots the autumn deep ploughing reached the depth of 25–30 cm. On the CS plots 8–10 cm deep disking was carried out and the surface coverage by plant residues reached 30%. The number of rounds was reduced by the application of combined machines.

Samples were collected from 3 soil layers along a catena, at 3–3 sites at the same time from both tillage types. Undisturbed soil samples were collected from 4 layers. Sampling occurred in the spring of 2005 and in the spring and autumn of 2010. The soil permeability tests were done simultaneously with sample collection. Bulk density, water infiltration capacity and hydraulic conductivity measurements were carried out on undisturbed samples of known volume (Buzás, 1993). Organic carbon content was determined by an NDIR spectrometer.

Soil Organic Carbon Research at the Ceglédbercel Site

The investigated area is located at Ceglédbercel SE from Budapest, Hungary. It is an intensively tilled arable field on sandy loess parent material.

The soil cover varies among the differently eroded and deposited types of Cambisol, Regosol, and Calcisol. The ridge and the upper third of the slope is an orchard with a very shallow solum. The slope steepness of the lower part changes between 5 and 17 %. The investigated part forms a valley with parent

material spots on the surface at the steepest points and 3 m deep sediments on the bottom. The elevation is 150 m a.s.l., mean temperature is 10.8 °C while the annual precipitation is around 600 mm (Dövényi, 2011).

For soil sample collection borings were constructed. Edelman augers were used to collect the samples from the lower horizons (Figure 1).

Figure 1. Sample collection from shallow drillings at the bottom of the catena, Ceglédbercel, Hungary, 2013.

Each boring was deepened to reach the parent material. Altogether 46 borings along downslope catenas were carried out on the summer of 2013. Samples were taken from different color and texture layers at varying intervals

from the surface to the parent material. In this study four soil profiles (along catena D) will be compared (Figure 2).

SOC content and total bound nitrogen (TNb) content were determined by Tekhmar-Dohrmann Apollo NDIR spectroscopy. $CaCO_3$ content was also analyzed by the Scheibler method. Organic matter compounds were compared on the basis of E_4/E_6, E_2/E_3 method (Tan, 2003) and the presence of aromatic factors.

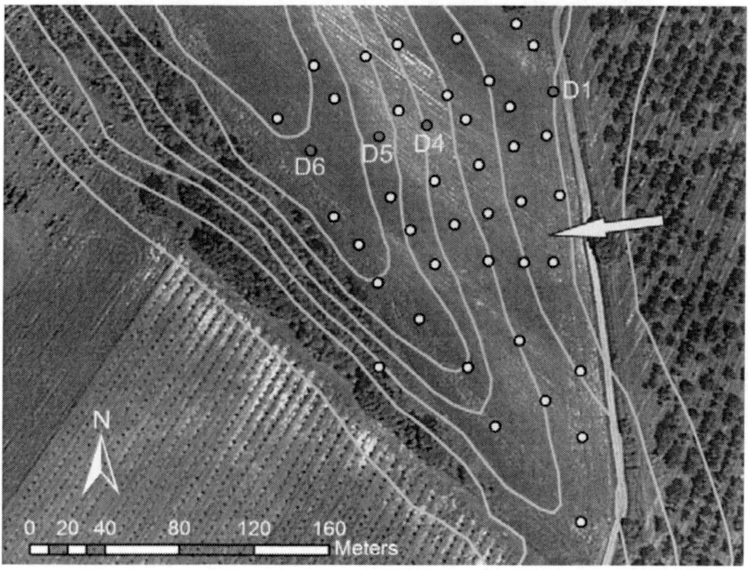

Figure 2. Sampling points at Ceglédbercel, Hungary. Dark dots indicate the profiles presented in this study (with 2.5m elevation lines, arrow presents slope direction).

According to Chin et al. (1994), the absorbance at 280nm is highly correlated with the aromatic character of SOC. That is why we also measured the absorbance in UV range by Shimadzu 3600 UV-VIS spectrophotometer.

The ultraviolet absorbance ratio index (URI, UVA_{210}/UVA_{254}) is another frequently used parameter to determine the polymerization of the dissolved organic matter (Her et al., 2008).

To compare the results of the different methods, spectra were recorded between the 800–180 nm wavelengths.

Research on Relation between Soil Organic Carbon and Water Erosion at Other Hungarian Sites

Soil organic carbon loss caused by soil water erosion was examined on various parts of Hungary. Slope sections were separated into 3 parts, upper third of the slope, middle third of the slope and lower third of the slope according to the methodology of the Soil Protection and Monitoring System of Hungary. In the present case we publish the comparison of the soil organic matter content of the upper and the lower third of the slope.

On the other hand, shallow drillings were made at the bottom of some slopes as far as the parent material was revealed. Drillings were made with spiral sampler (Figure 3).

There was one excavation site used for sampling in Maglód where an approximately 300 cm deep profile was opened for archeological examinations.

Samples were taken every 20 cm (0–20cm, 20–40cm... etc.). Soil organic matter was examined in laboratory with the Turin method.

Figure 3. Shallow drilling sampling point at Galgahévíz, Hungary.

RESULTS

Results from the Tillage Research

Change in SOM Content of the Fertile Layer

In 2005 the average thickness of the humic horizon was 80–90 cm in the entire study area. From the point of view of N supply, the SOM content was definitely low (1.25%). In 2010 the upper horizons already belonged to the intermediate N supply category (1.5–2%), while all but one of the other horizons remained in the low category of 2005 (Table 1).

Results of the studied five-year period showed a clear trend of increasing SOM values in the higher horizons for both tillage types.

Table 1. Bulk density (g cm^{-3}) and SOM (%) in 2005 and 2010

	2005				2010			
	Bulk density g cm^{-3}		SOM %		Bulk density g cm^{-3}		SOM %	
Horizon	CV	CS	CV	CS	CV	CS	CV	CS
Surface	1.35	1.59	0.93	0.93	1.35	1.39	1.55	2.04
Cultivated	1.48	1.68	0.81	0.79	1.56	1.58	1.18	1.37
Subsoil	1.65	1.60	0.56	0.60	1.59	1.58	0.57	0.56
Average (0-50 cm)	1.49	1.62	0.77[a]	0.77[b]	1.50	1.52	1.10[c]	1.32[d]
	Soil layer cm		SOM kg m^{-2}		Soil layer cm		SOM kg m^{-2}	
	CV	CS	CV	CS	CV	CS	CV	CS
Surface	10	10	1.26	1.47	10	10	2.09	2.84
Cultivated	20	10	2.41	1.32	20	10	3.68	2.17
Subsoil	20	30	1.85	2.90	20	30	1.80	2.64
In all	50	50	5.52	5.70	50	50	7.57	7.65
WA			1.10	1.14			1.51	1.53
Change in SOM content			100	103.2			137.1	138.5

WA: weighted average; SD: a) 0.14; b) 0.11; c) 0.09; d) 0.17, SOM: Soil Organic Matter, CV: conventional tillage, CS: conservation tillage.

The increase in the SOM content could be surprising, however, change in soil organic matter under CA is always reported in international literature (Diaz-Zorita and Grove, 2002; Wright et al., 2005; Sombrero and Benito, 2010; Plaza et al., 2012). The rate of increase depends on the tillage method

and the crop rotation (West and Post, 2002). The increase in SOM content in the upper horizons was more enhanced under CS tillage. In the deeper horizons this value was approximately half of the values measured in the CV areas. SOM values weighted by layer thickness and by bulk density values refer to the presence of the same amount of SOM under both tillage between 0 and 50 cm depth. In 2005 5.5–5.7 kg m^{-2} and in 2010 7.7 kg m^{-2} SOM were estimated in both cases, i.e. a 34–37% increase in 5 years (Figure 4). Consequently, the SOM accumulation was the major difference between the CV and CS tillage. Through the technology of cultivation (non-inversion shallow tillage or no-tillage) major part of organic matter is located in the upper soil layers. Particular feature of CA is that SOC and SOM accumulates in the very topsoil which leads to carbon vertical stratification (Hernanz et al., 2002; Moreno et al., 2006).

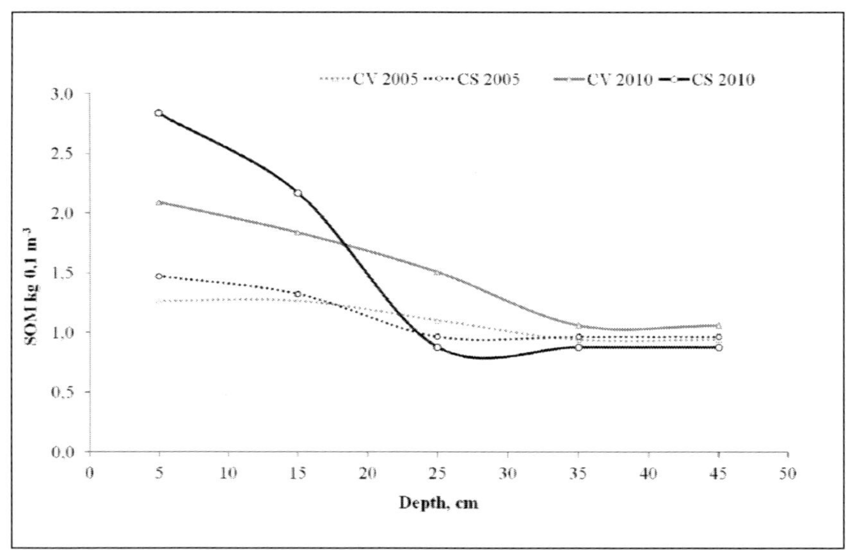

SOM: Soil Organic Matter, CV: conventional tillage, CS: conservation tillage.

Figure 4. SOM content by horizons and by treatments in 2005 and 2010.

This distribution effects soil structure, erosion, water holding capacity and biological activity of the soil as well. SOM redundancy in the topsoil layers increases the durability of surface aggregates (especially against water), thus protecting against erosion and facilitating water infiltration into the soil. The increase of SOM is correlated to the increase of aggregate stability (Hernanz et al., 2002). Organic matters in the soil are constantly transforming. Total SOM

content and proportion of different humus materials to each other are determined by the balance between humification and mineralization. Under changing conditions (e.g. starting cultivation of a soil or changing cultivation) strengths of accumulation and degradation of organic matter will change as well. Under CS cultivation organic matter accumulated in the uppermost horizon will tip the balance developed under former CV cultivation. Thus change (increase) of SOM content can be expected, till the new balance following the new circumstances evolves (Stefanovits 1999). This process occurs conversely as well; if CS tilled area will be ploughed again. Koch and Stockfish (2006) found that decrease can reach 2–5% in 0–45 cm depth already after one tillage operation, while 1.5–2.5 years after ploughing, losses from the 0–45 cm depth accounted for 6–10% of the initial total mass of SOC and SN.

Edaphon: The Key Factor

Main source of organic matter supply are plant residues. Transformation of plant residues is a consequence of complicated degrading and synthesizing microbial processes and connected biochemical reactions. Edaphon (flora and fauna of the soil) have huge role in the process of humification and in degrading organic materials.

Though they are very beneficial for us, members of edaphon are for the naked eye not, or only hardly perceptible. It has as much biomass in the upper 15 cm of an acre, as a gang of 40 cattle (20–25 t/ha).

All members of this abundance of spices has the optimal habitat connected to certain soil layer, structure (where temperature, pH, air-, water- and nutrient supply, presence of symbionts and other members of edaphon is acceptable for them).

This habitat will be formed according to their genetically determined demands by their movement, reproduction, and metabolism. They are though resistant and reproductive, a deep ploughing tumbles their habitat to such an extent, that it takes quite a while till they can pullulate much enough to be able to function again in accordance with their weight in the living society of the soil by processing and constructing materials.

To be able to achieve this, it is essential to get back to its specific habitat and pullulate there under optimal circumstances.

Earthworms have an exceptional role in this system. Their abundance indicates precisely the degree of soil life activity. Comparing with CV tillage, CS tillage creates much more favorable conditions for them (less disturbance,

better moisture conditions, more organic matter etc.), which is confirmed by the results of our examinations on the study area as well (Figure 5).

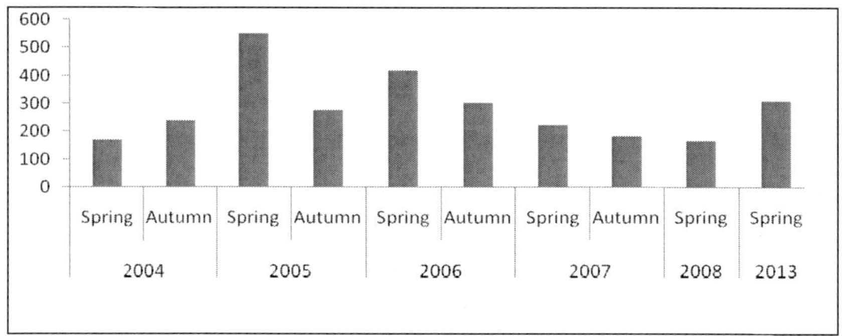

Figure 5. Average number of earthworms per m^2 as % of CV.

Earthworms constantly enlarge and maintain their vertical and horizontal tunnels to better their own circumstances. Macropores, the vertical tunnels help leading surface water of intensive precipitation load into the soil. During nutrition they carry organic materials from the surface into the deeper soil layers, facilitating humification of stem residues.

Dosing lime from their lime gland to their intestinal content, they regulate optimal pH for the microbial degradation of organic materials ingestioned together with soil. Cellulose degrading symbionts provide also their hosts (the earthworms) with sugar gained from degraded fibres.

Most useful elements of the soil are the tiny, waterproof aggregates of earthworm faeces cemented with stable humus containing calcium-humates. Earthworms' activity has a decisive role in the formation of macro- and microaggregates (Six et al., 2004).

These play an essential role in continuous and balanced nutrient supply of cultivated plants. Aggregates contribute to preserving spasmodically dosed chemical fertilizers in the root level and to distributing their availability in space and time as well. Earthworms and their surface runs can be observed quite easily, and they serve with useful information about the quality of edaphon and the intensity of soil life though their presence (abundance and tunnel frequency). Soil flora and fauna play essential role not only in degradation of organic matter and humification, but they also affect soil bulk density favorably owing to their activity.

Table 2. Main carbon related parameters of the investigated layers, Ceglédbercel, Hungary

Profile	Depth	pH			$CaCO_3$	URI	E_4/E_6	Abs at	E_2/E_3	SOC	TNb	C/N	OM	SIC	TSC
	cm	Dw	KCl	dpH	%			280 nm		g/kg	mg/kg		%	g/kg	g/kg
D1	0–10	7.70	7.24	0.46	3.0	1.59	13.20	1.21	3.26	14.0	148	95	2.4	3.6	17.6
D4	0–10	7.77	7.52	0.25	10.2	2.16	1.97	0.72	2.79	6.1	174	35	1.1	12.2	18.3
D5	0–10	7.73	7.37	0.36	6.8	1.71	2.24	0.94	2.81	11.9	260	46	2.0	8.1	20.0
D5	50–60	7.75	7.39	0.36	8.5	1.96	2.88	0.68	3.25	10.5	201	52	1.8	10.2	20.6
D5	80–90	7.68	7.04	0.64	20.7	2.12	4.45	0.52	3.78	7.7	164	47	1.3	24.9	32.6
D6/1	0–30	7.51	7.02	0.49	4.7	1.44	2.34	1.31	2.98	14.0	320	44	2.4	5.6	19.6
D6/2	30–80	7.80	7.00	0.80	0.9	1.59	3.90	1.12	3.17	14.5	307	47	2.5	1.0	15.5
D6/3	80–110	7.85	7.03	0.82	0.0	1.47	2.46	1.32	2.91	19.9	290	68	3.4	0.0	19.9
D6/4	110–125	7.88	6.65	1.23	0.4	1.56	3.71	1.09	2.97	17.4	284	61	3.0	0.5	17.9
D6/5	150–160	7.78	7.07	0.71	0.4	1.75	2.26	0.94	2.79	13.2	190	69	2.3	0.5	13.7
D6/6	180–190	7.68	6.84	0.84	0.0	1.93	2.70	0.76	2.73	7.1	187	38	1.2	0.0	7.1
D6/7	200–210	7.60	6.99	0.61	0.0	2.17	1.74	0.62	2.60	4.6	148	31	0.8	0.0	4.6
D6/8	210–220	7.51	6.98	0.53	0.0	2.83	2.61	0.29	3.21	2.8	110	25	0.5	0.0	2.8
D6/9	250	7.82	7.27	0.55	20.4	3.12	1.21	0.25	2.24	1.3	92	14	0.2	24.4	25.7

URI: UV absorbance ratio index, E_4/E_6: ratio between absorbance at 465 and 665 nm, Abs. 280 nm: Absorbance at 280 nm, E_2/E_3: ratio between absorbance at 254 and 365 nm, SOC: Soil organic carbon; TNb: Total bound Nitrogen; C/N: Carbon: Nitrogen ratio; OM: organic matter, SIC: Soil inorganic carbon, TSC: Total soil carbon.

In our investigations bulk density of the undisturbed CS soil samples from surface and from the cultivated layer was between 1.59–1.68 g cm^{-3}, which values are 15% higher than those for CT (between 1.35–1.48 g cm^{-3}) in 2005. However, the initial compaction of the CS soil stopped, probably because of the edaphon organisms and in 2010 the bulk density of the samples from the two tillage types was identical ($\sim$1.5 g cm^{-3}) (Table 1). This state guarantees good oxygen supply of the roots, even in the case of saturation of field water capacity.

Results of the Examination of Soil Erosion and Soil Organic Carbon from the Ceglédbercel Site

The solum thickness along the catena varies from 30 cm at D1 point to 230 cm D6 (Table 2). On the steepest part (D2-D3) the depth of the humic layer equals the tillage depth, i.e. 20 cm.

At the end of the steepest part (D4-D5) the thickness reaches one meter which presumably was the original, in situ depth of the non-eroded nor deposited haplic Cambisol in the region (Stefanovits 1971). During soil evolution under forest vegetation the total amount of $CaCO_3$ was leached out of the solum.

After the clearance ($\sim$300 years) arable farming was started followed by intensive erosion processes (Kertész 2004). As the volume of soil loss exceeded the whole profile on the upper parts loess appeared on the surface. From these parts with high $CaCO_3$ content on the surface the additional erosion started to spread lime on the deposited parts. However lime migration within the profile was observed as lime films on the aggregate surfaces the main driving force of $CaCO_3$ is definitely soil erosion. On the higher part each horizon contains more or less lime, while in the D6 profile only the uppermost and lowermost layers are calcariferous. The small amount of lime in the middle of the D6 profile is presumed to be the results of leaching which occurs only on the surfaces of megaaggregates, while the matrix is lime free. The lower part of this profile could be the in situ soil which is covered by early eroded lime free sediment, finally on the surface recently deposited calciferous sediment can be found. That means more than 1 m thick soil erosion and deposition since forest clearance what fits former results from the Somogy hills, Hungary (Jakab et al. 2010).

This considerable mass redistribution also affected the TOC dispersion. Originally the highest TOC value was on the soil surface which gradually

decreased downwards. Recently the highest values are in the 80-150 cm depth at the bottom (D6) profile. Presumably this part of the profile could be the original surface before the intensive erosion processes. Due to the frequent tillage operations and the lack of organic material bypass SOC generally decreases in the tilled horizon (Lal, 2005) as it is the case in the study site. On the surface the lower values situated on the steeper parts while the top and the bottom has more SOC. The OM compound also varies on the surface along the investigated catena.

In spite of the relatively high SOC content on the uppermost profile (D1) here the low molecular weight, more labile, acidic components dominate according to almost all the used indexes. The same tendency arises on the surface at the bottom, while the OM on the steeper parts of the surface seems to be more stable. The relative lack of the mobile components here could be the result of intensive erosion. On the other hand the volume of aromaticity is lower on the slopes and higher on the more flat areas. Although the differences are low the selective movement of the OM compounds is quite possible.

Comparing recent deposition processes remarkable differences can be observed between the different parts of the catena. Sedimentation on the higher parts creates very thin layers containing small, unaggregated particles (Figure 6). In this type of crust there is much less organic matter and nutrients occur because these materials are mainly delivered within macroaggregates (Yamashita et al. 2006). Crusts developed on the lower part of the catena differ morphologically from the one mentioned above (Figure 7). In there the layers are thicker and built of aggregates that cause changes in SOC content and OM compound as well. The difference between the two types of crust can justify the results of the buried sediment properties.

Concerning the vertical distribution of OM compounds it is clear that in the D5 profile the decreasing SOC content values with depth are together with molecular weight and aromaticity reduction (Table 2). This result suggests here a regular, in situ solum profile in spite of probable recent deposition and the high lime content. At the deepest profile the horizon around 1 m depth with the highest SOC content has the most polymerized and aromatized OM content. On the other hand the OM compound of this horizon is quite similar to that of the surface. This fact suggests two possibilities. (i) In the buried horizon there is no significant SOC carbon mineralization and if there is so; it concerns the whole spectrum of the OM. (ii) Recent deposition highly affected by the polymerized OM compounds while the lower molecular weight components are delivered or mineralized. And in the buried horizon there is also a slight special mineralization of the low weight components.

Figure 6. Thin sediment layers built by the deposition of individual soil particles at the top of the catena, Ceglédbercel, Hungary.

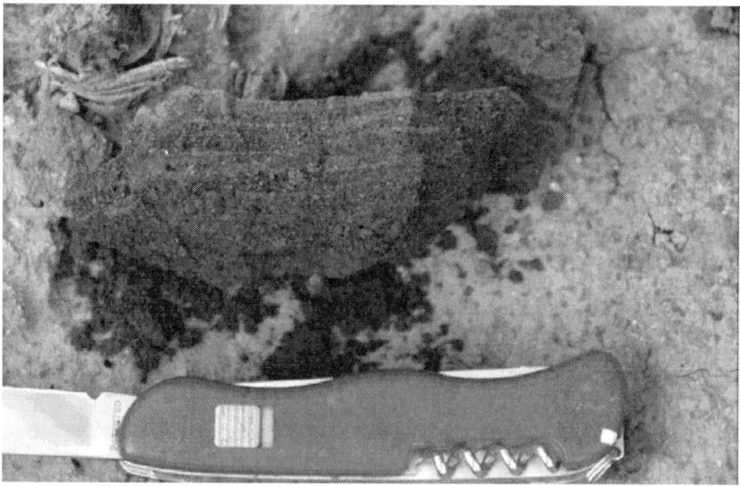

Figure 7. Coarse sediment layers built by the deposition of aggregates at the bottom of the catena, Ceglédbercel, Hungary.

Results of Soil Organic Matter Loss through Soil Water Erosion from Various Hungarian Sites

Examination of Different Slope Sections and Their Soil Organic Matter Content

Investigation of the slope thirds from various sites highlight some of the processes that contribute to the dynamism of soil erosion processes over the slope and carries valuable information on mineralization of soil organic carbon, anthropogenic effects on soil management and effects of land use on soil organic matter content and on its change, too.

Table 3. Comparison of soil organic matter content of different slope sections under different land use types on Hungarian sites

Situation of the site	Land use	UST	LST	SOMd
		SOM (%)		(%)
Galgahévíz, Sósi Creek	Arable field with alfalfa	1.51	2.22	+47
Galgahévíz, Sósi Creek	Forest 1.	3.07	2.83	-8
Galgahévíz, Sósi Creek	Arable field with maize	1.49	1.76	+18
Galgahévíz, Sósi Creek	Forest 2.	3.25	2.26	-30
Gömörszőlős	Arable field with maize	2.33	3.16	+36
Gömörszőlős	Grassland	3.91	4.45	+14
Alsószuha	Arable field	2.55	3.28	+29
Alsószuha	GFAF, 10 years after abandonment	3.01	2.37	-21
Alsószuha	GFAF, 40 years after abandonment	2.5	2.86	+14
Pilismarót (south)	Grassland	1.43	1.25	-13
Pilismarót (south)	Winter wheat	1.32	1.59	+20
Pilismarót (south)	Alfalfa	1.56	1.64	+5
Pilismarót (north)	Alfalfa	1.36	1.46	+7
Pilismarót (north)	Maize	1.72	1.45	-16
Nemesgulács	Pasture	4.25	4.74	+12

UST = Upper Third of the Slope, LST = Lower Third of the Slope, SOM = Soil Organic Matter, GFAF = grassland, former arable field, SOMd: change of soil organic matter compared to the upper slope third (100%).

Table 3 includes the results of soil organic matter measurements on different slope sections. However the slope is separated to 3 sections, here

(Table 3) only the lower and the upper third of the slope is presented, in order to show the accumulation effects of the erosion processes on soil organic carbon.

It is important to state that we only care about the organic compounds as majority of the examined soils are formed on loess rich sediments where $CaCO_3$ content can reach 25%, including a high amount of carbon. This carbon behaves differently as soil tillage brings it to the surface on the upper third of the slope so it will appear on the upper slope third and sometimes decreases to zero at the lower slope third.

Table 3 allows us to make multiple comparison based on a very simple methodology of comparing two sections of the slope.

First of all it is obvious that on intensive agricultural areas (arable fields) there is a bigger percentage of soil organic matter can be found at the lower third of the slope compared to its upper third counterpart (changes range between 5 and 47 %, taking the upper third of the slope as basis). The basic hypothesis is that water erosion moves the upper layer of the soil from the upper part of the slope and some of this removed material accumulates at the bottom of the slope.

It is known that accumulation takes place all over the slope and some of the material is going to other areas, leaving the given slope, e.g. arrives to a waterway or accumulates in a ditch if it exists at the bottom of the slope. In this case this basic hypothesis is well proved.

The question is what happened on the areas where a negative change took place?

There can be multiple explanations for that. On the one hand, extensive areas (grasslands and forests in the presented cases) protect the soil well enough to result this negative change between the slope thirds. From the soil protection point of view this change is positive and can serve as an example for farmers as soil protection measure.

On the other hand, there was one sample site where there was more soil organic matter at the upper third of the slope than at the lower third of the slope. This proves that there are always exclusions from the well-proven zero hypothesis. In this case a positive anthropogenic effect had to happen at the upper third of the slope or a negative effect had to happen at the lower third of the slope.

This one case leads us to the other hypothesis that suggests positive effects of extensive land use on soil quality and thus make is believe that there will not be soil organic matter accumulation at the lower third of the slope.

It is well-proven again as we can see in case of the forested areas and on grasslands except one where a former arable land was abandoned for 40 years and revegetated by grasses.

It is interesting because the other similar area – where abandonment was much younger being only about 10 years – followed the hypothesis, resulting 21% less soil organic matter at the bottom of the slope than at the upper third of the slope.

The reason of the above mentioned extremities (not following the hypothesis) is caused by the human factor, we lack data on exact crop rotations and other farming information (e.g. the amount of fertilizer, manure used etc.) from previous years, on the other hand there is lack on information if any of the areas were used for pasture with intensive fertilization of animals at any part of the slope. These two inputs could have been valuable and improve the quality of the output and the conclusions.

We can however expect that soil organic matter is enriching the lower part of the slope so investigation of the lower part more precisely can be quite helpful to understand the water erosion and sedimentation processes.

Examination of Sediments at the Bottom of the Slope

Various examined areas where shallow drilling took place resulted valuable figures related to soil organic matter movement (Table 4).

There is a huge variation in the distribution of soil organic matter in accumulated areas.

In the worst case (from the point of view of having the thickest sediment thickness) when the parent material was reached at the depth of 300–320 cm (in Galgahévíz, west of the cooperative), the amount of soil organic matter often becomes zero % (e.g. at the depth of 120–140cm) while it increases to 1.86 % again at the depth of 180–200 cm, decreasing to almost zero at the depth of 220–240 cm and stays like that until 300 cm. In case of the Maglód site (Figure 8), the soil organic matter content is high on the surface 5.1% at the depth of 0–20 cm and stays above 2 % in the whole examined profile. Normally – as in all other cases – the soil organic matter content decreases to 1 % and stays below that at the depth of 200–320 cm.

However the fluctuation of the soil organic matter content proves that water erosion carries different amount of soil organic matter in the sediments and as it accumulates at the bottom of the slope provides us valuable information again – as in case of the investigation of the slope thirds – on the water erosion process itself.

Table 4. Laboratory results of the shallow drillings in Hungary

Sampling depth	GW	GE	GP	PD	SD	M
	SOM (%)					
0–20	3.05	2.03	1.7	1.67	1.55	5.1
20–40	ND	ND	1.67	1.53	1.58	3.8
40–60	1.79	2.60	1.65	1.66	1.62	3.5
60–80	2.73	5.35	1.45	1.33	1.33	3.5
80–100	2.40	2.39	1.31	1.42	1.17	2.3
100–120	1.04	1.72	1.17	1.21	0.95	2.7
120–140	0.00	1.78	1.14	1.04	0.89	2.7
140–160	0.24	1.28	0.82	1.24	0.67	4.3
160–180	0.95	0.00		1.21	0.60	2.8
180–200	1.86	0.00		1.05		3.0
200–220	0.80	0.28		0.75		3.3
220–240	0.04	0.85		0.83		2.0
240–260	0.00	1.01				2.2
260–280	0.00					
280–300	ND					
300–320	0.17					

ND: no data, GW: Galgahévíz, west of the cooperative, GE: Galgahévíz, east of the cooperative, GP: Gerézdpuszta, PD: Pilledomb, SD: Somogydöröcske, M = Maglód.

Figure 8. The open profile near Maglód, Hungary.

The process of course is very complicated and we are not providing any solution or equations for the proper description of the process itself as it is varying according to the differences in the soil forming factors and especially along the differences in the parent material, the lope length and steepness, and soil management.

ACKNOWLEDGMENTS

Part of the research was co-financed by Kutató Kari Kiválósági Támogatás – Research Centre of Excellence – 17586-4/2013/TUDPOL.

The authors are grateful to the Hungarian Scientific Research Fund (OTKA) Ref. No: PD100929, PD OTKA 104899 and for Syngenta Ltd. for financial support and for Egegyumolcs Ltd. for providing the research field.

The authors are also thankful to E. Mészáros, M. Di Gléria and V. Újházi for their help during the project.

Reviewed by: Zita Zsembery. Leading Councilor, Dept. of Natura 2000, Ministry of Rural Development.

REFERENCES

Abbaszadeh Afshar, F.; Ayoubi, S.; Jalalian, A. Soil redistribution rate and its relationship with soil organic carbon and total nitrogen using 137Cs technique in a cultivated complex hillslope in western. *Iran Journal of Environmental Radioactivity* 2010, 101, 606–614.

Andarski, B. J.; Mueller, D H.; Daniel, T. C. Effects of tillage and rainfall simulation date on water and soil losses. *Soil Science Society of America Journal* 1985, 49, 1512–1517.

Bádonyi, K.; Madarász, B.; Kertész, Á.; Csepinszky, B. Talajművelési módok és a talajerózió kapcsolatának vizsgálata zalai mintaterületen. *Hungarian Geology Bulletin* 2008, 57, 147–167.

Barczi, A.; Tóth, T. M.; Csanádi, A.; Sümegi, P.; Czinkota, I. Reconstruction of the paleo-environment and soil evolution of the Csípő-halom kurgan, Hungary. *Quaternary International* 2006, 156-157: 49-59.

Birkás, M.; Jolankai, M.; Gyurica, C.; Percze, A. Tillage effects on compaction, earthworms and other soil quality indicators in Hungary. *Soil and Tillage Research* 2004, 78, 185–196.

Buzás, I. *Talaj- és agrokémiai vizsgálati módszerkönyv 1.* INDA 4231 Kiadó: Budapest, HU, 1993.

Chin, Y. P.; Aiken, G.; O'Loughlin, E. Molecular weight, polydispersity, and spectroscopic properties of aquatic humic substances. *Environmental Science Technology* 1994, 28, 1853–1858.

Díaz-Zorita, M.; Grove, J. H. Duration of tillage management affects carbon arid phosphorus stratification in phosphatic Paleudalfs. *Soil and Tillage Research* 2002, 66, 165–174.

Dövényi, Z. *Magyarország kistájainak katasztere MTAFKI Budapest*, Hungary. 2010 (In Hungarian)

Farsang, A.; Kitka, G.; Barta, K.; Puskás, I. Estimating element transport rates on sloping agricultural land at catchment scale (Velence Mts., NW Hungary). *Carpathian Journal of Earth and Environmental Sciences* 2012, 7(4), 15–26.

Fonseca, F., de Figueiredo, T. Impact of tree species replacement on carbon stocks in forest floor and mineral soil. In: Azevedo, J. C.; Feliciano, M.; Castro, J.; Pinto, M. A. (eds) *Proceedings of the "IUFRO Landscape Ecology Working Group International Conference".* Instituto Politécnico de Bragança, Bragança, Portugal 2010, 557–562.

Fonseca, F., de Figueiredo, T., Bompastor Ramos, M. A. Carbon storage in the Mediterranean upland shrub communities of Montesinho Natural Park, northeast of Portugal. *Agroforestry Systems* 2012, 86(3): 463–475.

Fonseca, F., de Figueiredo, T. Carbon in soils of Montesinho Natural Park, Northeast Portugal: preliminary map-based estimate of its storage and stability. *Spanish Journal of Rural Development* 2012, 3(1): 71–78.

Franzluebbers, A. J. Soil organic matter stratification ratio as an indicator of soil quality. *Soil and Tillage Research* 2002, 66, 95–106.

Gregorich, E. G.; Greer, K. J.; Anderson, D. W. Carbon distribution and losses: erosion and deposition effects. *Soil and Tillage Research* 1998, 47 (3-4), 291–302.

Her, N.; Amy, G.; Sohn, J.; Gunten, U. UV absorbance ratio index with size exclusion chromatography (URI-SEC) as an NOM property indicator. Journal of Water Supply: Research and Technology. *AQUA* 2008, 57(1), 35–44.

Hernanz, J. L.; López, R.; Navarrete, L.; Sánchez-Girón, V. Long-term effects of tillage systems and rotations on soil structural stability and organic

carbon stratification in semiarid central Spain. *Soil and Tillage Research* 2002, 66, 129–141.

Holland, J M. The environmental consequences of adopting conservation tillage in Europe: reviewing the evidence. *Agriculture, Ecosystems and Environment* 2004, 103, 1–25.

Jacinthe, P. A.; Lal, R.; Owens, L. B.; Hothem, D. L. Transport of labile carbon in runoff as affected by land use and rainfall characteristics. *Soil and Tillage Research* 2004, 77, 111–123.

Jakab, G.; Kertész, Á.; Madarász, B.; Ronczyk, L.; Szalai, Z. Az erózió és a domborzat kapcsolata szántóföldön, a tolerálható talajveszteség tükrében. *Tájökológiai Lapok* 2010, 8(1), 35–45. (The role of relief in soil erosion with special emphasis on tolerable soil loss) (In Hungarian with English abstract).

Kassam, A.; Friedrich, T.; Derpsch, R.; Lahmar, R.; Mrabet, R.; Basch, G.; González-Sánchez, E. J.; Serraj, R. Conservation agriculture in dry Mediterranean climate. *Field Crops Research* 2012, 132, 7–17.

Kertész, Á. Soil erosion and mass movement processes on the loess covered areas of Hungary. *Földrajzi Értesítő* 2004, 53(1-2), 13–20.

Kertész, Á.; Bádonyi, K.; Madarász, B.; Csepinszky, B. Environmental aspects of Conventional and Conservation tillage. In: Goddard, T.; Zoebisch, M.; Gan, Y.; Ellis, W.; Watson, A.; Sombatpanit, S. (eds) *No-till farming systems.* World Association of Soil and Water Conservation, Bangkok, 2007, 313–329.

Kertész, Á.; Madarász, B.; Csepinszky, B.; Benke, Sz. The role of conservation agriculture in landscape protection. *Hungarian Geology Bulletin* 2010, 59. 167–180.

Koch, H. J.; Stockfish, N. Loss of soil organic matter upon ploughing under a loess soil after several years of conservation tillage. *Soil and Tillage Research* 2006, 86, 73–83.

Lal, R. Soil erosion and carbon dynamics. *Soil and Tillage Research* 2005, 81, 137–142.

Lal, R.; Kimble, J. M.; Follet, R. F.; Cole, C. V. *The potential of US cropland to sequester carbon and mitigate the greenhouse effect.* Lewis Publishers: Boca Raton, US.

Li, L. L.; Huang, G. B.; Zhang, R. Z.; Bellotti, B; Li, G.; Chan, K. Y. Benefits of conservation agriculture on soil and water conservation and its progress in China. *Agricultural Sciences in China* 2011, 10, 850–859.

Lorenz, K.; Lal, R. The depth distribution of soil organic carbon in relation to land use and management and the potential of carbon sequestration in subsoil horizons. *Advances in Agronomy* 2005, 88, 35–66.

Madarász, B.; Bádonyi, K.; Csepinszky, B.; Mika, J.; Kertész, Á. Conservation tillage for rational water management and soil conservation. *Hungarian Geology Bulletin* 2011, 60, 117–133.

Malatinszky, Á.; Ádám, Sz.; Saláta–Falusi, E.; Saláta, D.; Penksza, K. Planning management adapted to climate change effects in terrestrial wetlands and grasslands. *International Journal of Global Warming* 2013, 5(3): 311–325.

Merino, A.; Pérez-Batallón, P.; Macias, F. Responses of soil organic matter and greenhouse gas fluxes to soil management and land use changes in a humid temperate region of southern Europe. *Soil Biology and Chemistry* 2004, 36(6): 917–925.

Merino, A.; Real, C.; Álvarez-González, J. G.; Rodríguez-Guitián, M. A. Forest structure and C stocks in natural *Fagus sylvatica* forest in southern Europe: The effects of past management. *Forest Ecology and Management* 2008, 250(3): 206–214.

Moreno, F.; Murillo, J. M.; Pelegrín, J. E.; Gíron, I. F. Long-term impact of conservation tillage on stratification ratio of soil organic carbon and loss of total and active $CaCO_3$. *Soil and Tillage Research* 2006, 85, 86–93.

Nadeu, E.; Van Oost, K.; de Vente, J.; Boix-Fayos, C. Calibration and application of an erosion and C redistribution model (SPEROS-C) to twelve small catchments in southeastern Spain. *Cuadernos de Investigacion Geografica* 2013, 39(2), 225–242.

Nagy, R.; Zsolt, Zs.; Papp, I.; Földvári, M.; Kerényi, A., Szabó, Sz. Evaluation of the relationship between soil properties and vineyards: a case study from a cool climate wine region of Hungary. *Carpathian Journal of Earth and Environmental Sciences* 2012, 7(1): 223–230.

Nie, X. J.; Zhao, T. Q.; Qiao, X. N. Impacts of soil erosion on organic carbon and nutrient dynamics in an alpine grassland soil. *Soil Science and Plant Nutrition* 2013, 59(4), 660–668.

Nie, X.; Wang, X.; Liu, S.; Gu, S.; Liu, H. Cs-137 tracing dynamics of soil erosion, organic carbon and nitrogen in sloping farmland converted from original grassland in Tibetan plateau. *Applied Radiation and Isotopes* 2010, 68(9), 1650–1655.

Olson, K. R.; Gennadiyev, A. N.; Zhidkin, A. P.; Markelov, M. V. Impacts of Land-Use Change, Slope, and Erosion on Soil Organic Carbon Retention and Storage. *Soil Science* 2012, 177(4), 269–278.

Pető, Á. Studying modern soil profiles of different landscape zones in Hungary: An attempt to establish a soil-phytolith identification key. *Quaternary International* 2013, 287, 149–161.

Plaza, C.; Courtier-Murias, D.; Fernández, M. J.; Polo, A.; Simpson, J. A. Physical, chemical, and biochemical mechanisms of soil organic matter stabilization under conservation tillage systems: A central role for microbes and microbial by-products in C sequestration. *Soil Biology and Biochemistry* 2012, 57, 124–134.

Post, W. M.; Emanuel, W. R.; Zinke, P. J.; Stangenberger, A. G. Soil carbon pools and world life zones. *Nature* 1982, 298, 156–159.

Six, J.; Bossuyt, H.; Degryze, S.; Denef, K. A history of research on the link between (micro)aggregates, soil biota and soil organic matter dynamics. *Soil and Tillage Research* 2004, 79, 7–31.

Skianis, G. A.; Vaiopoulos, D.; Evelpidou, N. Solution of the linear diffusion equation for modelling erosion processes with a time varying diffusion coefficient. *Earth Surface Processes and Landforms* 2008, 33(10): 1491–1501.

Sombrero, A.; de Benito, A. Carbon accumulation in soil. Ten-year study of conservation tillage and crop rotation in a semi-arid area of Castile-Leon, Spain. *Soil and Tillage Research* 2010, 107, 64–70.

Stefanovits, P. *Brown forest soils of Hungary*. Akadémiai Kiadó, Budapest; 1971; pp. 139–141.

Stefanovits, P.; Filep, G.; Füleky, Gy. *Talajtan*. Mezőgazda Kiadó: Budapest, HU, 1999.

Szatmári, G.; Barta, K. Csernozjom talajok szervesanyag-tartalmának digitális térképezése erózióval veszélyeztetett mezőföldi területen. *Agrokémia és Talajtan* 2013, 62:(1): 47–60. (in Hungarian with English abstract).

Szilassi, P.; Jordan, G.; van Rompaey, A.; Csillag, G. Impacts of historical land use changes on erosion and agricultural soil properties in the Kali Basin at Lake Balaton, Hungary. *Catena* 2006, 68(3): 96–108.

Tan, K. H. *Humic matter in soil and the environment principles and controversies*. Marcel Dekker Inc., New York 2003; pp. 176–181.

Tombácz, E.; Lámfalusi, E.; Szekeres, M.; Micheli, E. Humuszanyagok hatása a talajok felületi tulajdonságaira. *Agrokémia és Talajtan* 1996, 45(3–4), 238–248. (in Hungarian with English abstract).

Wang, Z.; Govers, G.; Steegen, A.; Clymans, W.; Van den Putte, A.; Langhans, C.; Merckx, R.; Van Oost, K. Catchment-scale carbon redistribution and delivery by water erosion in an intensively cultivated area. *Geomorphology* 2010, 124, 65–74.

West, T. O.; Post, W. M. Soil organic carbon sequestration rates by tillage and crop rotation: a global data analysis. *Soil Science Society of America Journal* 2002, 66, 1930–1946.

Wright, L. A.; Hons, M. F.; Matocha Jr. E. J. Tillage impacts on microbial biomass and soil carbon and nitrogen dynamics of corn and cotton rotations. *Applied Soil Ecology* 2005, 29, 85–92.

Yamashita, T.; Flessa, H.; John, B.; Helfrich, M.; Ludwig, B. Organic matter in density fractions of water-stable aggregates in silty soils: Effect of land use. *Soil Biology and Biochemistry* 2006, 38, 3222–3234.

INDEX

A

accounting, 6, 8, 9, 17, 67, 70
acetic acid, 26
acid, 20, 28, 43
acidic, 24, 44, 106
acidity, 95
adaptation, 5, 19, 20, 31, 36, 39, 40
adverse effects, 56
Africa, 93
agar, 22
age, 15, 18, 40
agriculture, vii, 51, 54, 55, 56, 57, 59, 67, 72, 89, 94, 114
Alaska, 37, 46
alfalfa, 67, 108
alters, 48, 49
AMF, 16
amino, 26, 28, 29, 43, 46
amino acid(s), 26, 28, 29, 43, 46
amoeboid, 37
anaerobic bacteria, 12
arable farming, ix, 92, 105
arbuscular mycorrhizal fungi, 10, 16
Argentina, vii, 51, 53, 81, 82, 83, 84, 85, 86, 87, 88, 89, 90
Arrhenius equation, 31
artificial intelligence, ix, 53
Asia, 93
assimilation, 2, 10, 28, 35

atmosphere, vii, viii, 2, 27, 30, 34, 36, 52, 56, 60, 63, 67, 87, 92
audit, 83
Austria, 20

B

bacteria, 2, 3, 5, 6, 7, 8, 9, 12, 13, 15, 16, 17, 21, 22, 23, 24, 25, 26, 28, 30, 35, 38, 41, 47
barriers, 53
base, 23, 26, 33
Belgium, 88
benefits, vii, 93
bioavailability, 43
biochemistry, 24
biodiversity, 22, 38, 41, 48, 93
biogeography, 21
biological activity, 101
biological consequences, 93
biomarkers, 5, 16, 25
biomass, viii, 2, 3, 5, 9, 10, 11, 13, 14, 15, 16, 17, 19, 20, 21, 22, 23, 24, 25, 26, 27, 29, 30, 35, 38, 41, 45, 48, 49, 52, 73, 74, 75, 80, 82, 102, 117
biosphere, 93
biotic, 5, 13, 36
biotic factor, 5, 13
birds, 94
boreal forest, 17, 20, 43, 46

Brooks Range, 25, 46
by-products, 116

C

calcification, vii, 1
calcium, 103
carbohydrate, 28
carbon content, viii, 29, 44, 52, 58, 59, 60, 70, 71, 79, 87, 89, 95, 96
carbon dioxide, 46, 47
carbonate, vii, viii, 1, 51, 57, 58, 59
carboxylic acid(s), 28
Caribbean, 16
case study, 115
catchments, 115
cation, 95
cattle, 102
C-C, viii, 27, 52, 56, 60, 63
Census, 58
chemical, 15, 23, 24, 93, 103, 116
China, 6, 15, 44, 45, 47, 89, 90, 114
chromatography, 23, 113
CIA, 60, 84
classes, 26
classification, viii, 52, 58
climate, viii, 2, 4, 8, 11, 19, 20, 21, 22, 24, 25, 27, 32, 34, 35, 36, 37, 38, 41, 43, 44, 47, 49, 52, 53, 54, 58, 61, 75, 88, 92, 115
climate change, viii, 3, 8, 11, 20, 21, 24, 27, 32, 34, 36, 37, 38, 41, 44, 49, 52, 115
clone, 4, 17
clustering, 5, 17
CO_2, vii, 1, 3, 5, 10, 11, 27, 30, 32, 33, 34, 36, 37, 38, 40, 42, 43, 48, 49, 82, 87
colonization, 4
color, 97
community(s), vii, 2, 3, 4, 5, 6, 7, 8, 9, 10, 11, 12, 14, 15, 16, 17, 18, 19, 20, 21, 22, 23, 24, 25, 26, 27, 28, 33, 35, 38, 40, 41, 42, 44, 45, 46, 49, 50, 113
compaction, 93, 105, 113
comparative analysis, 35
compensation, 81
competitors, 29

compilation, 73
complement, 3
complex interactions, 13
composition, 4, 5, 8, 9, 10, 12, 13, 14, 15, 16, 17, 18, 20, 21, 22, 23, 24, 25, 26, 27, 30, 31, 35, 40, 41, 42, 45, 48, 49, 50, 63
compounds, 26, 27, 28, 30, 31, 35, 40, 98, 106
conductivity, 96
conservation, 86, 89, 93, 94, 96, 100, 101, 114, 115, 116
consumers, 13, 26
consumption, 60, 67
contour, 96
control group, 20
controversies, 116
correlation(s), 13, 18, 23, 50
Costa Rica, 14, 43
cotton, 23, 117
covering, 18
crop(s), vii, viii, 13, 51, 52, 53, 55, 56, 58, 59, 60, 62, 63, 64, 67, 68, 69, 70, 71, 73, 75, 77, 78, 80, 81, 82, 85, 86, 87, 88, 90, 93, 101, 110, 116, 117
crop production, 55, 88
crop residue, viii, 52, 80
crop rotations, viii, 52, 55, 64, 71, 85, 110
crust, 7, 106
CT, 105
cultivation, viii, 52, 54, 55, 56, 58, 59, 81, 84, 85, 93, 101, 102
culture, 30
CV, 96, 100, 101, 102, 103
cycles, 12, 27, 32, 80, 86
cycling, 10, 11, 14, 16, 18, 43, 50
cyst, 36

D

data analysis, 117
data set, 73, 77
decomposition, vii, 1, 2, 17, 19, 20, 21, 23, 26, 39, 45, 48, 49, 50
degradation, ix, 53, 56, 85, 93, 102, 103
density values, 101

dependent variable, 75
deposition, 16, 19, 40, 42, 49, 50, 95, 105, 106, 107, 113
depth, viii, 51, 53, 56, 57, 58, 59, 60, 61, 76, 88, 96, 101, 102, 105, 106, 110, 111, 115
desiccation, 23
differential rates, 26
diffusion, 116
dispersion, 105
displacement, 67
distribution, 2, 5, 13, 18, 21, 22, 23, 39, 41, 47, 48, 50, 58, 61, 88, 101, 106, 110, 113, 115
distribution function, 88
diversity, 2, 3, 5, 6, 7, 8, 9, 10, 12, 13, 14, 16, 17, 18, 19, 21, 22, 25, 35, 38, 39, 43, 44, 45, 47, 49, 50
DNA, 8
DOC, 28, 29, 30, 39
dominance, 16, 26
drainage, 24
drought, 7, 41
drying, 12, 39, 42
durability, 101
dynamism, 108

E

earthworms, 94, 103, 113
ECM, 18, 19
ecology, 2, 4, 37, 38, 42
ecosystem, vii, 2, 3, 8, 11, 14, 15, 21, 26, 42, 45, 48, 49, 63, 66, 84
Ecuador, 16
egg, 36
El Niño, 37
electrical conductivity, 57
electrophoresis, 5, 8
emission, 63
energy, vii, 1, 11, 26
engineering, 5
enlargement, 94
environment(s), 5, 8, 11, 12, 14, 17, 21, 27, 30, 31, 49, 59, 66, 70, 93, 112, 116
environmental change, 14

environmental conditions, 14, 23, 95
environmental factors, 18, 21, 31, 47
environmental sustainability, 86
environmental variables, 23, 36
enzymatic activity, 13
enzyme(s), 3, 6, 17, 19, 31, 42, 50
erosion, 60, 86, 93, 94, 95, 99, 101, 105, 106, 109, 110, 113, 114, 115, 116
eukaryotic, 2, 3, 33
Europe, 24, 93, 114, 115
evidence, 4, 11, 12, 13, 14, 17, 19, 20, 21, 23, 25, 26, 31, 32, 34, 35, 36, 37, 114
evolution, 5, 66, 105, 112
examinations, 94, 99, 103
exclusion, 113
experimental design, 31
exploitation, 93
exposure, 12

F

farmers, 93, 109
farmland, 115
farms, 56
fatty acids, 3, 40
fertility, ix, 3, 7, 10, 27, 41, 44, 48, 53, 55, 73
fertilization, ix, 15, 18, 25, 53, 56, 67, 73, 75, 76, 77, 85, 110
fertilizers, 67, 68, 103
field nitrogen, ix, 53
films, 105
financial, 112
financial support, 112
fixation, vii, 1, 33, 67, 68, 81, 85
flooding, 58
flora, 94, 102, 103
flora and fauna, 94, 102, 103
food, ix, 10, 13, 26, 33, 35, 42, 43, 86, 89, 92
food chain, 27, 33, 35
food production, 86
food security, 89
food web, 10, 13, 26, 35, 42
force, 105

forest ecosystem, 41, 49
formation, 103
France, 12, 40
fungi, 2, 5, 6, 9, 10, 11, 13, 15, 16, 17, 18, 19, 20, 21, 22, 24, 25, 26, 28, 35, 38, 47

G

gel, 5
genetics, 3, 4
genus, 19, 43
gland, 103
global climate change, vii, 2, 26, 27
global scale, 5, 32, 33, 36, 40
global warming, vii, 2, 3, 20, 21, 24, 27, 31, 32, 33, 34, 36, 37
glucose, 16, 26, 28, 29, 30, 42, 45
gracilis, 13
grants, 37
grass(s), 10, 12, 67, 94, 110
grasslands, vii, 1, 2, 8, 9, 10, 11, 12, 13, 14, 15, 28, 32, 42, 59, 60, 88, 109, 110, 115
grazers, 16, 44
grazing, 16, 32, 55, 57
greenhouse, vii, 2, 3, 34, 44, 114, 115
greenhouse gas, 3, 115
groundwater, 67
growth, 15, 33, 49

H

habitat(s), 4, 5, 13, 19, 22, 24, 34, 42, 43, 48, 50, 94, 102
half-life, 29
halophyte, 5
hardwood forest, 42, 44, 50
health, 14
heterogeneity, 7, 17
history, 14, 60, 116
host, 19
human, 92, 93, 110
human activity, 92
humidity, 13
humus, 20, 39, 102, 103

Hungary, ix, 91, 92, 94, 96, 97, 98, 99, 104, 105, 107, 111, 112, 113, 114, 115, 116
hypothesis, 7, 15, 109, 110

I

identification, 41, 116
image, viii, 52, 58
improvements, 55, 56
in transition, 3
in vitro, viii, 52, 73, 81
incubation period, 23
India, 13
indirect effect, 24
individuals, 16
inorganic carbon, viii, 52, 57, 60, 90, 95, 104
integration, 59, 62
intervention, 93
inversion, 96, 101
Iran, 112
islands, 23
isomers, 46
isotherms, 54
isotope, 40, 46
Israel, 7, 8, 47
issues, 25, 28
Italy, 19

J

Japan, 17, 46
Jordan, 116

K

kinetics, 28, 34

L

labeling, 10, 33
laboratory studies, 31, 32

landscape, 5, 12, 18, 24, 28, 48, 50, 95, 114, 116
leaching, 95, 105
lead, 55, 56
legume, 67, 68
light, viii, 15, 52, 73, 74, 75, 76, 77, 78, 79
lignin, 13
linoleic acid, 16
lipids, 14
livestock, vii, 51, 55, 69, 86
logging, 32
longevity, 88
longitudinal study, 14
Luo, 41

M

magnitude, 7, 22, 74
majority, 6, 21, 109
Malaysia, 17, 46
management, viii, ix, 10, 43, 52, 64, 66, 70, 73, 81, 84, 85, 89, 92, 93, 108, 112, 113, 115
manipulation, 44
manure, 110
marine environment, vii, 1
Mars, 22
mass, 17, 23, 31, 66, 102, 105, 114
mass loss, 17
mass spectrometry, 23
materials, 23, 45, 53, 102, 103, 106
matrix, 105
matter, ix, 5, 9, 17, 24, 52, 87, 89, 95, 98, 101, 102, 110, 116, 117
measurements, 37, 48, 63, 94, 95, 96, 108
mechanistic explanations, 32
Mediterranean, 8, 113, 114
Mediterranean climate, 114
medium wheat, ix, 53
melt, 23
meta analysis, 87
meta-analysis, 64
metabolism, 2, 3, 27, 30, 31, 32, 33, 102
metabolized, 26, 28, 35
meter, 105
methodology, 62, 66, 99, 109
microbial community(s), vii, 1, 2, 3, 4, 5, 6, 7, 8, 9, 10, 11, 12, 14, 15, 16, 18, 19, 20, 21, 23, 24, 25, 26, 27, 28, 30, 31, 32, 33, 35, 36, 37, 38, 39, 40, 41, 42, 43, 44, 45, 46, 48, 49
microbiota, 2, 10, 28, 29, 31, 36, 37
microclimate, 19, 20
microcosms, 23, 29, 40
microorganisms, 4, 6, 16, 22, 29, 34, 44, 46, 94
migration, 95, 105
mineralization, ix, 12, 15, 16, 26, 29, 52, 57, 72, 74, 75, 77, 79, 80, 81, 83, 94, 95, 102, 106, 108
mixing, 93
modelling, 116
models, ix, 22, 37, 52, 67, 75
modifications, 34
moisture, 6, 7, 8, 9, 11, 12, 13, 22, 23, 27, 30, 32, 39, 41, 48, 49, 93, 103
moisture content, 14, 23, 32
molecular weight, 26, 28, 30, 35, 39, 95, 106
molecules, vii, 1, 3
Mongolia, 46
morphology, 19
multiple regression, 75

N

National Academy of Sciences, 42, 45
national parks, 93
natural resources, 93
nature conservation, 93
negative effects, 93
negative relation, 22
nematode, 13, 21, 49
neural network(s), viii, 52, 58, 61, 63, 70, 72, 79, 80, 83, 84, 85
neutral, viii, 4, 52, 70
New Zealand, 87
nitrification, 23
nitrogen, viii, 13, 14, 15, 16, 18, 39, 41, 43, 44, 45, 49, 52, 56, 57, 60, 61, 62, 66, 67,

68, 69, 70, 72, 73, 75, 77, 78, 79, 80, 81, 82, 83, 85, 86, 87, 88, 89, 98, 112, 115, 117
nitrogen fixation, viii, 52, 56, 67, 69
nitrogen stocks, viii, 52, 56, 81
nitrous oxide, 89
nutrient(s), 14, 16, 18, 19, 22, 25, 26, 27, 28, 30, 31, 33, 35, 36, 40, 43, 44, 47, 49, 55, 67, 88, 89, 94, 95, 102, 103, 106, 115
nutrition, 35, 73, 103

O

OECD, 67, 89
OH, 20
oil, 23, 32, 89
operations, 96, 106
organic compounds, 2, 25, 26, 27, 28, 35, 109
overlap, 59
oxygen, 105

P

Pampas, v, vii, ix, 51, 55, 56, 58, 59, 60, 64, 65, 66, 67, 68, 69, 70, 72, 73, 75, 76, 81, 84, 85, 86, 87, 89, 90
pasture(s), vii, viii, 28, 51, 52, 56, 58, 60, 62, 67, 68, 69, 71, 72, 81, 110
pathways, 45
PCR, 6
periodicity, 30
permeability, 96
permit, 23
pH, 8, 19, 22, 23, 24, 44, 54, 57, 60, 102, 103, 104
phosphate, 14
phosphorus, 15, 16, 45, 85, 94, 113
photosynthesis, vii, 1
Physiological, 44
physiology, 22, 89
plankton, 43
plant growth, 10

plants, vii, 1, 6, 7, 10, 22, 26, 27, 34, 38, 44, 103
pollution, 19, 41, 93
polydispersity, 113
polymerization, 98
polysaccharides, 13
pools, 15, 30, 63, 73, 82, 85, 116
population, 7, 13, 31, 47
population density, 13
Portugal, 113
positive correlation, 13
positive feedback, 34, 36
positive relationship, 22
potassium, 94
precipitation, 4, 7, 8, 9, 11, 27, 28, 30, 32, 35, 38, 41, 57, 80, 88, 97, 103
predators, 2, 13, 21
predictability, 80
priming, 17
principles, 31, 116
productivity rates, 72
project, 112
propagation, 94
protection, 4, 109, 114
proteomics, 3
Puerto Rico, 39

Q

quality control, 48
quartz, 4

R

radiation, 4, 115
rain forest, 16, 39, 44
rainfall, viii, 7, 52, 53, 55, 58, 62, 64, 65, 67, 68, 69, 112, 114
reactions, 102
reciprocal relationships, 37
recycling, 14
redistribution, 94, 95, 105, 112, 115, 116
redundancy, 101
regeneration, 94

regression, 48
relief, 53, 94, 114
replication, 30
reproduction, 102
reserves, 93
residues, 26, 28, 63, 94, 96, 102, 103
resistance, 72
resource availability, 9
resources, 12, 93
respiration, vii, 1, 10, 11, 14, 15, 19, 23, 24, 30, 31, 32, 33, 34, 35, 36, 37, 38, 39, 40, 41, 43, 44, 45, 46, 47, 48, 49, 63, 82
response, vii, 2, 11, 15, 19, 28, 31, 33, 35, 41, 44, 45, 47, 49, 87, 89
restoration, 10
Rhizopoda, 48
risk, 93
root(s), 5, 16, 28, 29, 30, 33, 34, 35, 36, 48, 63, 103, 105
rotations, viii, 52, 58, 60, 63, 113, 117
rules, 5
runoff, 114

S

samplings, 69
saturation, 31, 105
scientific papers, 93
scope, 2, 3, 9, 19, 21, 24, 31
sediment(s), 53, 97, 105, 106, 107, 109, 110
sedimentation, 96, 110
seeded surface, viii, 51
sensitivity, 30, 49
sequencing, 17
services, vii, 2, 27, 84
shape, 24, 38
shelter, 44
showing, 21, 23
shrubs, 5, 6, 7, 13, 26, 44, 47
SIC, 104
signals, 54
simulation, 112
society, vii, 2, 102

soil carbon, vii, viii, ix, x, 46, 52, 56, 58, 59, 60, 64, 73, 83, 84, 85, 88, 92, 95, 104, 117
soil erosion, 93, 94, 95, 96, 105, 108, 114, 115
soil particles, 36, 107
soil quality, ix, 53, 109, 113
soil samples, viii, 5, 7, 11, 19, 20, 23, 30, 39, 52, 96, 105
soil type, 10, 12
solution, 28, 29, 30, 95, 112
South America, 93
sowing, 78
soybean(s), vii, 51, 55, 56, 60, 63, 64, 67, 68, 69, 70, 81, 85
SP, 115
Spain, 114, 115, 116
species, 4, 5, 7, 9, 10, 13, 16, 18, 19, 21, 31, 42, 43, 44, 45, 53, 93, 113
species richness, 7
specific tax, 17
spectroscopy, 98
stability, 4, 27, 31, 89, 95, 101, 113
stabilization, 4, 116
stakeholders, 93
starch, 26, 45
state(s), ix, 31, 34, 88, 92, 93, 105, 109
sterols, 3
stimulation, 38
stock, viii, 52, 56, 58, 59, 60, 61, 62, 66, 69, 72, 85, 88
stoichiometry, 25
storage, 16, 19, 90, 113
stratification, 82, 89, 101, 113, 114, 115
stress, 11, 12, 41
strong interaction, 22
structure, 3, 7, 8, 9, 10, 13, 14, 16, 17, 18, 19, 20, 21, 22, 23, 25, 27, 31, 40, 42, 43, 45, 49, 93, 94, 95, 101, 102, 115
subgroups, 25
substrate(s), 6, 14, 19, 26, 28, 30, 36, 47
succession, 13
sulfate, 22
surface chemistry, 27
surplus, 67

survival, 35
susceptibility, 8
Sweden, 25
synthesis, 4

T

taxa, 2, 3, 4, 6, 10, 17, 36
taxonomy, 42
techniques, 3, 6, 8, 34, 56
technology(s), 93, 94, 101
temperature, ix, 7, 9, 22, 23, 26, 27, 30, 31,
    32, 35, 39, 40, 42, 43, 45, 48, 49, 53, 57,
    58, 62, 80, 97, 102
temperature dependence, 39, 45
temporal variation, 19, 25, 39, 46
terrestrial ecosystems, 28, 37, 46
testing, 58
texture, ix, 27, 53, 57, 61, 89, 97
tillage system, viii, 52, 56, 66, 82, 85, 86,
    113, 116
time periods, 36, 64
trade, 31
trade-off, 31
transformation, 23
translocation, 33
transplant, 14
transport, 34, 113
treatment, 11, 16, 29, 58, 73
Trinidad, 30, 45
tropical forests, 14, 15, 16, 17, 32, 35, 41,
    45
tropical rain forests, 18
tundra, 2, 21, 23, 24, 25, 29, 30, 32, 35, 38,
    39, 40, 42, 46, 47, 48, 49, 50
turnover, 16, 25, 26, 28, 30, 39, 47

U

UK, 37
United, 37, 42, 87
United States, 37, 42, 87
Uruguay, 88
USA, 38, 46, 86
UV, 4, 98, 104, 113
UV radiation, 4

V

variables, 19, 24, 35, 36, 47, 67, 70, 79, 85
variations, 4, 17, 19, 24, 31, 75
vegetation, 4, 5, 6, 7, 8, 9, 10, 13, 17, 19,
    20, 22, 24, 33, 34, 35, 43, 45, 47, 53, 56,
    59, 88, 94, 105

W

water, viii, 11, 12, 22, 31, 32, 45, 52, 56, 57,
    70, 71, 94, 95, 96, 99, 101, 102, 103,
    105, 109, 110, 112, 114, 115, 116, 117
watershed, 7
wavelengths, 98
web, 43
wetlands, 32, 40, 115
wetting, 11, 23, 33
wood, 93

Y

yield, ix, 34, 52, 53, 55, 64, 67, 75, 76, 77,
    78, 82, 84, 85, 87, 89, 90, 93

TRAITÉ THÉORIQUE ET PRATIQUE

DU

TRAVAIL DES VINS

PAR

E. MAUMENÉ

AUTEUR DE LA THÉORIE GÉNÉRALE DE L'ACTION CHIMIQUE
ANCIEN PROFESSEUR A REIMS,
A LA FACULTÉ (LIBRE) DES SCIENCES DE LYON,
LAURÉAT DE L'INSTITUT, DOCTEUR ÈS-SCIENCES,
MEMBRE DE PLUSIEURS SOCIÉTÉS SAVANTES, ETC.

TROISIÈME ÉDITION REVUE ET AUGMENTÉE

PREMIER VOLUME

PARIS

E. BERNARD & Cⁱᵉ, IMPRIMEURS-ÉDITEURS

LIBRAIRIE   IMPRIMERIE

53 bis, QUAI DES GRANDS-AUGUSTINS   71, RUE DE LA CONDAMINE, 71

1890

ENVIRONMENTAL HEALTH - PHYSICAL, CHEMICAL
AND BIOLOGICAL FACTORS

SOIL CARBON

TYPES, MANAGEMENT PRACTICES AND ENVIRONMENTAL BENEFITS

AQUILA MARGIT
EDITOR

New York

ENVIRONMENTAL HEALTH - PHYSICAL, CHEMICAL AND BIOLOGICAL FACTORS

Additional books in this series can be found on Nova's website
under the Series tab.

Additional E-books in this series can be found on Nova's website
under the E-book tab.

SOIL CARBON

TYPES, MANAGEMENT PRACTICES AND ENVIRONMENTAL BENEFITS

EXPOSÉ DES MOTIFS

DU LIVRE PREMIER

DU

CODE PÉNAL,

PRÉSENTÉS AU CORPS LÉGISLATIF

PAR MM. LES COMTES TREILHARD, FAURE ET GIUNTI,
Conseillers d'État.

Séance du 1er février 1810.

MESSIEURS,

Si la lecture des lois pénales d'un peuple peut donner une juste idée de sa morale publique et de ses mœurs privées, le Code pénal qui vous est annoncé, et dont nous vous portons le premier Livre, attestera les progrès immenses qu'ont faits parmi nous la raison et la philosophie.

Vous n'y trouverez que des peines nécessaires, des peines clairement énoncées, répressives, et jamais atroces; vous y verrez aussi des dispositions faites pour diminuer la masse des désordres, parce qu'elles placeront sous une surveillance active et salutaire les hommes dont les intentions perverses auront éclaté.

L'Assemblée constituante a dégagé notre législation pénale de plusieurs dispositions contre lesquelles l'humanité réclamait depuis long-temps ; elle a réduit la peine de mort à la simple privation de la vie ; elle a fait disparaître les supplices barbares du feu, de la roue, et d'être tiré à quatre chevaux. Toute mutilation est défendue, et les peines de lèvre coupée, de langue percée, et autres de cette nature, ne souillent plus le Code français. C'est déjà un grand pas vers la perfection ; mais cette assemblée célèbre, qui se distingua par tant de *conceptions utiles*, qui détruisit *tant d'abus*, qui avait sans contredit pour elle *la pureté des intentions*, ne se tint pas toujours en garde contre *l'enthousiasme du bien* : le flambeau de l'expérience qui lui manquait, a fait apercevoir depuis d'utiles améliorations, dont le Code de 1791 est susceptible.

L'Assemblée constituante crut devoir poser en règle qu'aucune peine ne serait perpétuelle ; celle des fers, la première après celle de mort, ne dut jamais être prononcée que pour un temps qui, dans aucun cas, n'excéderait vingt-quatre années.

La durée des peines fut déterminée pour chaque espèce de crime, d'une manière invariable ; la marque et la confiscation furent supprimées ; enfin un coupable qui avait subi sa condamnation fut lancé sans précaution dans la société pour y jouir de toute la liberté des autres citoyens.

Les bases du projet qui vous est soumis diffèrent, sur ces points importants, de celles posées par l'Assemblée constituante.

Nous avons pensé que, pour parvenir à une juste gradation des peines, il fallait en établir de perpétuelles.

Il nous a paru suffisant de régler la nature des peines

à appliquer, et de fixer les termes qu'elles ne pourraient excéder, sans déterminer la durée précise de celle qui serait prononcée contre chaque condamné ; les magistrats la régleront dans la latitude que la loi leur laisse.

Nous avons rétabli la peine de la marque.

La confiscation pourra être prononcée dans certains cas.

Enfin les condamnés, après avoir subi leur peine, seront placés sous une utile surveillance.

J'aurai occasion de remarquer dans la suite quelques autres différences moins importantes, entre la législation pénale de l'Assemblée constituante et celle qui vous est proposée.

Quant à présent, je dois me borner à exposer, en peu de mots, les motifs qui ont fait adopter nos nouvelles bases.

Et d'abord, pour peu qu'on veuille y réfléchir, on sera bientôt convaincu que la distance entre une peine temporaire et la mort est si immense, que, pour la combler, il faut nécessairement établir une peine perpétuelle ; sans elle, plus de gradation, et toute proportion entre la peine et certains crimes est absolument rompue.

On ne peut disconvenir, par exemple, qu'un fonctionnaire coupable de faux en écriture authentique, et dans l'exercice de ses fonctions, doit être puni beaucoup plus sévèrement qu'un particulier qui a commis le même crime ; et lorsque celui-ci subit une simple peine temporaire, si on ne prononce pas la peine de mort contre le premier, parce qu'il est dangereux de donner trop souvent au peuple le spectacle du sang versé, il mérite certainement de subir à perpétuité la peine prononcée temporairement contre l'autre.

Le faux monnayeur qui a altéré ou fabriqué des espèces

d'or où d'argent est puni de mort; convient-il d'appliquer la même peine à celui qui n'a altéré ou fabriqué que des espèces de cuivre? Si la gravité du crime et ses funestes conséquences ne permettent pas de se borner en ce cas à une simple peine temporaire, n'est-il pas plus convenable, dans l'alternative de la peine de mort ou d'une peine perpétuelle, de se borner à cette dernière?

La règle posée par l'Assemblée constituante, que nulle peine ne serait perpétuelle, détruit donc les proportions qui doivent exister entre les peines et les crimes; dans son système on est souvent exposé, ou à infliger au coupable une peine trop sévère, ou à lui faire grâce d'une partie de celle qu'il a encourue.

Vivement frappée de quelques erreurs graves reprochées aux tribunaux, l'Assemblée constituante ne crut pas pouvoir resserrer dans des bornes trop étroites la délégation de pouvoir faite à la magistrature; elle régla, en conséquence, avec une exacte précision, la durée de la peine qui devait être appliquée à chaque fait en particulier, et elle voulut qu'après la déclaration du jury la fonction du juge fût bornée à l'application mécanique du texte de la loi.

Sans doute le magistrat ne doit et ne peut prononcer que la peine de la loi; mais n'y a-t-il pas quelque distinction à faire entre deux hommes convaincus du même crime? Doit-on placer sur la même ligne le jeune homme séduit, que des conseils désastreux et son inexpérience ont précipité dans l'abîme, et l'homme dont la profonde corruption est manifeste, et dont la vie est souillée de crimes?

Ici nous avons pensé qu'une saine politique et la justice bien entendue appelaient sur la magistrature une marque honorable de confiance, non que les cours puissent changer la nature de la peine indiquée par la loi; mais la loi voudra que chaque espèce de peine puisse être prononcée pour un temps qui ne doit être moindre ni excéder les limites qu'elle prescrit. C'est dans cette latitude que les magistrats, après avoir présidé à toute l'instruction, pesant le degré de perversité de chaque accusé, connaissant parfaitement toutes les circonstances qui peuvent aggraver ou atténuer le fait; c'est, disons-nous, dans cette latitude que les magistrats fixeront la durée de la peine légale qu'ils doivent appliquer.

La peine de la marque ou de la flétrissure fut proscrite par l'Assemblée constituante, parce qu'elle offre un caractère de perpétuité que l'opinion d'alors repoussait; vous avez déjà vu que la perpétuité de quelques peines était nécessaire pour la perfection du système pénal; et l'on ne peut se dissimuler que l'apposition publique de la marque produit, et sur le coupable et sur les spectateurs, une impression qui ne peut être que vive et profonde.

Je pourrais ajouter que la marque est un des moyens les plus efficaces pour constater les récidives dont il est si important de s'assurer; mais je ne crois pas qu'il soit nécessaire de s'appesantir sur cet article, puisque déjà vous avez adopté le rétablissement de la peine de la marque pour certains crimes, et que l'expérience a démontré les bons effets de cette mesure.

La confiscation générale fut aussi écartée du Code de 1791; nous n'hésitons pas à en proposer le rétablissement.

Les intentions philantropiques de l'Assemblée constituante, quand elle rejeta la confiscation et la marque, étaient certainement louables; mais, ne craignons pas de le dire, cette Assemblée a trop souvent considéré les hommes, non tels qu'ils sont, mais tels qu'il serait à désirer qu'ils fussent ; elle était mue par un espoir de perfectibilité qui malheureusement ne se réalise pas ; et si, dans le mouvement rapide qui l'entraînait, cette erreur fut excusable, nous ne le serions pas, nous qui, éclairés par l'expérience, méditons dans le calme des passions ; nous ne serions, dis-je, pas excusables de persister à méconnaître l'efficacité incontestable de quelques moyens de répression qui ne furent pas bien appréciés en 1791.

On objecte que la peine de la confiscation réfléchit sur des enfants qui peuvent n'être pas complices du crime de leur père : mais qui donc souffrira pour les fautes des pères, si ce ne sont les enfants? Lorsqu'un homme a consumé tout son patrimoine par des spéculations insensées, ou par des voies souvent plus répréhensibles, ses enfants ne supportent-ils pas la peine des égarements de leur père?

Lorsque des réparations civiles prononcées en faveur d'une victime du crime absorbent toute la fortune du coupable, peut-on se récrier contre sa condamnation, sous le frivole prétexte que sa succession est ruinée?

Or, qu'est-ce que la confiscation prononcée pour des crimes qui ont pour but de renverser l'État, le Gouvernement et la fortune publique (car la confiscation n'est proposée que pour des crimes de cette nature); qu'est-ce, dis-je, que la confiscation dans des cas de cette espèce? C'est évidemment une indemnité légitime, toujours trop faible pour la réparation du tort que l'on a fait, et qui ne couvre presque jamais les dépenses qu'on a occasionnées ; la confiscation, qui doit être odieuse quand on l'appliquait sans choix et sans discernement, n'aura rien que de convenable, rien que de juste, lorsqu'elle sera appliquée avec mesure et discrétion.

Je ne vous dirai pas qu'en rejetant la confiscation pour des crimes contre la sûreté de l'État il serait souvent fort à craindre qu'on ne laissât aux ennemis de la chose publique des moyens de lui nuire ; je n'ai pas besoin de ces considérations secondaires pour justifier une mesure toute fondée sur un principe de justice ; déjà même la confiscation a été rétablie pour les crimes de fausse monnaie. Au reste, vous verrez dans la suite combien la rigueur de cette peine est adoucie dans l'exécution, et vous serez convaincus qu'on a su concilier ce que prescrivait la justice et ce que conseillait l'humanité.

Enfin, en nous occupant des voies de répression, nous n'avons pas négligé les moyens de prévenir le mal ; les condamnés, après avoir subi leur peine, demeureront, dans les cas prévus par la loi, sous la surveillance de la haute police.

Dans un petit État tout le monde est surveillé, parce qu'on est pour ainsi dire réuni sur un même point, et que personne ne peut se soustraire à l'œil vigilant de ses concitoyens ; dans un empire immense, il est nécessaire qu'une institution sage et active remplace cette surveillance respective qui ne peut pas y exister ; il faut que les hommes pervers ne soient jamais perdus de vue ; or, quelle dénonciation plus pressante que celle qui résulte d'un arrêt de condamnation ?

Je crois, Messieurs, que cette mesure sera vue avec reconnaissance par tous les amis de la paix publique. Je dirai dans la suite comment elle s'effectuera ; dans ce moment, je ne dois vous parler que des bases en général du projet qui vous est soumis.

J'ai justifié celles que nous avons adoptées en matière

criminelle ; j'ai peu d'observations a faire sur celles en matière correctionnelle.

L'Assemblée constituante punissait les délits par l'amende, la confiscation, en certains cas, de la matière du délit, et par l'emprisonnement.

Nous avons cru devoir ajouter à ces peines celle de l'interdiction, à temps, de certains droits civiques, civils, ou de famille, et même, dans quelques cas, le renvoi sous la surveillance spéciale du Gouvernement. Je n'ai rien à ajouter à ce que j'ai dit sur cette dernière. Quant à la privation temporaire de certains droits, je demanderai quelle peine plus convenable on peut infliger à celui qui, par exemple, aura troublé la paix et commis quelque délit dans une assemblée politique, que celle de lui en interdire l'entrée pendant un certain temps ? Au reste, on a dû prévoir l'abus et ne rien laisser à l'arbitraire du juge ; les peines de cette nature, ainsi que celle de la mise en surveillance, ne seront prononcées que dans les cas où elles seront autorisées par une loi précise.

Après avoir développé les nouvelles bases du projet du Code pénal, je dois vous donner une idée du plan que nous avons suivi.

L'ouvrage est divisé en quatre Livres : le premier énonce les peines établies par la loi ; il prescrit le mode de leur exécution, et il en règle les effets.

Le second a pour objet les personnes punissables, excusables ou responsables pour crimes ou pour délits.

Le troisième détermine la nature de la peine encourue pour chaque crime ou chaque délit commis, soit contre la chose publique , soit contre les particuliers.

Le quatrième enfin est destiné aux contraventions de police et aux peines dont elles sont susceptibles.

Cette division embrasse l'ensemble des matières criminelle et de police ; et vous verrez dans la discussion de ces différens Livres, que nous avons rempli plusieurs lacunes du Code de 1791.

Nous n'apportons aujourd'hui que le premier Livre : il expose, en général, les peines que les tribunaux pourront infliger, sans s'occuper, en aucune manière, de leur application aux faits particuliers. Il règle, comme je l'ai déjà annoncé, le mode d'exécution de ces peines et leurs effets : ces dispositions sont précédées d'un petit nombre d'articles préliminaires.

Le premier de ces articles définit les expressions de *crime*, *délit*, *contravention*, trop souvent confondues et employées indifféremment. Désormais le mot *crime* désignera les attentats contre la société qui doivent occuper les cours criminelles. Le mot *délit* sera affecté aux désordres moins graves qui sont du ressort de la police correctionnelle. Enfin le mot *contravention* s'appliquera aux fautes contre la simple police.

Le second article préliminaire punit des mêmes peines que le crime, les tentatives manifestées par des actes extérieurs, et suivies d'un commencement d'exécution, lorsque cette exécution n'a été suspendue ou n'a manqué son effet que par des circonstances fortuites, indépendantes de la volonté du coupable.

Il a commis le crime autant qu'il était en lui de le commettre ; il a donc encouru la peine prononcée par la loi contre le crime ; la sûreté publique avait déjà provoqué cette disposition, qui se trouve textuellement écrite dans une de nos lois. On peut même dire qu'elle est un développement nécessaire aux deux articles du Code pénal de 1791, qui infligent aux tentatives d'assassinat et d'empoisonnement les mêmes peines qu'au crime consommé.

Mais cette disposition ne peut pas être si généralement adoptée pour les délits, parce que les caractères n'en sont pas aussi marqués que les caractères du crime : leur exécution peut très-bien avoir été préparée et commencée par des circonstances et des démarches qui, en elles-mêmes, n'ont rien de répréhensible, et dont l'objet n'est bien connu que lorsque le délit est consommé; il a donc été sage de déclarer que les tentatives du délit ne seraient considérées et punies comme le délit même que dans des cas particuliers, déterminés par une disposition spéciale de la loi.

Le dernier des articles préliminaires retrace une maxime que l'on peut regarder comme la plus forte garantie de la tranquillité des citoyens : « Nulle contravention, « nul délit, nul crime, ne peut être puni de peines qui « n'étaient pas prononcées par la loi avant qu'ils fussent « commis. »

Un citoyen ne doit être puni que d'une peine légale; il ne doit pas être laissé dans l'incertitude sur ce qui est ou n'est pas punissable; il ne peut être poursuivi pour un acte qu'il a pu, de bonne foi, supposer au moins indifférent, puisque la loi n'y attachait aucune peine.

Vous pouvez, Messieurs, juger par la disposition de cet article, de l'esprit qui a présidé à la rédaction du Code pénal. Vous voyez que, si l'on s'est occupé efficacement de la recherche et de la poursuite des hommes qui se constituent en état de guerre avec la société, on n'a pas apporté moins de soin pour ne pas troubler la sécurité du citoyen paisible, qui ne transgresse les dispositions d'aucune loi.

Le premier Livre, dont vous entendrez bientôt la lecture, donne le tableau des peines que les tribunaux pourront prononcer.

Celles adoptées en matière criminelle, sont : la mort, les travaux forcés à perpétuité, la déportation, les travaux forcés à temps, la réclusion, le carcan, le bannissement, la dégradation civique, la marque, la confiscation, et le renvoi sous la surveillance de la haute police.

L'Assemblée constituante n'avait inséré dans son Code que les peines de mort, des fers, de réclusion, de la gêne, de la détention, de la déportation, de la dégradation civique et du carcan. Nous en avons conservé une partie, et nous avons apporté quelques modifications dans les autres.

Il nous a paru à propos de remplacer par la peine des travaux forcés celle des fers, qui, n'étant établie que pour les hommes, avait mis dans la nécessité d'introduire, particulièrement pour les femmes, la peine de la réclusion ; celle des travaux forcés, que nous substituons, peut être appliquée aux deux sexes, en donnant à chacun l'espèce de travail qui peut lui convenir. Ainsi les femmes ne pourront être employées à ces travaux que dans une maison de force ; les hommes pourront être employés à toute espèce de travaux pénibles, avec les précautions suffisantes pour prévenir leur révolte ou leur évasion.

La peine des travaux forcés étant commune aux deux sexes, nous avons fait de la peine de la réclusion, qui, dans le Code de 1791, est particulière aux femmes, une peine également commune, et nous avons pu supprimer la peine de la détention.

Nous avons aussi supprimé la peine de la gêne, qui consistait à être enfermé dans une maison de force, sans aucune communication à l'extérieur ni avec les autres prisonniers : cette peine était prononcée quelquefois pour vingt ans.

Nous avouerons que nous n'avons pas reconnu, dans cette occasion, les sentiments philantropiques de l'Assemblée constituante.

Quel est donc le sort d'un homme enfermé pour vingt ans, sans espoir de communication ni à l'intérieur ni à l'extérieur ? N'est-il pas plongé vivant dans son tombeau ? Quelle peut être d'ailleurs l'utilité de cette peine ? On ne peut pas dire qu'elle est établie pour l'exemple, puisque le condamné, soustrait à tous les yeux, est mort, pour ainsi dire, à la société ; d'ailleurs il est presque impossible qu'une disposition qui introduit une séquestration aussi sévère soit jamais exécutée ; nouveau motif pour faire disparaître du Code la peine de la gêne.

En supprimant cette peine, nous avons rétabli celle de la relégation ou du bannissement · elle nous a paru convenable pour certains crimes politiques qui, ne supposant pas toujours un dernier degré de perversité, ne doivent pas être punis des peines réservées aux hommes profondément corrompus.

Vous jugerez, Messieurs, dans la suite, si les peines que nous avons cru devoir adopter sont appliquées avec sagesse aux crimes et aux délits : le premier Livre du Code que nous vous présentons, ne s'occupe, je le répète, en aucune manière, de cette application ; les règles en seront tracées dans les autres Livres ; j'ai dû me borner aujourd'hui à vous faire connaître notre système pénal, et à vous donner une idée du mode d'exécution et des effets des peines qui pourront être infligées.

J'aurai peu d'observations à faire sur le mode d'exécution ; il s'éloigne peu du mode actuel, et les dispositions que nous vous présentons sont du nombre de celles qu'il suffit de lire pour les justifier.

L'Assemblée constituante a réduit la peine de mort à la simple privation de la vie; en applaudissant à cette mesure, nous avons cependant pensé qu'elle devait éprouver une légère dérogation pour un crime qu'on ne peut pas se dispenser de prévoir, puisqu'il ne nous est malheureusement pas permis de le regarder comme impossible, pour le parricide; le monstre aura le poing coupé: puisse notre siècle n'avoir jamais à rougir de cet horrible forfait !

Les condamnés à la peine des travaux forcés à perpétuité seront toujours flétris sur la place publique, par l'application d'une empreinte avec un fer chaud, sur l'épaule droite; les condamnés à d'autres peines ne subiront cette flétrissure que dans les cas où la loi l'aura attachée à la peine qui leur est infligée.

Ceux qui seront condamnés à la peine des travaux forcés à perpétuité ou à temps, et à la peine de réclusion, seront, avant de subir leur peine, attachés au carcan sur la place publique, pour y demeurer exposés aux regards du peuple, durant une heure.

La déportation s'effectuera par un transport dans un lieu déterminé par le Gouvernement, hors du territoire continental de l'Empire, et pour y demeurer à perpétuité.

Les condamnés au bannissement seront transportés hors du territoire de l'Empire; s'ils y rentrent avant le temps prescrit, ils seront punis de la peine de la déportation.

Si les déportés rentrent, ils subiront la peine des travaux forcés à perpétuité.

Celui qui aura été condamné à la réclusion sera renfermé dans une maison de force, et employé à des travaux dont le produit pourra être en partie appliqué à son profit.

La dégradation civique consistera toujours dans la destitution et l'exclusion des condamnés de toutes fonctions ou emplois publics; ces dispositions ne présentent rien de nouveau, rien qui exige une explication.

Quant à la durée des peines temporairement infligées, l'échelle en a été graduée de manière à correspondre à l'échelle des crimes, en sorte que la proportion entre le fait et la peine ne sera jamais rompue.

Vous avez vu dans le Code d'instruction criminelle, que les tribunaux de police ne pourront prononcer la peine d'emprisonnement que pour cinq jours; la peine d'emprisonnement, en matière correctionnelle, ne pourra être prononcée pour moins de six jours, ni pour plus de cinq ans, sauf les cas de récidive.

La durée de la peine du bannissement et de celle de la réclusion sera au moins de cinq ans, et de dix ans au plus.

La peine des travaux forcés ne pourra, comme les précédentes, être moindre de cinq années; elle ne pourra pas en excéder vingt.

Le projet règle au surplus avec précision le moment où commencera la peine, le lieu où seront faites les exécutions, les jours où il ne sera pas permis d'en faire.

Il serait superflu d'entrer dans des explications sur ces objets de détail; je passe aux effets des peines prononcées. Je crois pouvoir me dispenser de remarquer que toute peine en matière criminelle est infamante, et que les peines des travaux forcés à perpétuité et de la dégradation emportent la mort civile.

L'effet de la condamnation aux travaux forcés à temps, au bannissement, à la réclusion, ou au carcan, ne doit pas être aussi étendu; mais la tache d'infamie imprimée sur le front des condamnés ne permet pas que leur témoignage soit admis en justice, et surtout leur présence ne doit jamais souiller les rangs des braves qui ont porté si loin la gloire du nom français; ils sont, en conséquence, déclarés déchus du droit de servir dans les armées de Sa Majesté.

Ceux qui ont été condamnés à la peine des travaux forcés à temps et de la réclusion sont, de plus, pendant la durée de leur peine, dans un état d'interdiction légale; il ne faut pas, comme il est trop souvent arrivé, que des profusions scandaleuses fassent d'un séjour d'humiliation et de deuil un théâtre de joie et de débauche.

Le curateur qui administrera le bien du condamné ne pourra lui faire aucune remise de ses revenus pendant la durée de la peine; lorsqu'elle sera subie, le curateur rendra compte de son administration.

La confiscation ne pourra jamais porter le moindre préjudice aux droits acquis par des tiers sur les biens du condamné; si une sévérité juste et politique a nécessité l'adoption de cette mesure, l'humanité en tempérera la rigueur dans l'exécution; non-seulement les biens confisqués demeurent grevés des dettes légitimes, ce qui est de toute justice, mais les enfants et la famille du condamné éprouveront encore la bienfaisance du Gouvernement: les enfants recevront la moitié de la portion dont leur père n'aurait pu les priver dans sa succession; les parents qui pouvaient avoir droit à des aliments n'en seront pas déchus, et l'Empereur pourra encore disposer en tout ou en partie des biens confisqués en faveur des père, mère, enfants, ou des autres parents des condamnés. C'est ainsi qu'après avoir assuré la punition du coupable, la loi prépare le moyen de récompenser la bonne conduite des membres de sa famille.

Je passe aux effets du renvoi sous la surveillance de la haute police de l'État.

Nous devons attendre, comme je l'ai déjà observé, des résultats heureux de cette mesure; mais il a fallu prévoir les abus de l'exécution, et ne tolérer que la rigueur qui est indispensable.

Celui qui sera placé sous cette surveillance donnera une caution solvable de bonne conduite; on pourra exiger une caution de ses père, mère, tuteur ou curateur, s'il est en âge de minorité; toute personne pourra même être admise à fournir pour lui cette caution; à son défaut, le Gouvernement peut ordonner l'éloignement du condamné, même lui indiquer une résidence dans un lieu déterminé; et s'il n'obéit point à l'ordre qu'il aura reçu, le Gouvernement pourra le faire arrêter, et le détenir pendant tout le temps fixé pour l'état de surveillance.

Indépendamment des peines dont je viens de parler, les cours et tribunaux peuvent encore prononcer des restitutions, des amendes, des condamnations de frais; le projet pourvoit aussi au mode d'exécution de ces dispositions; mais les articles n'en sont susceptibles d'aucune observation particulière.

Il ne me reste plus actuellement qu'à vous faire connaître le dernier chapitre du premier Livre du Code pénal; il est relatif aux peines de la récidive pour crimes et délits.

Un premier crime ne suppose pas toujours nécessairement l'entière dépravation de celui qui s'en est rendu coupable; mais la récidive annonce des habitudes vi-

cieuses et un fond de perversité, ou au moins de faiblesse non moins dangereuse pour le corps social que la
perversité.

Un second crime doit donc être réprimé avec plus de
sévérité que le premier.

L'Asssmblée constituante n'a établi contre le second
crime que la peine prononcée par la loi, sans distinction
de la récidive ; mais elle a voulu qu'après la peine subie
les condamnés pour récidive fussent déportés ; disposition qui ne nous paraît pas conforme aux règles d'une
justice exacte, puisqu'elle ne fait aucune différence entre
celui dont le second crime entraîne la peine de la réclusion, et celui dont le second crime emporte la peine de
vingt-quatre années de fers, la plus grave du Code de
1791, après celle de mort.

Il nous a paru convenable de chercher une autre
règle plus compatible avec les proportions qui doivent
exister entre les peines et les crimes ; elle se présente naturellement : c'est d'appliquer au crime, en cas de récidive, la peine immédiatement supérieure à celle qui de-
vrait être infligée au coupable, s'il était condamné pour
la première fois.

Ainsi, si le second crime emporte la peine de la dégradation civique, le coupable sera puni de celle du carcan ; si le second crime emporte la peine du carcan, ou
celle du bannissement, le coupable sera condamné à
celle de la réclusion ; il sera condamné à la peine des travaux forcés à temps, si le second crime emporte la peine
de la réclusion ; à la peine des travaux forcés à perpétuité, si le second crime emporte celle des travaux forcés
à temps ou de la déportation ; et enfin il sera condamné
à la mort, si le second crime emporte la peine des travaux forcés à perpétuité.

2.

Lorsque le condamné pour un crime n'aura commis depuis qu'un délit de nature à être puni correctionnellement, il sera toujours condamné, dans ce cas, au *maximum* des peines correctionnelles, et même la condamnation pourra s'élever jusqu'au double, c'est-à-dire, jusqu'à dix ans.

Vous connaissez actuellement, Messieurs, toutes les bases sur lesquelles s'est élevé le nouveau Code ; nous le proposons avec confiance ; l'adoption que vous en ferez complétera notre législation criminelle.

Le Code d'Instruction, que vous avez sanctionné dans l'avant-dernière session, garantit que les méchants seront poursuivis, atteints et punis. Le Code pénal garantira les proportions qui doivent exister entre les peines et les crimes, ou les délits.

Nous n'avons jamais perdu de vue le but que nous devions atteindre, celui de concilier la sûreté publique qui réclame des peines répressives, et le vœu de l'humanité qui repousse toute rigueur qui n'est pas nécessaire.

J'ose dire que cet ouvrage porte l'empreinte de la sagesse profonde qui caractérise tous les Codes que Sa Majesté a donnés à la nation : le Code pénal méritera aussi la reconnaissance du peuple français, l'hommage des contemporains et le respect de la postérité.

MOTIFS DU LIVRE II,

PRÉSENTÉS

Par MM. le chevalier Faure, et les comtes Berlier et Portalis, Conseillers d'État.

Séance du 3 février 1810.

Messieurs,

Vous avez entendu, dans la dernière séance, l'exposé du système pénal qui forme la base du nouveau Code des délits et des peines.

Tel est l'objet du Livre premier.

Sa Majesté nous a chargés de vous présenter aujourd'hui le second Livre, qui contient plusieurs dispositions générales, destinées à faciliter l'application des cas particuliers, et à prévenir un grand nombre de difficultés qu'ils pourraient faire naître.

Cette partie regarde spécialement les complices et les personnes excusables ou responsables pour crimes ou délits.

Le Code pénal de 1791 ne parle que des complices de crime ; la loi rendue dans le cours de la même année sur les délits de police correctionnelle est muette à l'égard de la complicité. L'usage autorisé par la raison a rendu communes à cette dernière loi les règles établies par la première.

Comme le Code actuel ne s'occupe pas seulement de la répression des crimes, et que celle des délits est également l'objet de sa prévoyance, ses dispositions sur les complices s'appliquent aux uns et aux autres ; les expressions mêmes du Code ne permettraient pas d'élever le plus léger doute sur ce point.

Le Code établit d'abord pour règle générale que le complice d'un crime ou délit sera puni de la même peine que celui qui en est l'auteur. Cependant comme cette règle est susceptible de quelques exceptions, quoique très-rares, le Code permet ces exceptions, pourvu qu'elles soient le résultat d'une disposition de la loi ; elles trouveront leur place naturelle dans les articles relatifs aux cas pour lesquels elles seront jugées nécessaires.

La définition donnée par le Code, de ce qui constitue la complicité, est à peu près la même que celle de la loi de 1791 ; elle s'applique à toute personne convaincue d'avoir préparé ou facilité l'action par des moyens qu'elle savait devoir y servir.

Provocations faites, instructions données, armes fournies, peu importe le moyen ; c'est d'après le même esprit que le Code ajoute une disposition qui n'était point dans la loi de 1791 ; il veut que ceux-là soient déclarés complices, et punis comme tels, qui, connaissant la conduite criminelle des malfaiteurs, les logeront habituellement chez eux, ou souffriront qu'ils s'y réunissent habituellement. Car, dès qu'ils n'ignorent pas que ces hommes ne vivent que de crimes, ils ne peuvent se dissimuler que la retraite qu'ils leur donnent est un moyen de faciliter l'exécution de leurs desseins criminels ; la même observation s'applique aux recéleurs d'objets volés.

Nous remarquerons une distinction établie par le nouveau Code, et réclamée depuis long-temps par l'expérience. Lorsque le vol ne donne lieu qu'à des peines temporaires, il faut, quelque rigoureuses qu'elles soient, que le recéleur subisse la même peine ; il s'est soumis à ce risque dès qu'il a bien voulu recevoir une chose qu'il savait provenir d'un vol. Mais, lorsque le crime est accompagné de circonstances si graves, qu'elles entraînent la peine de mort, ou toute autre peine perpétuelle, on peut croire que si au temps du recélé ces circonstances eussent été connues du recéleur, il eût mieux aimé ne pas recevoir l'objet volé que de s'en charger avec un si grand risque; il convient donc, en pareil cas, pour condamner le recéleur à la même peine que l'auteur du crime, qu'il y ait certitude qu'en recevant la chose il connaissait toute la gravité du crime dont elle était le fruit. A défaut de cette certitude, la sévérité de la loi doit se borner à prononcer contre lui la peine la plus forte parmi les peines temporaires. C'est ce que décide le nouveau Code. L'absence d'une distinction si sage a souvent été cause que des recéleurs sont restés impunis. On a déclaré des recéleurs non convaincus de complicité, pour ne pas leur faire subir une peine dont l'excessive rigueur paraissait injuste.

Une autre règle commune à tous les prévenus, soit du fait principal, soit de complicité, est qu'on ne peut déclarer coupable celui qui était en état de démence au temps de l'action, ou qui, malgré la plus vive résistance, n'a pu se dispenser de céder à la force. Tout crime ou délit se compose du fait et de l'intention : or, dans les deux cas dont nous venons de parler, aucune intention criminelle ne peut avoir existé de la part des prévenus, puisque l'un ne jouissait pas de ses qualités morales, et qu'à l'égard de l'autre la contrainte seule a dirigé l'emploi de ses forces physiques.

Après cette disposition, le Code rappelle que nulle excuse ne peut être admise, à moins que la loi même ne déclare le fait excusable; ce principe est déjà consacré par l'article 339 du Code d'instruction criminelle.

Il ajoute que nulle peine ne peut être mitigée, ex-cepté dans les cas où la loi l'autorise formellement.

Ces deux dispositions ont pout but de prévenir l'arbitraire qui substitue les passions, toujours mobiles et souvent aveugles de l'homme, à la volonté ferme et constante de la loi.

Le Code détermine ensuite l'influence de l'âge des condamnés sur la nature et la durée des peines.

Il s'occupe d'abord de celui qui, au moment de l'action, n'avait pas encore seize ans. On se rappelle que l'article 340 du Code d'instruction criminelle a décidé, qu'à l'égard de l'accusé qui se trouverait dans cette classe, la question de savoir s'il a commis l'action avec discernement serait examinée. Les dispositions actuelles règlent ce qui doit être ordonné d'après le résultat de l'examen. Si la décision est négative, l'accusé doit nécessairement être acquitté; car il serait contradictoire de le déclarer coupable d'un crime, et de dire en même temps que ce dont il est accusé a été fait par lui sans discernement. Les juges prononceront donc qu'il est acquitté; mais ils ne pourront pas le faire rentrer dans la société, sans pourvoir à ce que quelqu'un ait les regards fixés sur sa conduite : ils auront l'option de le rendre à ses parents, s'ils ont en eux assez de confiance, ou de le tenir renfermé durant un espace de temps qu'ils détermineront. Cette détention ne sera point une peine, mais un moyen de suppléer à la correction domestique, lorsque les circonstances ne permettront pas de la confier à sa famille.

Sa plus longue durée n'excédera jamais l'époque où la personne sera parvenue à l'âge de vingt ans accomplis. Ces limites laissent un intervalle suffisant pour que les juges puissent proportionner la précaution au besoin : mais si la décision porte que l'action a été commise avec discernement, il ne s'agit plus de correction; c'est une peine qui doit être prononcée. Seulement ce ne sera ni une peine afflictive, ni une peine infamante. La loi suppose que le coupable, quoique sachant bien qu'il faisait mal, n'était pas encore en état de sentir toute l'étendue de la faute qu'il commettait, ni de concevoir toute la rigueur de la peine qu'il allait encourir. Elle ne veut point le flétrir, dans l'espoir qu'il pourra devenir un citoyen utile; elle commue, en sa faveur, les peines afflictives en peines de police correctionnelle : elle ne le soumet point à l'exposition aux regards du peuple ; enfin, elle consent, par égard pour son jeune âge, à le traiter avec indulgence, et ose se confier à ses remords.

Quant à la proportion établie pour la durée de ces peines, relativement à celles qu'eût subies le condamné, s'il avait eu plus de seize ans, nous nous abstiendrons d'entrer dans des détails qui seront suffisamment connus par la lecture des articles; ils sont d'ailleurs conformes à la loi de 1791.

Après avoir parlé de l'indulgence de la loi pour un âge où l'inexpérience atténue la faute, nous allons faire connaître son humanité pour une autre époque de la vie, où les forces du corps sont présumées n'être plus capables de supporter une peine très-rigoureuse. Le Code fixe cette époque à soixante-dix ans. Celui qui sera parvenu à cet âge, au moment de son jugement, ne sera condamné ni aux travaux forcés à perpétuité, ni à la déportation, ni même aux travaux forcés à temps; les juges prononceront contre lui la réclusion pour le temps qu'eût duré la peine qu'il aurait subie s'il n'eût pas été septuagénaire : lorsqu'il n'atteindra les soixante-dix ans que depuis sa condamnation, la peine de la réclusion doit remplacer aussi celle à laquelle il avait été condamné, et il subira cette nouvelle peine jusqu'à l'expiration du temps que portait le jugement.

On observera cependant que le dernier cas regarde seulement les condamnés aux travaux forcés à perpétuité ou à temps. Quant à celui contre qui la déportation a été prononcée, il est facile de sentir que, lorsqu'il ne devient septuagénaire qu'après avoir été transporté hors du territoire continental de l'Empire, et s'être fixé dans le lieu déterminé par le Gouvernement, sa nouvelle situation rend moins désirable pour lui cette commutation de peine, et qu'il ne trouverait pas assez d'avantage dans un retour dont l'unique effet serait une réclusion perpétuelle.

En rapprochant le mode proposé de celui qu'adopta l'Assemblée constituante, on aperçoit plusieurs différences. Suivant la loi de 1791, il faut, pour que le sort du septuagénaire soit adouci, qu'il ait atteint l'âge de soixante-quinze ans. Alors la durée de la peine est réduite à cinq années : ici la commutation n'est que pour la durée ; il ne s'en opère aucune dans la nature du châtiment. Si le crime emporte les fers, le coupable doit subir cette peine, quel que soit son âge, sauf la réduction du temps.

Pour nous, Messieurs, nous avons pensé qu'il serait plus convenable de ne rien changer à la durée de la peine, mais d'y substituer la réclusion, comme mieux appropriée à l'état d'un vieillard. Les travaux forcés seraient trop rigoureux pour la plupart des septuagénaires : il n'en est pas ainsi de la réclusion ; et comme le but de la loi ne peut être de faire rentrer dans la société le coupable qui a soixante-dix ans, plutôt qu'un autre coupable moins âgé, comme il s'agit uniquement d'empêcher qu'il ne succombe par l'effet de travaux et de fatigues excessives, on a donné la préférence au mode proposé.

Il nous reste à parler d'une espèce de responsabilité qu'il appartenait au Code pénal de consacrer dans ses dispositions; c'est celle des aubergistes et hôteliers qui n'auront pas inscrit sur leurs registres, le nom, la profession et le domicile des personnes qu'ils ont logées.

Si ces personnes ont, pendant leur séjour, commis un crime ou délit, ils seront responsables de tout dommage qui en sera résulté. Ils devront s'imputer d'avoir négligé de prendre ces précautions salutaires, qu'une sage police a prescrites dans tous les temps. On ne doit pas perdre de vue qu'ils ne seront soumis à cette responsabilité que lorsque le coupable qu'ils ont reçu dans leur maison y aura passé plus de vingt-quatre heures. Il eût été trop rigoureux, et même injuste, de leur appliquer la peine, quelque courte qu'eût été la durée de son séjour. Lorsqu'un voyageur ne s'arrête que pendant quelques heures dans une hôtellerie, et disparaît pour faire place à d'autres qui n'y restent pas plus long-temps, il serait le plus souvent impossible de remplir à l'égard du premier, comme à l'égard de ceux qui lui succèdent, toutes les formalités exigées par la loi. L'hôtelier ne doit répondre que de celui qu'il a été à portée de voir; mais il est inexcusable de ne s'être pas mis en règle lorsque la personne qu'il a logée n'a quitté sa maison qu'après les vingt-quatre heures.

Cette responsabilité est ajoutée aux différentes espèces prévues par le Code Napoléon. Nous nous contenterons de rappeler l'article 1384 de ce Code, qui porte qu'on est responsable, non-seulement du dommage que l'on cause par son propre fait, mais encore de celui qui est causé par le fait des personnes dont on doit répondre, ou des choses que l'on a sous sa garde. Les cas spécifiés dans ce même article et dans les articles suivants serviront d'appendice à cette partie du Code pénal.

Tels sont, Messieurs, les motifs sur lesquels repose le projet de loi soumis à votre sanction. Vous trouverez sans doute que les améliorations qu'il contient sont une nouvelle preuve des soins constants que Sa Majesté apporte à tout ce qui peut contribuer au perfectionnement des lois.

MOTIFS

Du Livre III, Titre I, Chapitres I et II,

PRÉSENTÉS

par MM. les comtes Berlier, Corsini et Pelet,
Conseillers d'État.

Séance du 5 février 1810.

Messieurs,

La nature des peines instituées par le nouveau projet de Code vous est déjà connue.

Il s'agit aujourd'hui d'en faire l'application aux diverses espèces de crimes et de délits qui affligent la société, et de commencer la nombreuse et triste nomenclature des actes qui portent ce caractère.

Ce tableau sera long, bien qu'il ne doive pas embrasser, d'une manière générale et absolue, tout ce qui est nuisible ou funeste; ainsi vous n'y verrez point figurer beaucoup d'actes qui, simplement contraires à la bonne foi ou à la délicatesse, peuvent être quelquefois réprimés par la seule voie civile; vous n'y verrez pas non plus retracer les trop nombreux générateurs des crimes, je veux dire les vices, redoutables fléaux qui échappent à l'empire des lois pénales, et dont il n'appartient qu'à d'autres institutions de prévenir ou de diminuer les ravages.

En ne traitant ici que des crimes et délits, et de leur punition, le sujet est vaste encore, et n'a que trop d'étendue.

Il n'y a sur ce point que bien peu de lumières à puiser dans les anciens usages de la monarchie. Qu'était-ce, en effet, que notre législation pénale jusqu'à l'époque où une Assemblée mémorable vint poser sur cet important objet des règles qui, reçues alors avec enthousiasme, doivent encore aujourd'hui être méditées avec respect, parce qu'elles émanaient de vues très-pures et de principes généralement vrais ?

Toutefois, malgré les lumières de cette assemblée, il était difficile qu'un si grand ouvrage atteignit, dès le début, toute la perfection dont il était susceptible.

Aussi le Code pénal de 1791 a-t-il déjà éprouvé d'assez importantes modifications.

L'on entreprend aujourd'hui de l'améliorer encore, et l'auguste chef de l'Empire qui a porté son active sollicitude sur les autres parties de la législation ne pouvait refuser à celle-ci ce vigilant et sage intérêt par lequel son règne sera illustré autant que par ses victoires.

Dans les détails qui vont, Messieurs, passer sous vos yeux, l'on n'a pas oublié que des lois qui statuent sur tout ce que les hommes ont de plus cher, la vie et l'honneur, ne doivent effrayer que les pervers ; but qui serait manqué, si elles imprimaient trop légèrement le caractère de crime à des actes qui ne sont pas essentiellement criminels.

L'on a soigneusement cherché à établir de justes proportions entre les peines et les délits.

L'on a enfin mis une extrême attention à n'omettre aucuns délits et à les bien préciser ; car, dans une société bien organisée, où les hommes sont placés sous l'égide de la loi, de telle sorte que nul ne peut être puni que des peines et pour les délits qui y sont exprimés, une juste inquiétude naîtrait dans l'âme de tous, si un seul pouvait être poursuivi criminellement pour des faits auxquels la loi n'aurait pas attaché ce caractère par une disposition formelle et non équivoque.

Ces idées fondamentales sont des guides dont on ne saurait, dans le travail qui nous occupe, s'écarter un seul instant.

Que dirai-je du plan et de la distribution des matières ? Deux grandes divisions s'y présentent; d'abord, *les crimes et délits contre la chose publique*; ensuite, *les crimes et délits contre les particuliers*.

Il eût sans doute été facile de multiplier les classes principales : un traité récent et estimé [1] donne un frappant exemple du vaste champ que la seule division des matières ouvrait aux combinaisons du législateur; mais s'il y a quelque fruit à recueillir de ces profondes méditations des jurisconsultes et des publicistes, c'est en les rattachant à la loi par des points imperceptibles. La métaphysique et la législation ont des formes et un langage différents.

Loin donc de multiplier les cadres principaux, le projet de loi resserre même ceux qui existent aujourd'hui.

Ainsi, dans l'état présent de notre législation, les crimes d'une part, et les délits de l'autre, sont classés séparément, et placés même dans deux Codes distincts.

Au premier aspect, cette division séduit et paraît utile, parce qu'elle s'applique à des faits qui n'ont pas la même gravité, et à des peines qui ne sont pas du même ordre.

Cependant les avantages de cette division ne sont qu'éphémères, et ses inconvénients sont réels; car tel

---

[1] Traité de Législation, par *Jérémie Bentham*.

délit de police correctionnelle peut, avec une circonstance de plus, s'élever à la qualité de crime; et tel crime peut, avec une circonstance de moins, n'être plus qu'un délit.

Un fait parfaitement identique, s'il est considéré sans acception de personnes, peut changer de classe, selon, par exemple, qu'il a été commis par un fonctionnaire public, ou par un simple particulier, ou suivant qu'il l'a été contre les ministres de la loi ou contre d'autres personnes.

Dans cette position, il a semblé convenable de ne point diviser en plusieurs tableaux les crimes et délits qui s'appliquent à des faits de même catégorie, quoique d'une intensité différente : pourquoi le même titre n'embrasserait-il pas le faux commis dans un testament, comme celui commis dans un passe-port? Ce qui est important et juste, c'est qu'un délit ne soit pas puni aussi sévèrement qu'un crime; mais ce qui est utile aussi, c'est que l'on puisse embrasser du même coup-d'œil tous les crimes et délits qui s'appliquent à la même catégorie de faits.

Unir ce qui a de tels rapports, ce n'est point confondre; et la confusion, ou du moins l'embarras, commencerait bien plutôt là où il faudrait, sur des questions analogues, recourir à des règles éparses.

Le nouveau projet de Code traitera donc à la fois des *crimes et délits* sur chaque matière, et des peines qui leur sont applicables.

Au surplus, si dans le langage ordinaire le mot *délit* a une double acception, et est pris tantôt pour le genre, tantôt pour l'espèce, il n'aura dans notre classification que cette dernière acception, et ne s'appliquera qu'à des infractions de moindre gravité que les crimes.

Le nouveau projet divise donc les crimes et délits en deux classes principales, les uns *contre la chose publique*, et les autres *contre les particuliers* : vaste division à laquelle viennent nécessairement aboutir toutes les infractions que l'imagination peut embrasser.

C'est en partant du même point que les lois romaines s'étaient bornées à la distinction des délits *publics*, pour lesquels le droit d'accusation était accordé à tout citoyen, et des délits *privés*, dont la réparation ne pouvait être poursuivie que par les parties lésées.

Si le droit d'accusation est chez nous soumis à d'autres règles, et si notre classification des crimes et délits diffère beaucoup dans les détails avec la classification romaine, la division principale en crimes et délits *publics* et *privés*, ou, ce qui est la même chose, en crimes et délits *contre la chose publique* et *contre les particuliers*, n'en a semblé ni moins juste ni moins utile; non, sans doute, qu'il n'existe entre l'État et ses membres une connexion intime et telle que les membres de l'association souffrent quand le corps de l'État est attaqué, et réciproquement : à Dieu ne plaise que la division proposée porte jamais à oublier ou méconnaître un principe d'une si haute utilité ! mais il est pourtant dans la nature des choses que l'atteinte directe regarde principalement quelquefois la chose publique, quelquefois les particuliers, et cette définition a pû être prise pour base première de la division des crimes et délits.

La loi qui vous est aujourd'hui proposée, Messieurs, et celle qui la suivra immédiatement, ne traitent que des crimes ou délits *contre la chose publique.*

Ces crimes ou délits sont sous-divisés en trois espèces, ceux *contre la sûreté de l'État*, ceux *contre les constitutions de l'Empire*, et ceux *contre la paix publique.*

Les crimes ou délits contre la sûreté de l'État sont eux-mêmes de deux sortes : ils attaquent la sûreté *extérieure*, ou compromettent la sûreté *intérieure*.

Sous l'un comme sous l'autre rapport, ils sont d'une extrême gravité : l'on va néanmoins, pour obtenir plus de clarté, retracer séparément les dispositions relatives à chacune de ces espèces, en commençant par les crimes ou délits dirigés contre la *sûreté extérieure de l'État*.

C'est ici que figureront ces Français dénaturés qui portent les armes contre leur patrie, qui entretiennent des intelligences avec l'ennemi, qui recèlent ses espions, ou qui lui livrent, soit des plans, soit le secret d'une négociation.

De si grands crimes n'admettent d'autre peine que la mort ; peine terrible que le législateur n'inflige qu'avec regret, mais qui, selon les expressions de Montesquieu (1), est *comme le remède de la société malade*.

Toutefois il convenait de bien caractériser les intelligences criminelles, pour qu'elles ne fussent point confondues avec des correspondances imprudentes.

Il convenait aussi de tracer une ligne de démarcation entre les communications données par les dépositaires eux-mêmes ou par d'autres personnes.

C'est ce qui a été fait en punissant toujours, mais en punissant moins ceux qui sont coupables à un moindre degré.

Ceux qui, par des actions hostiles ou des actes non approuvés par le Gouvernement, exposent l'État à une déclaration de guerre, compromettent sans doute la sûreté extérieure.

---

(1) Esprit des Lois, liv. XII, chap. IV.

La loi les proclamera donc coupables, bien que nul soupçon d'intelligence avec l'ennemi ne plane sur eux; mais comme, relativement à leurs actes, il n'est pas d'éléments susceptibles d'indiquer jusqu'à quel point les conséquences pouvaient en être connues de leurs auteurs, ceux-ci ne seront pas punis de la peine capitale, mais déportés ou bannis, selon les suites plus ou moins graves qu'auront eues leurs téméraires démarches.

En suivant l'ordre du projet, je dois maintenant vous entretenir des peines infligées aux crimes dirigés contre *la sûreté intérieure de l'État*.

Au premier rang de ces crimes, est celui de *lèse-majesté*. L'on a long-temps abusé de ce mot : plusieurs lois des empereurs romains déclaraient sacriléges, ou coupables de lèse-majesté, ceux qui avaient osé douter du mérite des personnes appelées par le prince à quelque emploi (1), ceux qui attentaient contre les ministres ou officiers du prince (2), et même les fabricateurs de fausse monnaie (3).

L'on admit aussi le crime de lèse-majesté divine, et l'on distingua le crime de lèse-majesté proprement dit, en plusieurs espèces; il fut, selon les circonstances, qualifié au premier ou au deuxième chef.

Cette législation diminuait, par de fausses applications, l'horreur que doit inspirer le crime de lèse-majesté.

---

(1) . . . . . *Dubitare an is dignus sit quem elegerit imperator.* Leg. 3, C. de Crim. sacril.

(2) . . . . . . *Nam et ipsa pars corporis nostri sunt.* Leg. 5, C. ad leg. Jul. majest.

(3) . . . . . *Majestatis crimen committunt.* Leg. 2, C. de falsâ Monetâ.

Ce crime est, par notre projet, réduit à des termes simples ; celui-là seul en est coupable, qui a eu part à un *attentat ou complot dirigé contre la personne ou la vie de l'Empereur*, et comme ce crime ainsi qualifié est le plus énorme de tous, il sera puni de la peine réservée au parricide ; c'est-à-dire, de la seule qui soumette le coupable à quelques mutilations avant qu'il reçoive la mort.

Si l'attentat ou le complot est dirigé, non contre la personne ou la vie du prince, mais contre l'autorité impériale ou contre les membres de la famille régnante, un tel crime, quelle que soit sa gravité, ne sera point assimilé au parricide, mais il n'entraînera pas moins la peine capitale, bien due, sans doute, à un forfait qui répand une si grande alarme dans la société.

Au surplus, ces mots mêmes, *attentat* et *complot*, avaient-ils un sens assez déterminé pour qu'il ne fût pas utile de les définir ? Si les définitions ne conviennent point aux faits dont le caractère est vulgairement fixé, et si alors elles sont plus dangereuses qu'utiles, il n'en est pas ainsi quand il s'agit d'imprimer un caractère spécial de crime à des projets qui, s'ils s'appliquaient à des délits ordinaires, seraient toujours odieux, mais ne seraient point alors considérés comme le délit même.

Deux hommes ont-ils le dessein de voler leur voisin ; cette horrible et funeste pensée ne sera pourtant pas reprimée comme le vol, si elle n'a été suivie d'aucun commencement d'exécution ; mais dans les crimes d'État, le complot formé est assimilé à l'attentat et au crime même.

Ainsi, dans cette matière, le crime commence et existe déjà dans la seule résolution d'agir, arrêtée entre plusieurs coopérateurs : le suprême intérêt de l'État ne permet pas d'attendre et de ne considérer comme criminels que ceux qui ont déjà agi.

La simple proposition non agréée de former un complot est punissable elle-même, mais à un moindre degré; car, bien qu'il n'ait manqué à celui qui a fait la proposition que de trouver des gens qui voulussent s'associer à ses desseins criminels, cependant le danger et l'alarme n'ont pas été portés au même point que si le complot eût réellement existé.

Hors la classe des attentats ou complots dirigés d'une manière spéciale contre le chef de l'État, sa famille ou son autorité, il est d'autres crimes qui compromettent encore la sûreté intérieure.

Ici se présentent les complots tendant à exciter la guerre civile, le massacre ou le pillage, soit des propriétés publiques, soit de celles qui appartiendraient à une généralité de citoyens; les enrôlements illicites, la rétention illégale du commandement de la force publique, l'emploi de cette force contre la levée des gens de guerre, la destruction des ports, arsenaux et autres établissements de cette espèce; crimes qui sont tous bien dignes du dernier supplice.

Mais quand quelques-uns de ces crimes, ou d'autres de même nature, seront commis ou tentés par des bandes séditieuses, il faudra infliger les peines avec la juste circonspection que commandent des affaires aussi complexes.

Dans cette multitude de coupables, tous ne le sont pas au même degré, et l'humanité gémirait si la peine capitale était indistinctement appliquée à tous, hors les cas où la sédition serait dirigée contre la personne ou l'autorité du prince, ou aurait pour objet quelques crimes approchant de cette gravité.

Les chefs et directeurs de ces bandes, toujours plus influents et plus coupables, ne sauraient être trop punis; en déportant les autres individus saisis sur les lieux, on satisfera aux besoins de la société sans alarmer l'humanité.

L'on pourra même user d'une plus grande indulgence envers ceux qui n'auront été arrêtés que depuis, hors des lieux de la réunion séditieuse, sans résistance et sans armes.

La peine de la sédition sera, sans inconvénients, remise à ceux qui se seront retirés au premier avertissement de l'autorité publique : ici la politique s'allie à la justice; car, s'il convient de punir les séditieux, il n'importe pas moins de dissoudre les séditions.

Nous venons, Messieurs, de fixer votre attention sur les principales dispositions ayant trait aux crimes et complots qui attaquent la sûreté de l'État : mais comment, en cette matière, traitera-t-on les provocateurs?

Quelque grave que soit la peine que le projet leur destine, puisqu'il les considère comme complices, quand la provocation a été suivie d'effet, ce n'est point sans doute ce qui peut alarmer, si d'ailleurs la provocation est bien caractérisée; or, elle ne pourra résulter que de discours tenus en lieux ou réunions publics, ou d'écrits placardés ou imprimés.

A ces premiers caractères il faut en ajouter un autre; la provocation devra être directe.

Ainsi quelques vœux insensés, ou quelques rêves criminels, couchés sur un papier manuscrit et non colporté, ne constitueront pas la provocation que la loi assimile au crime même; et, s'ils sont découverts et de nature à appeler la surveillance de l'autorité publique, ce sera sans excéder les bornes posées par une sage prévoyance : un gouvernement fort et juste ne relevera ni l'échafaud de *Sidney*, ni celui de ce malheureux Syracusain qui, ayant rêvé qu'il avait tué *Denys* le tyran, fut condamné à mort parce que ses juges trouvèrent, dans son rêve même, la preuve qu'il s'était occupé de cet objet pendant ses veilles : une telle extension du droit de punir est trop loin de nos mœurs et de la justice.

Parmi les peines qui seront infligées à certains crimes d'Etat, je n'ai point nommé encore *la confiscation*, qui, en cette matière, suivra ordinairement la peine de mort.

La confiscation ! ce mot, qui laisse de si tristes souvenirs, sera, dans son application actuelle, facile à justifier.

Il ne s'agit point, comme on vous l'a déjà annoncé, de faire revivre ce système de confiscation qui, s'appliquant à une foule de délits communs, semblait n'exister que pour l'avantage du fisc ou des seigneurs haut-justiciers.

C'est avec raison sans doute que de graves écrivains (1) ont censuré ce déplorable usage ; ils s'étonnaient justement que la législation punît les enfants du crime de leurs pères, et que le fisc s'enrichît du malheur des familles.

De si puissantes considérations ne pouvaient manquer de partisans dans le conseil d'un prince qui, lui-même, y rappellerait les idées libérales, si elles cessaient d'y ré-

----

(1) Esprit des Lois, tome I, liv. V, chap. XI.
Beccaria, passim, et Commentaires à la suite, §. II.
Jérémie Bentham, troisième partie, chap. IV.
Voyez aussi le parallèle du Code pénal de l'Angleterre avec les lois pénales de France, par *Bexon*, chap. XIX.

gner : mais, odieuse lorsqu'elle s'étend à une multitude de délits communs, la confiscation n'est plus que juste quand, restreinte, comme dans notre Code, aux principaux crimes d'Etat et à la fabrication de la fausse monnaie, et ne s'exerçant d'ailleurs qu'après de fortes et nombreuses déductions au profit des familles, elle ne saurait plus être considérée que comme une faible et très-insuffisante représentation de l'indemnité due à l'Etat pour le vaste et inappréciable dommage qu'il a souffert.

Observons d'ailleurs qu'en admettant, dans des cas peu nombreux et très-graves, la peine de confiscation, qui eût pu recevoir un autre nom, s'il s'en fût présenté un qui eût été jugé propre à ce remplacement, le projet de loi se garde bien d'en étendre les effets au-delà des biens que le condamné possédait lors de sa condamnation, et ne consacre point cette barbare fiction de la corruption du sang, qui rend, en Angleterre, le fils d'un homme frappé de confiscation inhabile à succéder à son aïeul (1).

Une telle disposition, évidemment dirigée contre les descendants du coupable, ne pouvait trouver place dans notre législation ; et nous ne saurions admettre non plus cette loi romaine (2), qui vouait les enfants des criminels d'Etat à un tel degré d'abjection et de pauvreté, que la vie fût pour eux un supplice et la mort un bienfait : *Mors solatium et vita supplicium.* Leur condition est assez malheureuse pour ne point l'aggraver par un tel anathême : ah ! laissons-leur plutôt l'espoir de recouvrer

____

(1) Des Lois de police et criminelles de l'Angleterre, ouvrage traduit de *Blackstone,* par *Ludot,* chap. XII.

(2) *Leg. quisquis* 5. C. *ad leg. Jul. majest.*

comme un bienfait du prince, ce qu'ils ont perdu par le crime de leurs pères. Cette expectative, consolante pour eux, deviendra aussi un moyen politique de les rattacher par la reconnaissance au gouvernement de leur pays.

Je vous ai rendu compte, Messieurs, de la partie du projet qui regarde les crimes d'Etat, et fixe les peines qui leur sont applicables.

Mais ici se présente un nouveau sujet de discussion. En matière de complots ou crimes contre l'Etat, remettra-t-on la peine à ceux d'entre les coupables qui révéleront ce qu'ils savent, ou procureront l'arrestation de leurs complices? Infligera-t-on des peines à ceux qui, instruits d'un complot, même non approuvé par eux, ne l'auront point révélé?

De ces deux questions, la première, quoique fort controversée dans les assemblées législatives qui ont précédé la constitution de l'an VIII, ne devait pas donner naissance à tant d'hésitation. Si les peines sont instituées dans l'intérêt de la société, comment le même intérêt ne porterait-il pas à en faire la remise, quand la révélation peut procurer de grands avantages à l'Etat, ou le soustraire à de grands dangers?

La deuxième question offrait plus de difficulté.

Elle ne saurait être résolue par la loi que le sombre et farouche Louis XI porta contre ceux qui, sachant qu'il existait une conspiration, ne la dénonçaient pas.

L'application qui fut faite de cette loi, dans le procès du grand écuyer *d'Effiat Cinq-Mars*, au malheureux *Augustin de Thou*, l'a depuis long-temps marquée d'un juste sceau de réprobation.

Tout le monde sait que, loin d'approuver le complot plus exactement tramé contre le cardinal de Richelieu que contre le roi Louis XIII, *de Thou* avait cherché lui-même à en dissuader le grand écuyer; l'instruction en fournissait la preuve : il n'y avait donc nulle complicité à lui imputer : mais il avait eu connaissance du complot, et ne l'avait point révélé; il fut, pour cette réticence, condamné à mort.

L'opinion publique, plus forte que les arrêts, s'est depuis long-temps prononcée contre cette terrible exécution : mais qu'est-il arrivé? que l'énormité de la peine appliquée dans cette malheureuse circonstance n'en a plus laissé apercevoir d'applicable; des hommes éclairés (1) ont même écrit qu'on ne pouvait obliger personne à devenir délateur, ni à s'exposer aux peines de la calomnie en révélant des complots dont ils seraient rarement en état de fournir la preuve.

Ne nous laissons point aveugler par le prestige des mots; le délateur odieux est celui qui crée des complots imaginaires ; mais puisque notre législation invite partout les citoyens à faire connaître aux magistrats les délits et leurs auteurs, comment ne pourrait-elle point le leur prescrire sous de certaines peines , relativement aux crimes qui attaquent la sûreté de l'État? Si la patrie n'est pas un vain mot, ceci ne saurait être un vain devoir.

Mais si c'est un devoir, il faut le remplir, lors même qu'il en résulterait des embarras ou dangers personnels ; la loi d'ailleurs protégera toujours le révélateur véridique.

Qu'y a-t-il donc dans cette matière de sage et utile? C'est qu'en introduisant une peine contre la non-révélation des crimes d'État elle ne soit point effrayante par

---

(1) Voyez notamment le Commentaire sur le livre des Délits et des Peines, §. XV.

son énormité ; par-là l'on servira mieux, non-seulement l'autorité publique, mais encore l'humanité, que par un silence absolu sur cette espèce de délit : car que pourrait-il arriver, surtout sous un gouvernemeet qui serait faible et soupçonneux ? qu'au lieu de peines justes et modérées, il porterait, dans son inquiétude, des lois de colère, et irait peut-être jusqu'à frapper la non-révélation de propos simplement indiscrets ou vagues, aussi-bien que celle d'un complot réel.

Les peines qu'introduit le projet de Code au sujet de la non-révélation seront d'un ordre différent, selon que le complot non révélé regardera ou non la personne du chef de l'Empire.

Au cas de l'affirmative seulement, il y aura lieu à une peine afflictive ; la réticence relative aux autres crimes d'État ne sera punie que de peines de police correctionnelle.

Au surplus, le projet de loi a respecté les liens de la nature, en n'imposant pas aux proches parents l'obligation qu'elle a tracée pour les autres citoyens. L'intérêt qu'a l'État de connaître et de prévenir les complots dirigés contre lui ne le portera jamais à exiger d'un père qu'il lui livre son fils, ou d'un frère qu'il lui livre sa sœur.

Vous connaissez maintenant, Messieurs, les principales dispositions du projet sur les crimes et délits contre la sûreté de l'État.

Ici va commencer l'examen d'une autre classe de crimes et délits ; je veux dire de ceux qui sont dirigés contre les constitutions de l'Empire.

C'est par ces constitutions que les citoyens jouissent de certains droits politiques dont l'exercice est une propriété sacrée.

Toutes personnes qui troublent ou empêchent cet exercice se rendent donc coupables ; mais leur délit s'aggrave, et peut même s'élever au rang des crimes, s'il est le résultat d'un plan concerté pour être en même temps exécuté dans divers lieux : dans ce dernier cas, l'ordre public, plus grièvement blessé, réclame aussi une plus sévère punition.

Cette espèce d'infraction sera rare sans doute ; et si la loi a dû s'en occuper, elle n'a pas moins dû prévoir les délits plus communs peut-être qui auront lieu dans l'exercice même des droits dont il s'agit, et principalement dans les scrutins.

Il y a délit toutes les fois que le vœu des citoyens est dénaturé par des falsifications, soustractions ou additions de billets ; et ces coupables manœuvres acquièrent un nouveau degré de gravité lorsqu'elles sont l'ouvrage des scrutateurs eux-mêmes ; car il y a, dans ce cas, violation du dépôt et abus de confiance. Mais malgré tout ce qu'a d'odieux une telle infraction, l'on a dû craindre d'ouvrir une issue trop facile à de tardives et téméraires recherches pour des faits qui ne laissent plus de traces quand le scrutin est détruit, et qu'on a terminé les opérations qui s'y rapportent.

Combien, dans cette matière surtout, les espérances trompées, les prétentions évanouies, et l'amour-propre blessé, ne feraient-ils pas naître d'accusations hasardées, s'il était permis de les recevoir après coup, et hors les cas où le coupable est surpris, pour ainsi dire, en flagrant délit !

Notre projet de loi, en s'occupant des délits commis dans l'exercice des droits civiques, ne pouvait rester muet sur la turpitude de ceux qui achètent ou vendent des suffrages.

Laissons aux Anglais le scandaleux privilège de briguer les suffrages de leurs concitoyens à prix d'argent et à force de dépenses; l'honneur français repousse un tel moyen; et la peine qu'encourront chez nous ceux qui achètent ou vendent des suffrages est tracée par la nature même de leur délit : ils ont méconnu la dignité de leur caractère, ils ont profané l'un de leurs plus beaux droits; que l'exercice de ces droits leur soit donc retiré pendant un temps suffisant pour l'expiation d'un pacte honteux, et qu'il leur soit infligé une amende comme supplément de peine due à l'esprit de corruption et de vénalité qui les a conduits.

La loi, qui pourvoit à ce que l'exercice des droits civiques ne soit ni entravé ni souillé, ne pouvait omettre de s'expliquer sur la garantie due constitutionnellement aussi à la liberté civile, sans laquelle tous les autres droits ne seraient eux-mêmes qu'un vain mot.

Protecteurs nés de cette liberté, les magistrats, qui, étant formellement requis de faire cesser ou de constater une détention illégale ou arbitraire, ne le font point, ne sont pas moins coupables que s'ils l'avaient ordonnée eux-mêmes.

L'ordre du fonctionnaire supérieur donné à des fonctionnaires subordonnés pour effectuer une détention illégale, ne deviendra même pour ceux-ci un légitime sujet d'excuse, qu'autant qu'il sera relatif à des objets pour lesquels il était dû obéissance hiérarchique; et dans ce cas la responsabilité pesera toute entière sur le supérieur qui aura donné l'ordre.

Mais si cet ordre émanait d'un ministre même, comment la réparation en serait-elle poursuivie? Le sénatus-consulte du 28 floréal an XII a prévu cette infraction; et s'il n'en a point indiqué la peine, c'est un soin qu'il a évidemment laissé à la loi organique, et un devoir qu'il faut remplir en ce moment.

Quelque grave, au surplus, que paraisse d'abord cet objet à raison de l'élévation des personnes qu'il concerne, il ne peut résulter de la répression de tels actes aucun trouble pour la société; car, d'une part, si la signature du ministre lui avait été surprise au milieu de ses nombreux travaux, il sera à l'abri de toutes poursuites en faisant cesser l'acte arbitraire, et en dénonçant les auteurs de la surprise; et, d'un autre côté, quand cet acte serait réellement son ouvrage, le ministre ne sera pas immédiatement sujet aux poursuites des personnes qui se prétendraient lésées.

Le recours préalable à la commission sénatoriale, créée pour la protection de la liberté individuelle, et la nécessité d'en obtenir une décision, ne peuvent manquer d'obvier à tous les inconvénients qui résulteraient d'une action brusque et rapide dirigée contre un si haut fonctionnaire.

Si la réclamation est mal fondée, la commission sénatoriale n'y aura aucun égard; mais, si elle l'accueille, le ministre devra réparer le grief; sinon il se rendra évidemment coupable.

Sans doute, grâces à l'harmonie qui règne entre les grands pouvoirs politiques, nous ne serons pas témoins de pareils débats; mais, s'ils devaient éclater jamais, il convient de leur donner dès à présent des règles qui vaudront d'autant mieux, qu'elles auront été posées dans un temps plus calme.

Hors le cas de désobéissance qui vient d'être prévu et qui sera puni du bannissement, la peine de la dégradation civique est celle qui a paru généralement la plus convenable à la matière.

Ce sera donc celle que l'on proposera d'infliger et aux officiers de police judiciaire qui, au mépris des prérogatives constitutionnelles de certains fonctionnaires, auraient concouru à les poursuivre, sans les autorisations requises, et aux juges et officiers publics qui auraient retenu ou fait retenir un individu hors des lieux destinés à cet usage; car les lois ne veillent pas seulement pour la liberté des citoyens, elles ne permettent pas de vexer ceux qui ont mérité de la perdre.

A l'égard des gardiens et concierges qui auront reçu un prisonnier sans mandat, ou auront refusé, soit de le représenter, soit d'exhiber leurs registres aux magistrats chargés de cette surveillance, c'est une peine autre que la dégradation civique qui convient à une telle classe de coupables, et ils seront punis d'emprisonnement et d'amende.

Je viens de retracer les principales dispositions contenues dans le projet de loi sur les atteintes portées à la liberté; je vais parler d'une classe d'infractions qui n'appellent pas moins toute la sollicitude du législateur; ce sont les *coalitions de fonctionnaires.*

Ces coalitions, inquiétantes de leur nature, pourraient souvent devenir funestes; elles sont toujours un mal, mais elles peuvent varier d'intensité, selon l'objet qu'elles ont.

Si donc une peine de police correctionnelle a semblé suffisante pour réprimer un simple concert de mesures contraires aux lois, quand nulle circonstance plus grave n'y est jointe, une peine d'un ordre plus élevé a paru nécessaire, quand ce concert est dirigé contre l'exécution même des lois ou contre les ordres du Gouvernement.

Ce crime acquiert un nouveau degré d'intensité quand

la coalition a lieu entre des autorités civiles et des corps militaires.

Il devient énorme, quand il dégénère en complot contre la sûreté de l'Etat.

Des peines graduées d'après ces idées obtiendront sans doute votre assentiment.

Mais il ne suffisait pas d'atteindre les coalitions dirigées vers des mesures actives; il est une espèce de coalition qui se présente, au premier aspect, comme passive dans ses moyens d'exécution, et dont les résultats troubleraient la société à un haut degré; ce sont les démissions combinées, et dont l'objet ou l'effet serait d'empêcher ou de suspendre la justice ou tout autre service public.

Des fonctionnaires qui répondraient aussi mal à la confiance du Gouvernement et aux besoins de la cité, seront justement punis, quand on leur enlevera par la dégradation civique des droits qu'ils ont abdiqués de fait.

Il reste, Messieurs, une autre classe de crimes et délits contre les constitutions de l'Empire.

C'est par ces constitutions qu'existent, avec des pouvoirs distincts et indépendants, l'autorité judiciaire et l'autorité administrative; si l'une empiète sur l'autre, l'ordre constitutionnel est troublé; et il ne l'est assurément pas moins lorsque l'une ou l'autre de ces autorités ose s'arroger la puissance législative.

Ainsi, ni les juges ni les administrateurs ne peuvent suppléer par des règlements à des lois ou à des décrets.

Ils ne sauraient non plus, sans devenir coupables, délibérer sur la question de savoir si les lois seront ou non publiées; le temps est passé où les parlements exerçaient cette prérogative : aujourd'hui cette prétention, contraire à toute l'économie de nos pouvoirs constitués, ne serait pas un simple blasphême politique, elle serait le renversement de tout le systême constitutionnel.

Nos constitutions, et l'ordre public, s'opposent aussi à ce qu'un tribunal défende d'exécuter les ordres d'une administration, ou à ce qu'une administration intime des ordres ou défenses à un tribunal.

Il n'y aurait qu'anarchie dans un Etat où de pareilles prétentions seraient tolérées, et où chaque autorité se croirait en droit de se faire ainsi justice à elle-même; c'est à un pouvoir supérieur, à un régulateur commun, qu'il faut recourir en cas de dissentiment sur les attributions respectives; et tout juge ou administrateur qui franchit cette limite devient coupable et encourt la dégradation civique.

Une amende réprimera suffisamment le délit des juges qui auraient procédé au jugement d'affaires revendiquées par l'autorité administrative, ou d'administrateurs qui, après une réclamation légale, auraient retenu la connaissance d'affaires du ressort des tribunaux : hors les cas où les juges ou administrateurs sont avertis par un conflit ou acte équivalent, leurs jugements ou arrêtés, même incompétents, pourront être cassés; mais la loi ne punira point comme des délits ce qui peut n'être que des erreurs.

J'ai mis sous vos yeux, Messieurs, les principales dispositions du projet relatives aux deux premières classes *de crimes et délits contre la chose publique.*

Parmi ces crimes, vous avez pu en remarquer plusieurs qui sont hors du ressort des tribunaux ordinaires, et dont le jugement appartiendra, soit à la haute-cour, soit à des tribunaux spéciaux; mais notre projet, qui ne change rien aux règles générales ou particulières sur la compétence ou la procédure, aura atteint le seul but qu'il se proposait, si, avec les améliorations que lui ont procurées les judicieuses observations de votre commission, il est parvenu, quels que puissent être les magistrats chargés d'appliquer ses dispositions, à éclairer et alléger leur ministère, en traçant les délits avec clarté et en graduant les peines avec sagesse.

MOTIFS

Du Livre III, Titre Ier, Chapitre III,

PRÉSENTÉS

PAR MM. LES COMTES BERLIER, CORSINI ET PELET,
Conseillers d'État.

Séance du 6 février 1810.

MESSIEURS,

Lorsque, dans votre dernière séance, nous vous avons entretenus des principales dispositions portées au nouveau projet de Code pénal sur les deux premières subdivisions *des Crimes et Délits dirigés contre la chose publique,* nous n'avons rempli qu'une partie de la tâche qui nous était imposée.

Pour compléter ce tableau, nous venons aujourd'hui mettre sous vos yeux la troisième subdivision, intitulée : *Des Crimes et Délits contre la paix publique.*

Ce texte est vaste, et il ne saurait être oiseux de bien déterminer son acception ; car, exactement et rigoureusement appréciés, il n'est aucuns crimes ni délits qui n'altèrent la tranquillité publique à un degré quelconque ; mais il en est pourtant, et même un grand nombre, qui lèsent plus spécialement le corps de l'État que les particuliers.

Du faux.

C'est à ce caractère que l'on s'est arrêté pour qualifier les crimes et délits contre la paix publique ; et vous ne

5

serez point surpris, Messieurs, d'y voir figurer au premier rang *le crime de faux.*

Fausse monnaie.

L'on ne peut prononcer ce mot sans songer d'abord à la fausse monnaie, à cause de la gravité de ce crime et des alarmes qu'il répand dans la société.

Si l'Assemblée constituante réduisit aux fers la peine de ce crime, jusque-là puni de mort, l'on sait que cet essai philantropique ne fut point heureux, et que peu après il fallut rétablir la peine capitale.

Notre projet a maintenu cette peine, et y assujettit egalement ceux qui contrefont ou altèrent les monnaies d'or et d'argent ayant cours légal dans l'Empire, et ceux qui les distribuent, exposent ou introduisent en France.

Cette disposition avait d'abord alarmé quelques esprits (1), qui auraient désiré qu'on établît une distinction entre le fabricateur et le distributeur : mais toute inquiétude à ce sujet était vaine ; car, d'une part, le distributeur qui ignore le vice de la chose ne commet ni crime ni délit ; et, d'un autre côté, ceux qui ont remis en circulation des pièces qu'ils savaient être fausses, mais qu'ils avaient reçues pour bonnes, ne seront punis que d'une amende, attendu que la loi doit compatir à leur position, et ne voit en eux que des malheureux cherchant à rejeter sur la masse la perte dont ils étaient personnellement menacés.

Cela posé, qu'est-ce que peut être un distributeur ou introducteur qui connait la fausseté des pièces, et n'a

---

(1) Voyez les observations de quelques-unes des cours consultées sur le projet de Code pénal.

pas pour lui l'excuse de les avoir reçues pour bonnes? Qu'est-il, sinon le facteur volontaire, et conséquemment le complice du fabricateur? Il subira donc la même peine.

Mais cette peine si grave sera-t-elle appliquée à toute espèce de fausse monnaie, à celles de billon ou de cuivre, par exemple, et aux monnaies étrangères? La valeur exiguë des premières ne cause pas le même degré d'alarme, et la valeur purement commerciale des secondes en rend aussi la circulation moins dangereuse pour la multitude, qui, le plus souvent, ne connaîtra point ces signes monétaires, et qui, d'ailleurs, ne sera pas tenue de les accepter : la peine capitale ne sera donc point appliquée à ces deux classes de faux, qui seront suffisamment punis par les travaux forcés.

Au surplus, le crime de fausse monnaie, sans être précisément de la catégorie de ceux qui sont dirigés contre la sûreté de l'État, a plusieurs points de communs avec eux.

Vous ne serez donc point surpris, Messieurs, de voir appliquer à ce crime, et la remise de la peine en cas de révélation, et la peine de réticence, comme pour les crimes d'État. Le suprême intérêt qu'a la société d'écarter ou de faire cesser un tel fléau rend cette application légitime et nécessaire.

Vous ne serez pas étonnés non plus d'y trouver la confiscation unie à la peine capitale. *Les pertes de l'État*, a dit un orateur, pour le cas que nous examinons (1), *peuvent être immenses; elles sont vagues et inappréciables; c'est alors qu'à titre de dommages et inte*

---

(1) Voyez le Discours préliminaire de M. Target, sur le Code, pag. 21 et 22.

rêts, il est juste et nécessaire qu'elles soient réparées par la confiscation générale des biens du condamné.

C'est d'ailleurs notre législation actuelle; et une explication bien simple vient la justifier.

Dans les crimes et délits ordinaires, où il n'y a que peu de parties lésées, et où la mesure du dommage est connue ou susceptible de l'être, les réparations civiles suffisent à tout ce qui regarde l'intérêt privé ; mais peut-il en être ainsi quand le dommage est disséminé sur des milliers de personnes? et si le fruit du crime devait, à défaut de parties civiles, passer nécessairement des mains du coupable à celles de ses enfants, ne serait-ce pas une espèce de prime accordée aux faux monnayeurs sur tous les autres criminels ?

En adoptant la confiscation pour ce cas, vous apercevrez aisément, Messieurs, qu'elle n'a point l'odieux objet de dépouiller les familles, mais pour but unique de ne les point gratifier des dépouilles d'autrui : la justice et l'intérêt de l'État réclamaient cette disposition.

Vous trouverez sans doute également juste et convenable que les mêmes règles et les mêmes peines soient applicables aux effets émis par le trésor public avec son timbre, et aux billets de banque, qui ont tant d'affinité avec la monnaie même dont ils sont en quelque sorte le supplément et dont ils remplissent l'office.

Contrefaction des sceaux, timbres, poinçons, etc.

Mais si la peine capitale convient à de tels crimes, et peut être appliquée aussi à la contrefaction des sceaux de l'État, des peines inférieures devront être infligées à la contrefaction des autres sceaux, timbres, poinçons et marques, en graduant ces peines selon l'importance de la destination qu'avait l'instrument contrefait.

L'on a aussi distingué la fabrication d'un faux timbre d'avec le faux emploi d'un timbre vrai : cette disposition manquait dans notre législation.

Faux en écritures.

Jusqu'ici, Messieurs, dans les diverses especes de faux dont on vient de donner l'analyse, c'est l'État ou le corps social qui est principalement attaqué ou lésé : dans le faux appliqué aux écritures publiques ou privées, l'intérêt individuel joue un plus grand rôle, et peut-être eût-on pu renvoyer cette partie au chapitre *des crimes contre les particuliers,* s'il n'eût semblé nuisible de scinder cette matière.

Le faux en écritures est matériel quand il s'est opéré par fausses signatures, par altération ou intercalation d'écritures, par supposition de personnes; mais il est aussi une autre espèce de faux moins facile à caractériser, et qui a lieu quand un officier public écrit des conventions autres que celles qui lui ont été tracées ou dictées, et constate comme vrais des faits faux, ou comme avoués des faits qui ne l'étaient pas.

Toutefois il faut prendre garde de réputer crime ce qui ne serait qu'un malentendu ou une méprise : le rédacteur d'un acte peut mal saisir la volonté des parties, et pourtant n'être pas criminel; il ne le sera, aux termes du projet, que quand il aura *frauduleusement dénaturé la substance ou les circonstances de l'acte.* D'après ce caractère, il ne reste rien qui puisse alarmer l'innocence.

Le faux en écritures privées sera puni de la réclusion, et le faux en écritures publiques, des travaux forcés; mais dans cette dernière espèce de faux, si la peine n'est que temporaire à l'égard du simple particulier contrefac-

5.

teur d'écritures authentiques, elle sera perpétuelle à l'égard de l'officier public qui commettrait ce crime; celui-ci est doublement coupable, il a trahi la foi due à son caractère.

Les faux commis en écritures de commerce et de banque ont mérité une mention spéciale sans laquelle ils eussent été confondus avec les faux en écritures privées; l'extrême faveur due au commerce a donné lieu d'assimiler ces faux à ceux commis en écritures publiques.

Faux commis dans les passe-ports, feuilles de route et certificats.

Mais il est une autre espèce de faux qui, dans le silence des lois, a souvent embarrassé les tribunaux; c'est le faux commis dans les passe-ports, feuilles de route et certificats.

Sans doute ce serait blesser la justice que d'assimiler la contrefaction d'un passe-port à celle d'une lettre de change, ou la fabrication d'un certificat de maladie à celle d'une obligation que l'on créerait à son profit sur un tiers.

Des peines de police correctionnelle suffiront ordinairement pour la répression des faux passe-ports, si ce n'est à l'égard des officiers publics qui auraient participé au faux; car ils sont plus criminels que de simples particuliers quand ils abusent ainsi du pouvoir qui leur a été confié.

Les mêmes vues ont semblé applicables aux fausses feuilles de route, mais en prenant de plus en considération la lésion que le trésor public aurait pu recevoir par le paiement de sommes non dues; car alors il y a vol joint au faux, et lieu d'appliquer des peines plus fortes.

À l'égard des certificats de maladie ou d'infirmités, fabriqués dans la vue d'affranchir quelqu'un d'un service public ; ou s'il s'agit d'attestations d'indigence ou de bonne conduite, fabriquées pour procurer à celui qui y est désigné ou qui en est porteur, des secours, du crédit ou des places ; un tel délit a semblé n'appeler que des peines de police correctionnelle : mais on a dû éviter de confondre avec des certificats de cette espèce ceux qui auraient eu pour objet de se faire donner ou payer des sommes dues ou des effets appartenant à un tiers ; car, en ce cas, c'est la peine ordinaire du faux qui devra être infligée.

Dans les actes que l'on vient de désigner, il convenait de classer non-seulement ceux qui étaient matériellement faux, mais encore ceux qui, originairement véritables, auraient été altérés pour servir à d'autres personnes.

Le projet prévoit et embrasse ces différentes espèces : il y a lieu d'espérer qu'elles seront plus efficacement réprimées par des dispositions mieux adaptées au caractère particulier de chacune d'elles.

Dispositions communes à toutes les espèces de faux.

Quelques dispositions communes à toutes les classes ou espèces de faux terminent cette partie du projet.

Ainsi l'usage d'une pièce fausse étant partout puni comme sa fabrication même, il convenait de dissiper toutes les inquiétudes en exprimant que ce terrible anathème ne regarde que ceux qui ont eu connaissance du faux.

La marque, rarement applicable à des peines temporaires, sera pourtant infligée à tout faussaire condamné aux travaux forcés à temps, ou à la réclusion ; c'est l'état actuel de la législation ; et il était difficile de le changer pour un crime qui inspire à la société de si vives alarmes, et dont les auteurs ne sauraient être trop signalés.

Enfin, dans tous les cas où le faux n'entraînera ni la peine capitale ni la confiscation générale, une amende sera jointe à la peine prononcée : *il est raisonnable, il est utile* (1) *que les crimes qui ont eu pour principe une vile cupidité soient réprimés par des condamnations qui attaquent et affligent cette passion même par laquelle ils ont été inspirés.*

Vous connaissez maintenant, Messieurs, les principales dispositions relatives aux faux : la peine de faux témoignage sera placée au chapitre des crimes contre les particuliers.

Crimes et délits des fonctionnaires publics dans leurs fonctions.

Parmi les crimes et délits qui compromettent le plus la paix publique, il était impossible de ne pas accorder aussi un rang principal à ceux que commettent les fonctionnaires publics dans l'exercice de leurs fonctions : l'ordre est manifestement troublé quand ceux que la loi a préposés pour le maintenir sont les premiers à l'enfreindre.

Tout crime commis par un fonctionnaire dans l'exercice de ses fonctions le constitue en forfaiture, et la dégradation civique est la moindre peine qui y soit attachée; mais la peine peut s'élever selon la nature et l'intensité du crime.

---

(1) Voyez le discours de M. Target, pag. 21.

Crime de soustraction.

Ainsi la peine des travaux forcés à temps est infligée au fonctionnaire public qui détruit ou soustrait les actes ou titres dont il est dépositaire; et il a paru convenable d'appliquer aussi cette peine aux soustractions de deniers publics commises par les personnes chargées de leur perception.

Cependant l'on a cru devoir admettre une modification pour le cas où la somme soustraite serait si modique, qu'il deviendrait vraisemblable que le percepteur avait le dessein de s'en servir pendant quelque temps, plutôt que celui d'en frustrer le trésor public.

Lors donc que le *déficit* sera moindre du tiers de la recette d'un mois, ou ne surpassera pas le montant du cautionnement fourni, et qu'en même temps il sera inférieur à 3000 fr., un emprisonnement de deux à cinq ans a paru une peine suffisante envers d'imprudents percep-teurs qui sont coupables sans doute, mais pourtant beaucoup moins que ceux qui seraient partis avec le dé-pôt tout entier.

Rejeter toute distinction, dans cette conjoncture, se-lon quelques opinions sévères, et placer sur le même rang deux actes qui diffèrent dans leurs circonstances comme dans leurs résultats, ce n'eût pas été seulement blesser la justice, mais encore les vues saines d'une bonne administration.

Qu'arriverait-il, en effet, si un léger *déficit* et une soustraction totale étaient frappés de la même peine? Ne serait-ce pas, dès que le dépôt serait entamé pour la plus légère partie, une invitation au percepteur de soustraire le tout, puisqu'il trouverait dans ce simple et funeste calcul de plus grands bénéfices, sans s'exposer à une plus grande peine? Des dispositions pénales mal combinées seraient plus nuisibles qu'utiles à la société.

Crime de concussion.

Les concussions commises par les fonctionnaires publics ne pouvaient manquer d'appeler aussi l'attention du législateur.

Ce crime existe toutes les fois qu'un fonctionnaire exige ou reçoit ce qu'il sait ne lui être pas dû, ou excéder ce qui lui est dû; et l'on conçoit aisément que, s'il importe de poser des barrières contre la cupidité, c'est surtout quand elle se trouve unie au pouvoir (1).

La peine de réclusion, toute grave qu'elle est, sera donc infligée au fonctionnaire coupable de concussion; et les simples commis ou préposés seront, pour le même fait, punis de peines correctionnelles.

Je n'ai pas besoin sans doute de justifier cette différence dans la peine, quoiqu'il s'agisse du même délit : investi d'un plus haut caractère, celui qui doit aux autres citoyens l'exemple d'une conduite pure et sans tache, est bien plus répréhensible quand il tombe en faute; il doit donc être puni davantage; et cette idée, ainsi que ses applications, se reproduiront souvent dans le cours de cette discussion.

Délit des fonctionnaires qui s'immiscent dans des affaires incompatibles avec leur qualité.

La position spéciale des fonctionnaires publics peut aussi et doit même, en plusieurs circonstances, leur faire interdire ce qui serait licite à d'autres personnes.

---

(1) *Lege Juliâ* 3., *ff. De Leg. Jul. repetundarum.*

Ainsi un fonctionnaire devient coupable lorsqu'il prend, directement ou indirectement, intérêt dans les adjudications, entreprises ou régies, dont sa place lui donne l'administration ou la surveillance : que deviendrait en effet cette surveillance, quand elle se trouverait en point de contact avec l'intérêt personnel du surveillant ? et comment parviendrait-on, sans blesser l'honneur et la morale, à concilier ce double rôle de l'homme public et de l'homme privé ?

Tout fonctionnaire qui se sera souillé d'une telle turpitude sera donc justement puni d'emprisonnement, et déclaré indigne d'exercer désormais des fonctions dans lesquelles il se serait avili.

La sollicitude de la loi a pu et dû aussi embrasser, dans ses dispositions, des défenses aux commandants militaires et aux chefs d'administrations civiles, de s'immiscer dans le commerce des principaux comestibles, sous certaines peines de police correctionnelle.

Si l'ordre public s'oppose à ce que de tels fonctionnaires puissent, à la faveur de leur caractère, exercer, pour leur avantage particulier, une influence dangereuse sur le prix des principaux comestibles, l'interdiction d'un tel commerce est juste et convenable, même envers les administrateurs qui n'auraient pas la criminelle pensée d'en abuser.

En effet, il faut écarter tout ce qui pourrait inspirer aux citoyens de justes sujets d'inquiétudes ou d'alarmes ; il serait fâcheux que la masse des citoyens craignît l'abus, et encore plus qu'elle y crût : la considération qui environne les fonctionnaires naît principalement de la confiance qu'ils inspirent ; et tout ce qui peut altérer cette confiance ou dégrader leur caractère, doit leur être interdit.

De la corruption des fonctionnaires publics.

Que dirons-nous de la corruption?

Le fonctionnaire corrompu est celui qui met son autorité à prix, soit pour faire un acte de sa fonction non sujet à salaire, soit pour ne pas faire un acte qui entre dans l'ordre de ses devoirs.

De tels hommes sont de vrais fléaux, et la société serait bientôt dissoute s'ils étaient nombreux. La république romaine était bien près de sa ruine quand Cicéron se plaignait de ce qu'il y était passé en maxime qu'un homme riche, quelque coupable qu'il fût, ne pouvait pas être condamné (1).

Le crime de corruption, isolé de toutes autres circonstances, ne sera jamais puni d'une peine moindre que le carcan, et d'une amende double des promesses agréées ou des présents reçus.

Mais si le fonctionnaire public qui retire de ses fonctions un lucre illicite devient criminel par ce seul fait, ce crime peut s'aggraver beaucoup quand il est commis pour arriver à un autre, et que celui-ci a été suivi d'exécution.

C'est surtout dans les jugements criminels que cette aggravation peut se faire remarquer : l'on sent combien serait déplorable la corruption qui rendrait un criminel à la société, et combien serait énorme et atroce celle qui ferait succomber un innocent.

Jamais donc il ne sera, pour corruption pratiquée et soumise dans les jugements criminels, appliqué une

______

(1) *Pecuniosum hominem, quamvis sit nocens, neminem posse damnari.* Cic. act. 1 in Verr. n. 1.

peine moindre que la réclusion : mais si la corruption a eu pour résultat de faire condamner un innocent à une peine plus forte, cette peine, quelle qu'elle puisse être, deviendra le juste châtiment du fonctionnaire corrompu. La loi du talion ne fut jamais plus équitable ni plus exempte d'inconvénients.

Dans tous les cas, la même peine sera subie par le corrupteur et par le fonctionnaire qui se sera laissé corrompre, et jamais le prix honteux de la corruption ne deviendra l'objet d'une restitution ; la confiscation en sera prononcée au profit des hospices, et ce qui était destiné à alimenter le crime tournera quelquefois du moins au soulagement de l'humanité.

D'autres peines seront infligées à d'autres délits.

Abus d'autorité.

Les abus d'autorité, dont je vais actuellement vous entretenir, sont, par le projet de loi, divisés en deux classes, savoir : *contre les particuliers* et *contre la chose publique.*

Abus d'autorité contre les particuliers.

Les fonctionnaires abusent de leur autorité *contre les particuliers,* quand ils s'introduisent illégalement dans leurs domiciles ; quand ils dénient de leur rendre justice après une réquisition des parties et un avertissement de leurs supérieurs ; enfin, quand ils portent atteinte au secret de la correspondance.

Dans ces cas divers, le fonctionnaire sera puni d'une simple amende.

L'on a, dans cette matière, cherché plutôt une peine efficace qu'une peine sévère.

L'espèce de délit qu'on examine ne tire point sa source

de passions viles et basses, comme les concussions ou la corruption ; un zèle faux ou mal entendu peut produire assez souvent des abus d'autorité; et il importe de les réprimer, mais avec modération, si l'on veut que ce soit avec succès.

Une amende d'ailleurs a sa gravité relative aux personnes qui en sont l'obj t ; un fonctionnaire qui n'a point abdiqué tous les sentiments d'honneur sera plus qu'un autre sensible à cette peine, et ne s'y exposera plus.

Toutefois l'abus d'autorité qui aurait été porté jusqu'aux violences envers les personnes sera spécialement puni d'après la nature de ces violences ; car il n'y aurait plus de sûreté pour les citoyens, s'il en était autrement.

Au reste, si le plus fréquent abus du pouvoir est, par la nature des choses, celui que l'on se permet envers des personnes subordonnées, l'abus d'autorité peut aussi être dirigé contre la chose publique.

Abus d'autorité contre la chose publique.

C'est ce qui aurait lieu, si des fonctionnaires publics se permettaient de requérir ou ordonner l'emploi de la force publique pour empêcher l'exécution d'une loi , ou la perception d'une contribution légale, ou l'effet d'un ordre émané de l'autorité légitime.

Cet abus d'autorité est d'une nature fort différente de celui que nous avons examiné d'abord; c'est une espèce de révolte, qui sera d'autant plus grave et susceptible de peines d'autant plus fortes, qu'elle aura eu plus de développements et d'effets.

Nous avançons, Messieurs, dans le détail des crimes et délits des fonctionnaires publics, et nous en avons retracé les principaux.

Il en reste pourtant de deux espèces encore.

De quelques délits des officiers de l'état civil.

Des officiers de l'état civil inscrivent-ils leurs actes sur des feuilles volantes, ou procèdent-ils à des mariages sans s'être assurés des consentements nécessaires pour leur validité, ou admettent-ils une femme qui a déjà été mariée à un nouveau mariage avant le terme indiqué par le Code Napoléon?

Dans ces cas divers, ils compromettent l'état civil des personnes; ils se rendent coupables au moins de négligence, et le besoin de régulariser une partie aussi importante justifiera aisément les peines de police correctionnelle qui leur sont infligées.

De l'exercice de l'autorité publique illégalement anticipé ou prolongé.

C'est aussi pour régulariser l'exercice même de l'autorité publique que l'on réprimera par des peines de cette nature toutes personnes qui seraient entrées en fonctions sans avoir prêté le serment requis, ou qui s'y seraient maintenues après révocation ou remplacement.

Ces deux délits ne seront cependant pas confondus : le dernier est le plus grave, et n'est jamais susceptible d'excuse; le premier peut être excusé par l'absence des fonctionnaires entre les mains desquels le serment devait être prêté, et par le besoin de pourvoir au service. Les poursuites, dans ce cas, dépendront donc des circonstances ; et il eût été imprudent de poser à cet égard une règle inflexible.

Je ne puis, Messieurs, terminer l'exposé de la partie relative aux crimes et délits des fonctionnaires publics, sans appeler votre attention sur une disposition finale qui a paru aussi importante que juste.

Toujours relative aux fonctionnaires, et à eux seuls, cette disposition ne les considère plus comme délinquants dans l'exercice ou à l'occasion de l'exercice de leurs fonctions, mais comme délinquants dans l'ordre commun, et se rendant eux-mêmes coupables de quelques-uns des crimes ou délits dont la surveillance ou la répression leur étaient confiées par la loi.

Dans cette fâcheuse hypothèse, n'infligera-t-on que les peines de l'ordre commun? Et si, par exemple, un officier de police judiciaire a commis un vol, ne sera-t-il puni que comme un voleur ordinaire?

Il est difficile de ne pas considérer comme plus coupable celui qui, chargé par la loi de réprimer les crimes et délits, ose les commettre lui-même; et il a paru convenable d'élever la peine à son égard.

Si donc il s'agit d'un délit de police correctionnelle, le fonctionnaire qui l'aura commis subira toujours le *maximum* de la peine attachée à l'espèce de ce délit; et s'il s'agit de crimes, il subira la peine immédiatement supérieure à celle qu'eût méritée tout autre coupable; gradation qui ne cessera qu'au point où elle atteindrait la peine de mort.

Cette disposition toute morale ne saurait qu'honorer notre législation.

Je viens de parler des crimes et délits *des fonctionnaires publics*, classe dans laquelle n'entrent pas les ministres des cultes, à qui nulle autorité temporelle n'est départie, mais dont l'influence et la conduite ne sauraient être étrangères à la paix publique

Crimes et délits des ministres des cultes.

Le projet de loi s'occupe donc, dans un chapitre particulier, des troubles qui seraient apportés à l'ordre public par ces ministres dans l'exercice de leur ministère.

Cette matière est grave sans doute ; et autant la société doit de reconnaissance et d'égards à ces pasteurs vénérables dont les discours et l'exemple sont un constant hommage à la religion, aux mœurs et aux lois, autant elle doit s'armer contre ces hommes fanatiques ou séditieux qui, au nom du ciel, voudraient troubler la terre, et n'invoqueraient la puissance spirituelle que pour avilir ou entraver l'autorité des lois et du Gouvernement.

Les crimes et délits des ministres des cultes dans l'exercice de leur ministère sont, par notre projet, divisé en plusieurs classes.

Des contraventions propres à compromettre l'état civil des personnes.

Les ministres qui procèdent aux cérémonies religieuses d'un mariage sans qu'il leur ait été justifié de l'acte de mariage reçu par les officiers de l'état civil, compromettent évidemment l'état civil des gens simples, d'autant plus disposés à confondre la bénédiction nuptiale avec l'acte constitutif du mariage, que le droit d'imprimer au mariage le sceau de la loi était naguères dans les mains de ces ministres.

Il importe sans doute qu'une si funeste méprise ne se perpétue point ; et ce motif est assez puissant pour punir d'une amende les ministres de cultes qui procèdent aux cérémonies religieuses d'un mariage sans justification préalable de l'acte qui le constitue réellement.

Cette peine, légère d'abord, s'aggravera en cas de ré-cidive, et entraînera à la seconde récidive, ou, en d'autres termes, à la troisième infraction, la peine de déporta-tion; parce que celui qui a failli trois fois se place évi-demment dans un état de désobéissance permanente et de révolte contre la loi.

Critiques, censures ou provocations contre l'autorité publique.

Les critiques, censures ou provocations dirigées par ces ministres contre l'autorité publique sont d'une im-portance qui ne permettait point le silence et appelait des mesures répressives.

L'on a distingué la critique ou censure simple d'avec la provocation directe à la désobéissance : dans ce der-nier cas, la culpabilité plus forte entraîne une plus grande peine.

L'on a distingué aussi les censures et provocations faites dans un discours public d'avec celles consignées dans un écrit pastoral; et ces dernières sont punies da-vantage, comme étant le produit plus réfléchi de vues perverses, et comme susceptibles d'une circulation plus dangereuse.

Correspondance avec des cours ou puissances étrangères sur des matières de religion.

Enfin, le projet de loi proclame comme infraction de l'ordre public toute correspondance que des ministres de cultes entretiendraient sur des questions ou matières re-ligieuses avec une cour ou puissance étrangère, sans l'au-torisation du ministre de l'Empereur chargé de la sur-veillance des cultes.

Cette disposition, d'une haute importance, ne saurait

alarmer que les artisans de troubles, et les hommes, s'il en est encore, assez insensés pour croire, ou assez audacieux pour dire que l'*Etat est dans l'Eglise, et non l'Eglise dans l'Etat*.

Cette maxime ultramontaine, qui put prevaloir lorsqu'un pontife étranger disposait des empires et déposait les rois, a été d puis long-temps reléguée dans la c asse des erreurs qu'enfantèrent les siècles d'ignorance.

Il ne s'agit pas, au reste, de rompre les rapports légitimes d'aucun culte avec des chefs même étrangers; il n'est question que de les connaître; et ce droit du Gouvernement, fondé sur le besoin de maintenir la tranquilité publique, impose aux ministres des cultes des devoirs que rempliront avec empressement tous ceux dont les cœurs sont purs et les vues honnêtes. Si cette obligation gêne les autres, son utilité n'en sera que mieux prouvée.

Nous ne sommes point au terme de la longue et penible nomenclature des crimes et délits qui attaquent la paix publique.

Les crimes ou délits qui blessent l'autorité publique avec un caractère spécial de résistance ou de désobéissance n'ont point encore passé sous vos yeux; et ils sont nombreux, puisqu'ils se divisent en huit classes; la rebellion, les outrages et violences envers les dépositaires de l'autorité, le refus de service, l'évasion des détenus et le recèlement des criminels, les bris de scellés, les dégradations de monuments, l'usurpation de titres, et enfin les entraves au libre exercice des cultes.

Je vais parcourir ces diverses espèces, sans m'arrêter particulièrement à chaque disposition, mais de manière à indiquer les vues principales du projet relativement à chaque classe.

Rebellion.

Le crime de *rebellion* est plus ou moins grave, d'après certains caractères qui sont devenus la base de la distribution des peines en cette matière.

Les rebelles étaient-ils nombreux ou non, armés ou sans armes? L'intensité de la rebellion dépend essentiellement de ces circonstances.

La qualité des rebelles peut aussi n'être pas sans importance : étaient-ce des ouvriers attachés à des ateliers publics, des personnes admises dans des hospices, des prisonniers même? Entre personnes de cette espèce, les rebellions ont un caractère d'autant plus dangereux, qu'il y a plus de tendance et d'occasions pour s'y livrer.

Les peines de la rebellion, établies et graduées d'après ces idées, seront quelquefois correctionnelles, quelquefois afflictives.

Mais, pour en faire une juste application, et ne point confondre surtout les réunions armées ou non armées, il convenait de bien fixer le caractère de celles qui, au premier aspect, semblent mixtes, et où les rebelles sont en partie armés et en partie sans armes.

Ces cas sont fréquents; et le projet de loi règle que la réunion armée sera celle où trois personnes au moins porteront des armes ostensibles.

Cette règle est juste; et les individus non armés ont au moins à s'imputer de s'être placés sous la protection ou la bannière de ceux qui avaient des armes.

Il convient, au surplus, de remarquer que, si la rebellion dont on traite en ce moment, dirigée contre les agents de la force publique en fonctions, a un objet différent de celui des bandes et attroupements séditieux dont je vous ai entretenus dans votre dernière séance, une telle rebellion pourra néanmoins, comme dans les cas de sédition, n'être suivie d'aucune peine envers ceux des rebelles avec attroupement qui se seraient retirés au premier avertissement de l'autorité : c'est le même motif, c'est la même alliance de l'indulgence avec la politique.

Pareillement, dans l'espèce présente, comme on l'a déjà observé dans l'autre, les crimes individuels commis dans le cours de la rebellion seront distingués du crime même de rebellion, et pourront donner lieu à de plus fortes peines contre ceux qui s'en seraient personnellement rendus coupables : mais ces peines spéciales ne s'étendront pas aux autres rebelles ; car si, dans le tumulte qui accompagne ordinairement de telles scènes, il s'est commis sur l'un des points un crime plus grave que celui de la rebellion même, ne serait-ce pas une rigueur poussée jusqu'à l'injustice que d'en appliquer sans distinction la peine à tous les rebelles ?

Sans doute ils doivent tous être punis ; mais le crime de rebellion est le seul qui soit commun à tous, et ceux qui n'ont pas pris part à d'autres crimes spéciaux n'en sauraient être considérés comme complices.

Après le crime de rebellion, le projet de loi s'occupe des outrages et violences envers les dépositaires de l'autorité et de la force publique.

Outrages et violences envers l'autorité.

Ici s'est offert un sujet de discussion assez grave, mais dont la solution pourtant a été facile : convenait-il de punir les outrages commis, même hors de tout exercice de fonctions, de peines de différents ordres, graduées d'après la simple considération du rang plus ou moins élevé que les personnes outragées tiennent dans la société ?

En agitant cette question, l'on n'a pas tardé à reconnaître que l'application d'une telle idée serait impraticable ; qu'en tarifant les peines selon le rang de l'offensé, cela irait à l'infini ; qu'il faudrait aussi prendre en considération le rang de l'offenseur ; enfin, l'on a reconnu que cela était moins utile que jamais dans un système qui, assignant à chaque classe de peines temporaires un *maximum* et un *minimum*, laissait à la justice une suffisante latitude pour varier la punition des outrages *privés* d'après la considération due aux personnes.

Il ne sera donc ici question que des seuls outrages qui compromettent la paix publique, c'est-à-dire de ceux dirigés contre les fonctionnaires ou agents publics, dans l'exercice ou à l'occasion de l'exercice de leurs fonctions : dans ce cas, ce n'est plus seulement un particulier, c'est l'ordre public qui est blessé ; et dans un grand intérêt les peines peuvent changer de classe et de nature, parce que le délit en a changé lui-même, et que l'outrage dirigé contre l'homme de la loi, dans l'exercice de ses fonctions ou de son ministère, quoique conçu dans les mêmes paroles ou les mêmes gestes, est beaucoup plus grave que s'il était dirigé contre un simple citoyen.

La hiérarchie politique sera, dans ce cas, prise en considération : celui qui se permet des outrages ou violences envers un officier ministériel est coupable sans doute ; mais il commet un moindre scandale que lorsqu'il outrage un magistrat.

L'offense envers celui-ci peut même varier d'intensité, selon qu'elle est commise dans le sanctuaire même de la justice, ou ailleurs, mais toujours à l'occasion de ses fonctions.

Dans la classification de ces outrages on a placé au moindre degré de l'échelle ceux qui sont commis par gestes ou par menaces.

Les paroles outrageantes, qui ont ordinairement un sens plus précis et mieux déterminé que de simples gestes ou menaces, ont paru être un délit supérieur à celui-ci.

Au sommet de l'echelle viennent les coups, qui, punissables envers tout citoyen, sont le comble de l'irrévérence envers les dépositaires de l'autorité.

D'après ces idées générales, le projet distribue des peines quelquefois correctionnelles, quelquefois afflictives.

A ces peines il pourra s'en joindre d'un ordre particulier, telles que les réparations par écrit ou à l'audience; l'éloignement, pendant un temps donné, du lieu où siége le magistrat offensé; et, en cas d'infraction de cette mesure, le bannissement.

Dans toutes ces dispositions, on a cherché, en observant d'ailleurs une juste gradation dans les peines, à faire respecter les organes de la justice et ses agents.

Refus d'un service dû légalement.

Le paragraphe qui traite du refus de remplir un service dû légalement n'est pas susceptible d'observations.

Les témoins, les jurés et les dépositaires de la force publique, requis par l'autorité civile et ne répondant point à ses ordres, seront punis de peines correctionnelles, seules convenables pour réprimer une désobeissance qui ne dégénère point en révolte.

Évasion de détenus, recèlement de criminels.

Mais parmi les actes de désobéissance à l'autorité publique l'on peut classer aussi l'évasion des détenus et le recèlement des criminels.

Le délit de recèlement ne s'appliquera point aux proches parents, qui trouvent dans les affections naturelles une excuse que la loi sait apprécier et admettre; mais nulles autres personnes ne pourront, sous prétexte d'humanité, soustraire le coupable à sa punition, ou le prévenu aux recherches de la justice.

L'évasion constitue un délit d'une autre espèce : considérée dans la personne des détenus eux-mêmes, elle ne saurait être traitée avec rigueur. Le désir de la liberté est si naturel à l'homme, que l'on ne saurait prononcer que celui-là devient coupable qui, trouvant la porte de sa prison ouverte, en franchit le seuil : le délit ne commence à son égard que lorsqu'il a employé des moyens criminels, tels que le bris de prison ou la violence.

A l'égard de ceux que la loi a préposés à sa garde, la position est toute différente, et la simple évasion du détenu constitue ses gardiens en délit.

Ce délit sera plus ou moins grave, selon qu'il résultera de connivence, ou simplement de négligence. La gravité sera aussi mesurée d'après celle du crime ou du délit pour lequel la détention avait eu lieu ; car, si la peine doit être proportionnée au préjudice que reçoit la société, il est certain que l'évasion d'un homme détenu pour une rixe ne répand point le même degré d'alarme que l'évasion d'un incendiaire ou d'un assassin.

Bris de scellés, dégradation de monuments, usurpations de titres.

Je n'arrêterai point votre attention, Messieurs, sur les bris de scellés, dégradations de monuments et usurpations de titres.

Les dispositions qui regardent ces diverses espèces d'attentats contre la paix publique se justifient d'elles-mêmes.

J'observerai seulement que la peine du bris de scellés est graduée elle-même sur l'importance des objets qui étaient sous le scellé, et d'après les caractères auxquels la loi attache plus ou moins d'importance.

C'est sans doute une chose utile et juste que d'appliquer cette gradation toutes les fois qu'elle est praticable ; et les dispositions dont je vous ai déjà donné connaissance ont pu vous convaincre que nulle occasion tendant à ce but n'a été négligée.

Entraves au libre exercice des cultes.

Je vais maintenant vous entretenir des peines que l'on propose d'appliquer aux entraves mises au libre exercice des cultes.

Ce libre exercice est l'une des propriétés les plus sacrées de l'homme en société, et les atteintes qui y seraient portées ne sauraient que troubler la paix publique.

Nulle religion, nulle secte n'a donc le droit de prescrire à une autre le travail ou le repos, l'observance ou l'inobservance d'une fête religieuse ; car nulle d'entre elles n'est dépositaire de l'autorité ; et tout acte qui tend à faire ouvrir ou fermer des ateliers, s'il n'émane du magistrat même, est une voie de fait punissable.

Les désordres causés dans l'intérieur d'un temple, ou dans des lieux actuellement servant aux exercices d'un culte, sont aussi un délit qu'il importe de réprimer ; l'auteur du trouble est également coupable, soit qu'il appartienne au culte dont les cérémonies ont été troublées, soit qu'il lui soit étranger ; car respect est dû à tous les cultes qui existent sous la protection de la loi.

Le perturbateur sera donc puni, et la peine s'aggravera si le trouble a dégénéré en outrages contre les objets du culte, et si ces outrages ont été commis *dans les lieux destinés ou servant actuellement à l'exercice ou au service d'un culte.*

Mais ces expressions mêmes indiquent la limite dans laquelle le législateur a cru devoir se renfermer : la juste protection due aux différents cultes pourrait perdre cet imposant caractère, et dégénérer même en vexation ou tyrannie, si de prétendus outrages faits à des signes placés hors de l'enceinte consacrée pouvaient devenir l'objet de recherches juridiques : chacun de nous se rappelle la condamnation prononcée, dans le siècle dernier, contre le jeune et malheureux *Delabarre* ; et nul ne voudra que le jet imprudent d'une pierre lancée au milieu des rues ou des champs puisse fournir matière à une accusation de sacrilège.

Renfermée dans ses vraies limites, la loi n'en sera que plus respectée; elle prononcera une peine sévère et prise dans l'ordre des peines infamantes contre quiconque oserait porter une main téméraire sur le ministre du culte en fonction ; mais à moins qu'il n'y ait des circonstances aggravantes, elle ne punira les autres troubles que de peines correctionnelles, graduées d'après le scandale qui aura pu en résulter : ce ne sont pas, surtout en matière de troubles de cette espèce, les peines les plus sévères qui seraient les plus efficaces.

Après avoir retracé les crimes et délits qui compromettent la paix publique sous le rapport d'une résistance plus ou moins directe à l'action de l'autorité, le projet qui vous est soumis s'occupe des dispositions relatives aux associations de malfaiteurs, aux vagabonds et aux mendiants : je viens en trois mots d'indiquer trois classes d'individus dont le nom seul est un sujet d'alarme pour la société.

Remarquons, au reste, que les malfaiteurs dont il s'agit en ce moment ne sont pas ceux qui agissent isolément, ou même de concert avec d'autres pour la simple exécution d'un crime : sous ce rapport, il est déjà beaucoup de malfaiteurs dont la peine a été déterminée selon la nature de leurs crimes.

Associations de malfaiteurs.

Ce que le projet de loi considère plus particulièrement ici, ce sont les bandes ou associations de ces êtres pervers qui, faisant un métier du vol et du pillage, sont convenus de mettre en commun le produit de leurs méfaits.

Cette association est en soi-même un crime qui, lorsqu'il n'aurait été accompagné ni suivi d'aucun autre, entraînera la peine des travaux forcés à temps contre les chefs, et celle de la réclusion contre tous les autres individus de la bande.

Vagabondage.

Mais ces bandes sont ordinairement recrutées par les vagabonds; et tout ce qui touche au vagabondage trouve naturellement ici sa place. Le projet de loi définit le vagabondage; il l'érige en délit, et lui inflige une peine correctionnelle. Toutefois il ne s'arrête point là. Que serait-ce, en effet, qu'un emprisonnement de quelques mois, si le vagabond était ensuite purement et simplement replacé dans la société à laquelle il n'offrirait aucune garantie ?

Celui qui n'a ni domicile, ni moyens de subsistance, ni profession ou métier, n'est point en effet membre de la cité; elle peut le rejeter et le laisser à la disposition du Gouvernement, qui pourra, dans sa prudence, ou l'admettre à caution, si un citoyen honnête et solvable veut bien en répondre, ou le placer dans une maison de travail jusqu'à ce qu'il ait appris à subvenir à ses besoins, ou enfin le détenir comme un être nuisible ou dangereux, s'il n'y a nul amendement à en espérer.

Mendicité.

Les mendiants ne sont pas dignes de beaucoup plus de faveur, aujourd'hui surtout que la bienfaisante activité du Gouvernement réalise le vœu philantropique de tant d'écrivains distingués, et ouvre, sous le nom de dépôts de mendicité, des asiles où les pauvres infirmes sont nourris aux frais de l'État, qui ne leur demandera d'ailleurs que le travail dont ils seront capables.

Quand de tels établissements existeront partout, il ne restera plus de prétexte ni d'excuse à la mendicité; mais jusque-là la crainte de frapper le malheur et l'indigence exigera quelques ménagements en faveur des mendiants invalides.

D'après ces idées, le projet de loi assujettit, sans distinction, à des peines correctionnelles toutes personnes qui mendient dans les lieux pour lesquels il y a des dépôts de mendicité.

Dans les autres lieux on distinguera; et la mendicité, toujours punissable à l'égard des individus valides, ne deviendra un délit à l'égard des autres qu'autant qu'ils feindraient des plaies, qu'ils mendieraient en réunion, ou qu'ils seraient entrés dans une maison sans permission des personnes qui y demeurent.

Dans sa prévoyance, le projet de loi a posé aussi quelques règles communes aux vagabonds et aux mendiants.

Dispositions communes aux vagabonds et aux mendiants.

Tout individu de cette qualité appelle une répression plus spéciale, s'il a été saisi travesti ou muni d'armes, de limes ou de crochets; s'il a été trouvé porteur d'effets d'une certaine valeur, ou s'il a exercé des violences, quelque légères qu'elles soient.

De la part des hommes dont on s'occupe en ce moment, il n'est aucun des signes indiqués qui ne soit propre à porter l'alarme et n'atteste un délit consommé ou prêt à l'être.

L'ordre public doit s'armer plus fortement contre ceux qui le menacent davantage; et c'est aussi dans ces vues que la marque sera infligée à tout vagabond ou mendiant qui aura encouru la peine des travaux forcés à temps, et qu'après toute espèce de condamnation à des peines afflictives, ou même simplement correctionnelles, les vagabonds et mendiants seront mis à la disposition de la haute police.

Réflexions générales sur les mises à la disposition de la haute police.

Cette attribution à la haute police est d'une grande importance; restreinte, par les dispositions générales du projet, aux gens sans aveu et aux individus condamnés à des peines afflictives ou au bannissement, ne s'exerçant au-delà qu'en vertu de condamnations spéciales et pour des cas bien déterminés, c'est une véritable institution dont le nom, quelque sévère qu'il puisse paraître au premier aspect, doit rassurer et non alarmer les bons citoyens.

La société n'a-t-elle donc en effet aucunes précautions

à prendre, lorsque les hommes qui l'ont grièvement troublée rentrent dans son sein ? et s'ils ne peuvent trouver sur toute la surface de l'Empire un seul citoyen solvable qui veuille cautionner leur conduite future, n'est-ce pas un nouveau degré de suspicion qui s'élève contre eux et autorise, soit à les éloigner d'un lieu désigné, soit à leur prescrire l'habitation d'un autre, soit enfin à les arrêter et détenir s'ils désobéissent ?

Eh ! quand cette restriction des droits individuels du condamné pourrait être considérée comme une aggravation de la peine principale, elle serait juste encore, puisqu'elle complète la garantie sociale.

Chez un peuple voisin dont la législation, en matière criminelle surtout, a été peut-être trop vantée, quoique souvent digne d'éloges, l'obligation de fournir cette caution a sans doute été portée trop loin quand la loi a permis de l'imposer, selon les circonstances, à tout particulier, même domicilié et non repris de justice, sur l'affirmation assermentée d'un autre citoyen touchant le péril auquel celui-ci se prétendrait exposé (1) par suite de paroles ou démarches menaçantes

Mais s'il y a de graves inconvénients à armer ainsi les citoyens les uns contre les autres, et si une telle législation semble plus propre à répandre du trouble et des inquiétudes qu'à les calmer, la scène change lorsque la surveillance légale, spécialement dirigée contre des gens sans aveu ou repris de justice, a été remise par l'autorité judiciaire, qui a déjà usé du droit de punir, à l'autorité

____

(1) Des Lois de police et criminelles de l'Angleterre, ouvrage traduit de l'anglais de *Blackstone*, par *Ludot*, chap. I.

administrative chargée du soin de prévenir de nouveaux crimes.

Dans ce système tout se trouve en harmonie ; et si cette heureuse innovation n'arrête pas toutes les récidives, elle en préviendra beaucoup, et assurera du moins, par le cautionnement même, une indemnité aux parties qui seraient lésées par un nouveau délit.

Parmi les innovations heureuses du projet de loi, nous espérons que l'on pourra compter aussi les dispositions qu'il a adoptées, dans l'intérêt de la paix publique, contre les distributions d'écrits, images ou gravures que l'on ferait paraître sans nom, soit de l'auteur, soit de l'imprimeur ou graveur.

Distribution d'écrits, images ou gravures sans noms d'auteur, imprimeur ou graveur.

Sans rien préjuger sur les mesures d'un autre ordre que l'on pourrait prendre contre certains ouvrages dont la circulation serait dangereuse, il est, dès ce moment, et il a toujours été reconnu que l'émission d'un ouvrage entraîne une juste responsabilité, toutes les fois qu'il nuit, soit à l'ordre public, soit à des intérêts privés.

Mais l'on n'a pas jusqu'à présent tiré de ce principe toutes les conséquences qui en dérivaient naturellement ; la première sans doute est que celui qui imprime ou fait imprimer doit se faire connaître ; car que deviendrait sans cela la responsabilité, dans tous les cas où il pourrait échoir de l'appliquer ?

Dans tout système qui ne dégénérera point en licence, l'on ne saurait se plaindre d'une telle obligation : si l'ouvrage est bon, ce n'est point une gêne sensible ; s'il est dangereux ou nuisible, cette obligation devient un frein utile.

Disons donc que la société a de justes et grandes raisons pour connaître celui qui est responsable : si l'auteur, timide ou modeste, n'a pas voulu se nommer, le même motif n'existe pas pour l'imprimeur. L'alternative laissée sur ce point répond à toutes les objections que l'on pourrait élever dans l'intérêt des lettres.

Ce qui importe surtout ici, c'est qu'il y ait au moins une personne responsable, qu'elle soit connue, et que par ce moyen l'on puisse, le cas échéant, exercer toutes les actions ou poursuites que réclamerait l'ordre public.

Ainsi, puisqu'il est utile que tout ouvrage littéraire porte le nom de son auteur ou de l'imprimeur, la loi peut l'ordonner ; et, par une juste et immédiate conséquence de cette première disposition, elle pourra prohiber la distribution de tous ouvrages qui ne seraient point revêtus de ce caractère.

Si donc on colporte un ouvrage sans nom d'auteur ni d'imprimeur, le colporteur pourra être immédiatement saisi, et, pour cette seule contravention, puni de peines correctionnelles, réductibles toutefois à des peines de simple police, s'il révèle les personnes qui l'ont chargé de la distribution.

Par cette voie l'on remontera ordinairement jusqu'à l'imprimeur, et de celui-ci même jusqu'à l'auteur, sur lequel pesera toujours la plus forte peine, lorsqu'il sera découvert.

Cette peine cependant variera selon la nature de l'ouvrage distribué en contravention aux lois ; ordinairement correctionnelle, elle pourra devenir afflictive, si l'écrit anonyme contient provocation à des crimes.

Dans ce dernier cas, la peine de complicité restera irrévocablement applicable à l'imprimeur, justement considéré comme ayant connu les caractères pernicieux de l'ouvrage auquel sa criminelle complaisance aura donné cours ; et l'atténuation de la peine, pour cause de révélation, se bornera aux simples distributeurs : ceux-ci, aveugles instruments d'écrivains pervers, ont paru susceptibles de cette modération de peines, qui d'ailleurs profitera même à l'ordre public en intéressant les colporteurs à révéler ce qu'ils savent, pour n'être pas traités comme complices.

Dans la combinaison des mesures que je viens de vous exposer, Messieurs, il n'y a rien (vous vous en convaincrez facilement) qui soit dirigé contre le sage emploi des lettres, mais seulement contre les productions clandestines : or, tout auteur qui veut porter ses coups dans l'ombre mérite bien qu'on le suive à la trace ; et si, comme nous l'espérons, le projet de loi atteint ce but, il aura beaucoup fait pour le maintien du bon ordre.

Des sociétés ou réunions illicites.

Il me reste à vous parler, Messieurs, des sociétés ou réunions ayant pour but de s'occuper journellement ou périodiquement d'objets religieux, politiques ou littéraires.

Je me garderai bien de traiter ce sujet avec l'importance qu'on eût pu y mettre il y a quelques années : tout ce qui fut dit et écrit alors dérivait d'idées et de principes qui ne peuvent plus recevoir d'application sous la forme de Gouvernement qui a été depuis adoptée en France.

Le droit absolu et indéfini qu'aurait la multitude de se réunir pour traiter d'affaires politiques, religieuses, ou autres de cette nature, serait incompatible avec notre état politique actuel.

Mais si le gouvernement monarchique doit être assez

fort pour repousser ce qui pourrait lui nuire, il est aussi dans son essence de n'admettre aucune rigueur inutile : il n'interviendra donc point, hors les cas qui l'intéresseraient spécialement, dans ces petites réunions que les rapports de famille, d'amitié ou de voisinage peuvent établir sur tous les points d'un si vaste empire ; et lorsqu'il ne se passera dans ces petites réunions rien de contraire au bon ordre, l'autorité publique, qui ne saurait être tracassière, ne leur imposera aucune obligation spéciale, eussent-elles pour objet la lecture en commun de journaux ou autres ouvrages.

Cette obligation spéciale de se faire connaître de l'autorité et d'obtenir son assentiment commencera là seulement où le nombre des sociétaires serait tel, qu'il pût devenir un juste sujet de surveillance plus particulière.

C'est alors que de telles associations ne pourront exister qu'avec l'autorisation du Gouvernement et sous les conditions qui leur seront imposées : c'est alors aussi qu'en cas d'infraction ces associations pourront être dissoutes, et leurs chefs et directeurs condamnés à des amendes, et même à l'emprisonnement.

Les dispositions du projet de loi, conformes à ces idées, vous paraîtront sans doute avoir atteint le but qu'elles se proposaient.

Ici, Messieurs, se termine le tableau des crimes et délits *contre la paix publique*; tableau qui n'est lui-même que le complement du chapitre *des crimes et délits contre la chose publique.*

Cet exposé, bien que restreint aux dispositions principales, a été long, parce qu'il embrassait une multitude de matières dont plusieurs, dérivant de sources un peu abstraites, avaient besoin d'être ramenées à des termes simples, précis et tels qu'ils convinssent à une législation pénale.

Je me suis au surplus abstenu d'en comparer les détails avec ceux du Code de 1791.

Semblables sur plusieurs points, plus ou moins différentes sur d'autres, souvent ajoutées, les dispositions du nouveau projet de loi sont le résultat de méditations dans lesquelles nous nous sommes efforcés de mettre à profit les travaux mêmes de nos devanciers, et les leçons fournies par l'expérience des derniers temps.

Un travail de cette nature offrait de grandes difficultés ; la plus grave sans doute était de bien graduer les peines et d'en faire une juste application aux divers crimes ou délits.

Cet effet s'obtiendrait exactement, s'il existait une progression de peines parfaitement correspondante à la progression des délits, et *si* (selon les expressions de *Beccaria*) *la géométrie était applicable à toutes les petites combinaisons obscures de nos actions* (1).

En l'absence d'un tel guide, le législateur doit consulter son cœur au moins autant que son esprit : il doit aussi reconnaître et respecter les limites que la nature des choses a mises à sa puissance.

Dans l'application de la peine capitale, et même des peines perpétuelles, la gravité nécessairement énorme des crimes qui y donnent lieu ne laisse pas apercevoir de nuances propres à entraîner la modification de la peine.

Il en est autrement à l'égard des crimes inférieurs, et dont la peine n'est que temporaire ; plus on descendra dans cette classe, plus il deviendra évident que chaque

____

(1) Traité des Délits et des Peines, §. VI, p. 31.

espèce est susceptible de varier d'intensité ; une sage circonspection commandait donc de laisser sur ce point une suffisante latitude aux juges ; et ce parti adopté par le projet, en même temps qu'il satisfait à la justice, a paru propre à rassurer aussi la conscience du législateur.

Puisse ce nouveau travail obtenir votre approbation et répondre aux vues bienfaisantes de l'auguste chef de cet Empire ! Puisse le nouveau Code, dont plusieurs parties essentielles vous sont actuellement connues, obtenir bientôt une place honorable à côté de ceux qu'a déjà tracés et donnés à la France le Héros législateur du dix-neuvième siècle !

MOTIFS

Du Livre III, Titre II, Chapitre I,

PRÉSENTÉS

PAR MM. LE CHEVALIER FAURE, LES COMTES RÉAL
ET GIUNTI, Conseillers d'État.

Séance du 7 février 1810.

MESSIEURS,

Le projet de Code pénal offert à vos méditations vous
a présenté, dans le titre premier du Livre III, le tableau
des crimes et délits contre la chose publique.

Ce titre II du même Livre a pour objet les crimes et
délits contre les particuliers.

Cette seconde partie est aussi d'une extrême importance : elle embrasse un grand nombre d'attentats, dont
la répression est indispensable pour garantir à chacun
des membres de la société la jouissance paisible de tous
les avantages qu'il a droit d'attendre du pacte social. En
vain les meilleures lois civiles auraient été faites, si la
violence ou la fraude, l'intérêt ou la méchanceté pouvaient se jouer impunément de la vie, de la liberté, de
l'honneur et de la fortune des citoyens ; ou si le vice
livré aux excès les plus honteux pouvait impunément
outrager les mœurs.

Ce titre se divise en deux chapitres. L'un est relatif
aux attentats contre les personnes ; le second concerne
les attentats contre les propriétés. Le premier forme la
matière dont nous aurons l'honneur de vous entretenir
aujourd'hui.

Nous parlerons d'abord des actes attentatoires à la vie.

Attentats à la vie.

On attente à la vie d'une personne, soit en lui donnant la mort, soit en exerçant sur elle des actes de violence. Ceux-ci, quoiqu'ils n'entraînent pas sur-le-champ la perte de la vie, peuvent cependant en abréger le cours, ou donner lieu à des maladies ou infirmités.

Pour que l'homicide soit un crime, il faut qu'il soit volontaire. S'il est tel, il est qualifié meurtre. Mais si le meurtre est commis avec préméditation ou guet-apens, la loi le qualifie assassinat.

L'assassinat est donc un plus grand crime que le meurtre, et le meurtre n'emporte la même peine que l'assassinat que dans des cas particuliers où l'assimilation est nécessitée par l'atrocité du crime résultant soit de la qualité de la personne homicidée, soit d'autres circonstances aggravantes.

La peine de l'assassinat est la mort : c'est celle du talion. Toute autre peine, quelque rigoureuse qu'elle fût, ne serait pas assez répressive, et le plus souvent produirait l'impunité. Sans cette peine, la haine ou la vengeance d'un lâche pourrait se satisfaire en jouant, si je puis parler ainsi, un jeu trop inégal contre le citoyen dont il méditerait la mort : l'un ne mettrait au jeu que sa liberté, et l'autre y mettrait sa vie.

Après avoir dit que le nouveau Code porte la peine de mort contre les assassins, nous n'avons pas besoin d'ajouter que l'homicide par poison sera puni de la même peine.

Le crime d'empoisonnement est un véritable assassinat; car il suppose nécessairement un dessein antérieur. Il est d'ailleurs, de tous les crimes, le plus lâche parmi les plus atroces.

Le nouveau Code le définit ainsi : « Est qualifié em-
« poisonnement tout attentat à la vie d'une personne par
« l'effet de substances qui peuvent donner la mort
« plus ou moins promptement, de quelque manière que
« ces substances aient été employées ou administrées, et
« quelles qu'en aient été les suites. »

Cette définition est plus complète que celle adoptée
par la loi de 1791, en ce qu'elle comprend tout moyen
dont on aurait fait usage pour commettre ce crime, et ne
borne pas les tentatives au cas particulier où le poison
aurait été présenté ou mêlé avec des aliments ou breuvages : il est tant de moyens que la scélératesse peut in-
venter, et dont l'histoire offre l'exemple, qu'il était indispensable de recourir à des termes généraux.

D'un autre côté, il était inutile d'ajouter la disposition de cette même loi de 1791, qui porte que si,
avant que l'empoisonnement ait été effectué, ou avant
que l'empoisonnement des aliments et breuvages ait été
découvert, l'empoisonneur arrêtait l'exécution du crime,
soit en supprimant les aliments et breuvages, soit en
empêchant qu'on en fasse usage, l'accusé sera acquitté.

Cette disposition était nécessaire lorsqu'elle fut adoptée, parce qu'alors il n'existait aucune loi contre les tentatives de crime ; mais l'article 2 du nouveau Code, qui
les prévoit et les définit, annonce assez qu'aucune de ces
tentatives ne sera considérée comme le crime même,
lorsqu'elle aura été arrêtée par la volonté de l'auteur, et
non par des circonstances fortuites et indépendantes de
sa volonté.

Quant au parricide, qui consiste dans le meurtre des
pères ou mères légitimes, naturels ou adoptifs, ou de
tout autre ascendant légitime, ce crime, même commis
sans préméditation ni guet-apens, révolte tellement la

nature, que, loin de pouvoir être puni d'une peine moindre que l'assassinat, il mérite une peine plus forte. Aussi est-il dit, dans le premier Livre du nouveau Code, qu'avant d'être exécuté à mort il aura le poing droit coupé. Nous ne répéterons point les observations qui vous ont été présentées à cet égard.

On sait que chez les Romains le coupable de parricide était condamné au supplice le plus affreux.

Vous remarquerez, Messieurs, que le nouveau Code assimile les pères et mères adoptifs aux pères et mères légitimes. Le Code Napoléon a consacré cette assimilation par ses diverses dispositions. Suivant l'article 349, « l'obligation naturelle qui continuera d'exister entre « l'adopté et ses père et mère, de se fournir des ali- « ments dans les cas déterminés par la loi, sera consi- « dérée comme commune à l'adoptant et à l'adopté l'un « envers l'autre. »

Ajoutons que l'article 350 accorde à l'adopté, sur la succession de l'adoptant, les mêmes droits que ceux qui appartiennent à l'enfant né en mariage.

Le meurtre d'un enfant nouveau-né, crime que le projet qualifie infanticide, sera puni de la même peine que l'assassinat. On se rappelle que la qualification d'assassinat est donnée à tout meurtre commis avec préméditation. Or, il est impossible que l'infanticide ne soit pas prémédité : il est impossible qu'il soit l'effet subit de la colère ou de la haine, puisqu'un enfant, loin d'exciter de tels sentiments, ne peut inspirer que celui de pitié. Il est hors d'état de se défendre, hors d'état de demander du secours ; et par cela seul il est plus spécialement sous la protection de la loi. Des hospices sont établis pour recevoir ceux dont on ne peut prendre soin. L'infanticide est donc, sous tous les rapports, un acte de barbarie atroce ;

et quand il serait quelquefois le fruit du dérèglement de mœurs, une telle cause ne peut trouver d'indulgence dans une législation protectrice des mœurs.

La peine de l'assassinat sera aussi celle du meurtre qui aura été précédé, accompagné ou suivi de quelque crime ou délit. Ce concours de circonstances qui s'aggravent réciproquement est d'une nature si effrayante, qu'une peine inférieure ne suffirait pas pour tranquilliser la société.

Enfin, le Code assimile aux assassins, et punit comme tels, tous malfaiteurs, quelle que soit leur dénomination, qui, pour l'exécution de leurs crimes, emploient des tortures ou commettent des actes de barbarie. Ces individus, à qui les moyens les plus horribles ne coûtent rien, pourvu qu'ils arrivent à leurs fins, et qui portent la terreur et la désolation partout où ils existent, ne peuvent être retenus que par la crainte du dernier supplice.

Quant au meurtre dénué de toute espèce de circonstances aggravantes, il sera puni de la peine qui suit immédiatement celle de mort, c'est-à-dire de la peine des travaux forcés à perpétuité. Dès que ce crime n'est point le résultat d'un dessein formé avant l'action, dès qu'il ne présente aucun des caractères dont nous avons parlé, il est sans contredit moins grave que l'assassinat, et dès-lors ne doit pas emporter la même peine; autrement cette juste proportion qu'on ne saurait observer avec trop de soin entre les délits et les peines, et cette gradation qui en est la suite nécessaire, ne subsisteraient plus.

Le nouveau Code ne se borne pas à établir des peines contre les coupables des divers crimes dont nous venons de parler; il en établit aussi contre ceux qui se permettent des menaces d'attentats contre la vie des per-

sonnes, lorsque ces attentats, s'ils étaient commis, seraient punis d'une peine capitale, ou au moins égale à
celle des travaux forcés à temps.

De telles menaces, lorsqu'elles sont écrites, annoncent
un dessein prémédité de faire le mal. Le plus souvent
l'écrit où elles se trouvent contient un ordre quelconque;
par exemple, l'ordre de déposer une somme d'argent
dans un lieu indiqué. Quel que soit l'ordre, la loi punit
le crime de la même peine que le vol avec violence.
N'est-ce pas en effet un crime semblable? La personne
menacée est dans une situation d'autant plus critique,
qu'elle ne peut pas se mettre continuellement en garde,
et qu'elle craint toujours que, si elle n'obéit point à
l'ordre, tôt ou tard, et au moment où elle y songera
le moins, elle ne finisse par être victime du crime dont
elle est menacée. La terreur que ces menaces inspirent
ne nuit pas seulement à la tranquillité de la personne
qui en est l'objet, elle est partagée par beaucoup d'autres
qui redoutent pour eux le même sort.

Ce que nous venons d'observer trouve également son
application, si l'écrit, au lieu de contenir l'ordre de déposer une somme, contient celui de remplir une condition
quelconque; en ce dernier cas, il y a toujours violence,
et violence préméditée avec dessein d'obtenir ce qu'on
n'a pas le droit d'exiger.

Lorsque la menace écrite n'a été accompagnée d'aucun
ordre ou condition, on ne peut l'attribuer qu'au désir de
répandre l'effroi, sans aucun but de s'approprier le bien
d'autrui. Le coupable doit être puni; mais il ne le sera
que des peines de police correctionnelle. Ce délit est en
effet bien moins grave que le premier.

Le Code veut aussi que des peines de police correctionnelle soient prononcées, quoique les menaces soient

verbales, toutes les fois qu'elles seront accompagnées d'un ordre ou condition. Les menaces verbales seront moins punies que les menaces écrites, parce que le coupable agissant plus à découvert, il est moins difficile de se mettre en garde contre lui; que dès-lors elles excitent une crainte moins forte; que, d'un autre côté, la préméditation n'est pas nécessairement attachée aux menaces verbales comme elle l'est aux menaces écrites.

A l'égard des menaces verbales qu'aucun ordre ni condition n'auront accompagnées, nulle peine n'est établie : on a considéré qu'étant dénuées de tout intérêt, elles peuvent être le résultat d'un mouvement subit produit par la colère, et dissipé bientôt par la réflexion.

Nous observerons ici que, dans les deux cas où la menace est punie correctionnellement, les coupables peuvent être mis sous la surveillance de la haute police. Cette faculté laissée aux juges leur impose le devoir d'examiner jusqu'à quel point ces individus sont dangereux, soit par leur vie habituelle, soit par leurs liaisons.

Passons maintenant à l'examen des attentats qui ne portent point le caractère de meurtre, mais qui cependant présentent des actes de violence que la loi doit sévèrement réprimer. Ainsi des coups auront été portés, ou des blessures auront été faites, et la personne blessée ou frappée aura essuyé une maladie, ou se sera trouvée dans l'incapacité absolue de se livrer à aucun travail personnel; si la maladie ou l'incapacité de travail a duré plus de vingt jours, le coupable sera puni de la réclusion. Le même crime emportera la peine des travaux forcés à temps, lorsqu'il y aura eu préméditation ou guet-apens; et comme les juges, en appliquant la loi, auront une latitude de cinq ans jusqu'à dix pour la réclusion, et de cinq ans jusqu'à vingt pour les travaux forcés à temps, il leur sera facile de proportionner la peine à la gravité du fait. C'est par cette raison qu'il n'a pas été jugé nécessaire de faire entrer dans le nouveau Code les distinctions qui se trouvent dans la loi de 1791, sur les différentes espèces de mutilations.

Si les blessures ou les coups sont d'une nature moins grave que ceux qui doivent donner lieu à la réclusion ou aux travaux forcés à temps, ils ne seront punis que des peines de police correctionnelle. Mais la durée de l'emprisonnement et la quotité de l'amende dépendent des circonstances dont la preuve aura été acquise. Il suffira que les juges se renferment dans les limites tracées par la loi à l'égard de cette espèce de délit.

Enfin, quelle qu'ait été la nature du crime ou délit, le Code veut que la peine soit plus forte, si la personne maltraitée est le père ou la mère légitime ou adoptif, ou tout autre ascendant légitime. Cette différence dérive du même principe que la disposition relative au parricide. La lecture de l'article fera voir que la peine est élevée dans une juste proportion, comparativement à celle que le coupable aurait subie, si le crime ou délit eût été commis envers tout autre.

On doit observer que, lorsque les blessures ou les coups seront susceptibles d'être qualifiés tentatives d'assassinat, les dispositions qui viennent d'être analysées ne seront plus applicables : il faudra se reporter à l'article du Code relatif aux tentatives de crime ; et si le cas d'attaque à dessein de tuer a été l'objet d'une disposition spéciale dans la loi de 1791, c'est parce que cette loi ne contenait aucune disposition générale sur les tentatives.

L'article du nouveau Code relatif à l'avortement offre aussi plusieurs modifications importantes. La nécessité de punir ce crime n'a pas besoin d'être démontrée ; la loi de

1791 ne l'a pas oublié; mais elle punit de la même peine indistinctement toute personne coupable de ce crime. Cette confusion n'existera point dans la nouvelle loi. La femme coupable du crime d'avortement sera punie de la réclusion. Mais une peine plus rigoureuse, celle des travaux forcés à temps, aura lieu contre les médecins, chirurgiens et autres officiers de santé qui auront procuré à la femme les moyens de se faire avorter. Ils sont en effet plus coupables que la femme même, lorsqu'ils font usage, pour détruire, d'un art qu'ils ne doivent employer qu'à conserver. Le chancelier Daguesseau rapporte, à ce sujet, qu'Hippocrate, dans le serment qu'on trouve à la tête de ses ouvrages, promet solennellement de ne jamais donner à une femme grosse aucun médicament qui puisse la faire avorter. Son serment, dit-il, est suivi d'imprécations qui prouvent que ce crime était considéré comme un des plus grands qu'un médecin pût commettre. En effet, si la femme ne trouvait pas tant de facilité à se procurer les moyens d'avortement, la crainte d'exposer sa propre vie en faisant usage de médicaments qu'elle ne connaitrait pas, l'obligerait souvent de différer son crime, et elle pourrait ensuite être arrêtée par ses remords. La disposition relative aux médecins ne se trouve point dans la loi de 1791.

Je ne m'arrêterai point à la partie du Code qui concerne l'homicide, les blessures et les coups involontaires résultant du défaut d'adresse ou de précaution : ces délits sont punis de peines de police correctionnelle, et les termes généraux dans lesquels ces articles sont conçus embrassent toutes les espèces.

Je passe aux crimes ou délits qui, quoique volontaires, sont susceptibles d'être excusés. On se rappelle que le Code d'instruction criminelle porte qu'aucun fait propose pour excuse par l'accusé ne sera, quelque prouvé qu'il soit, pris en considération par le juge, s'il n'est déclaré excusable par la loi.

C'est ici que le Code détermine les divers cas où des crimes et délits commis envers les personnes peuvent être excusés; il n'admet point l'excuse sans une provocation violente, et d'une violence telle, que le coupable n'ait pas eu, au moment même de l'action qui lui est reprochée, toute la liberté d'esprit nécessaire pour agir avec une mûre réflexion. Sans doute il a commis une action blâmable, une action que la loi ne peut se dispenser de punir; mais il ne peut être, aux yeux de la loi, tout-à-fait aussi coupable que si la provocation qui l'a entraîné n'eût pas existé.

Cette provocation, nous ne pouvons trop le redire, doit être de nature à faire la plus vive impression sur l'esprit le plus fort.

Le Code renferme plusieurs dispositions sur les faits qui sont susceptibles d'être déclarés excusables. Je me contenterai d'en citer une seule. « Dans le cas d'adultère, « porte le Code, le meurtre commis par l'époux sur son « épouse, ainsi que sur le complice, à l'instant où il les « surprend en flagrant délit dans la maison conjugale, « est excusable. » Cet outrage fait au mari est une de ces provocations violentes qui appellent l'indulgence de la loi. On remarquera que la loi n'excuse ce meurtre que sous deux conditions : 1° si l'époux l'a commis au même instant où il a surpris l'adultère; plus tard, il a eu le temps de réfléchir, et il a dû penser qu'il n'est permis à personne de se faire justice à soi-même; 2° s'il a surpris l'adultère dans sa propre maison. Cette restriction a paru nécessaire : on a craint que, si ce meurtre, commis dans tout autre lieu, était également excusable, la tranquillité

des familles ne fût troublée par des époux méfiants et in-
justes, qu'aveuglerait l'espoir de se venger des prétendus
égarements de leurs épouses.

Il est certains meurtres à l'égard desquels la loi n'admet point d'excuse, quoiqu'il y ait eu provocation
violente.

Par exemple, aucune provocation, quelque violente
qu'elle soit, ne peut excuser le parricide : le respect religieux qu'on doit à l'auteur de ses jours, ou à celui que la
loi place au même rang, impose le devoir de tout souf-
frir plutôt que de porter sur eux une main sacrilége.

A l'égard du meurtre commis par l'époux envers son
épouse, dans tout autre cas que celui dont nous venons
de parler au sujet de la femme adultère, ou du meurtre
commis par l'épouse envers son époux, le crime n'est
excusable que lorsqu'au moment même où il a été commis la vie de l'auteur du meurtre a été mise en péril par
l'époux ou l'épouse homicidée. C'est en effet la seule excuse qui puisse être admise à l'égard de personnes obli-
gées par état de vivre ensemble, et de n'épargner aucun
sacrifice pour maintenir entre eux une parfaite union.

Lorsque la loi déclare un fait excusable et que ce fait
est prouvé, les juges ne peuvent prononcer des peines
afflictives ou infamantes : il y aurait de la contradiction à
déclarer infâme, en vertu de la loi, celui qu'elle recon-
naît digne d'excuse. Les peines de police correctionnelle
sont donc les seules qui doivent être prononcées. Le
Code établit sur ce point une échelle de proportion relative à la peine que le coupable eût dû subir si l'excuse
n'avait pas existé.

Il est des circonstances où l'homicide, les blessures et
les coups ne sont susceptibles d'aucune peine ; en un mot,
où il ne résulte de ces actes aucun crime ni délit.

Le cas arrive, soit lorsque ces actes étaient ordonnés par la loi et commandés par l'autorité légitime, soit lorsqu'ils étaient commandés par la nécessité actuelle de la légitime défense de soi-même ou d'autrui.

Ces mots *nécessité actuelle* prouvent qu'il ne s'agit que du moment même où l'on est obligé de repousser la force par la force. Après avoir vu la loi défendre d'exercer des violences, on la voit ici permettre de les repousser. Elle veut que les hommes écoutent et respectent cette défense dans le commerce paisible qu'ils ont ensemble; mais elle les en dispense lorsque l'on commet contre eux des actes hostiles : elle ne leur commande pas d'attendre alors sa protection et son secours, et de se reposer sur elle du soin de leur vengeance, parce que l'innocent souffrirait une mort injuste avant qu'elle eût pu faire subir au coupable le juste châtiment qu'il aurait mérité.

J'ai terminé mes observations sur la partie du Code relative aux attentats contre la vie des personnes.

Le Code s'occupe ensuite des attentats contre les mœurs.

Attentats contre les mœurs.

« Les peines qui sont de la juridiction correctionnelle,
« dit l'auteur de l'Esprit des Lois, suffisent pour répri-
« mer ces sortes de délits : en effet ils sont moins fondés
« sur la méchanceté que sur l'oubli ou le mépris de soi-
« même. Il n'est ici question, ajoute-t-il, que des crimes
« qui intéressent uniquement les mœurs, non de ceux
« qui choquent aussi la sûreté publique, tels que l'enlè-
« vement et le viol. »

La distinction établie par Montesquieu a été suivie dans le Code.

Le viol sera puni de la réclusion. Il en sera de même de tout autre attentat à la pudeur, consommé ou tenté avec violence contre des personnes de l'un ou de l'autre sexe. La loi de 1791 n'a parlé que du viol; elle s'est tue sur d'autres crimes qui n'offensent pas moins les mœurs; il convenait de remplir cette lacune. Celui qui aura commis l'un de ces attentats envers une personne âgée de moins de quinze ans accomplis encourra la peine des travaux forcés à temps. Il est même des circonstances qui, réunies au crime, attireront sur le coupable la peine des travaux forcés à perpétuité. Ces circonstances, spécifiées par le Code, résulteront, soit de la qualité du coupable, soit des moyens qu'il aura employés.

Le Code prononce aussi des peines de police correctionnelle contre les personnes convaincues d'avoir débauché ou corrompu la jeunesse : il est, en ce point, conforme à l'ancienne loi; mais, de plus, le coupable sera interdit de toute tutelle et curatelle, et de toute participation au conseil de famille pendant un temps déterminé. Si c'est le père ou la mère, il sera, indépendamment des autres peines, privé de tous les droits et avantages qu'il aurait pu réclamer, en vertu du Code Napoléon, sur la personne et les biens de l'enfant. Cette dernière disposition vengera les mœurs outragées par ceux qui devaient en être les plus fidèles gardiens.

Parmi les attentats aux mœurs est comprise la violation de la foi conjugale, soit que ce délit ait été commis par la femme, soit qu'il l'ait été par le mari. L'adultère de la femme est un délit plus grand, parce qu'il entraîne des conséquences plus graves, et qu'il peut faire entrer dans la famille légitime un enfant qui n'appartient point à celui que la loi regarde comme le père. Le Code pénal, en établissant la peine qui doit être prononcée contre la

9

femme, n'a fait que se conformer à l'article 298 du Code Napoléon; de ce Code où l'on remarque partout le respect le plus religieux pour les mœurs : il porte un emprisonnement par voie de police correctionnelle, de trois mois au moins et de deux ans au plus.

On a rappelé dans le projet l'article 309 de ce même Code, qui laisse le mari maître d'arrêter l'effet de cette condamnation, en consentant à reprendre sa femme : en effet, la femme n'est coupable qu'envers son mari; il doit donc avoir le droit de lui pardonner.

Si la femme n'est coupable qu'envers le mari, lui seul est en droit de se plaindre; l'action doit être interdite à tout autre, parce que tout autre est sans qualité et sans intérêt.

Bien plus, le mari serait privé de cette action s'il avait été condamné lui-même pour cause d'adultère. Alors la justice le repousserait comme indigne de sa confiance; et n'ayant pu, comme on va le voir, être convaincu d'adultère que sur la plainte de sa femme, il serait trop à craindre qu'il n'agît par récrimination.

Le complice de la femme sera condamné à la même peine, et de plus à l'amende.

À l'égard de la poursuite contre le mari pour cause d'adultère, elle ne peut avoir lieu que sur la plainte de la femme, parce qu'elle seule est intéressée à réclamer contre l'infidélité de son époux; et la femme ne peut intenter cette plainte que lorsqu'il a entretenu sa concubine dans la maison conjugale. Dans tout autre cas, les recherches dégénéreraient souvent en inquisition; mais dans celui prévu par la loi le délit est notoire. C'est d'après le même esprit que le Code Napoléon n'admet la femme à demander le divorce pour cause d'adultère de son mari, qu'en rapportant la même preuve à l'égard de la concubine Quant au délit, il sera puni d'une amende.

La loi de 1791 avait gardé le silence sur la violation de la foi conjugale de la part de l'époux ou de l'épouse : les dispositions du nouveau Code rempliront cette lacune.

La loi proposée prévoit, comme celle de 1791, le crime commis par la personne qui a contracté un nouveau mariage avant la dissolution du premier. La peine sera celle des travaux forcés à temps, et remplacera celle des fers. Le crime est très-grave en effet ; il renferme tout à la fois l'adultère et le faux ; car le coupable a déclaré faussement devant l'officier de l'état civil, et même attesté par sa signature, qu'il n'était point engagé dans les liens du mariage. Nous ne parlerons point des conséquences qui résultent de ce crime pour la seconde femme et pour les enfants. Ces détails n'entrent point dans notre sujet.

Nous arrivons maintenant à la partie du Code relative aux arrestations illégales et séquestrations de personnes.

Arrestations illégales.

Il ne s'agit point ici de celles commises par des fonctionnaires publics : cette matière est réglée par le titre 1er du troisième Livre. Les dispositions actuelles n'ont trait qu'aux attentats à la liberté, commis par des particuliers. On peut être arrêté par toute personne, lorsqu'on est surpris commettant un crime ou délit que toute personne a le droit de dénoncer. On peut aussi être arrêté par celui qu'une loi autorise à cet effet, ou qui est porteur d'ordre de l'autorité compétente. Hors ces cas, celui qui se permet de faire une arrestation est coupable de crime.

Prêter un lieu pour séquestrer la personne arrêtée est un acte de complicité : ce crime appelle un châtiment rigoureux; il porte atteinte à l'une des jouissances les plus précieuses que la société garantit à chacun de ses membres. Le Code prononce la peine des travaux forcés à temps contre l'auteur et son complice. Il se relâche cependant de sa rigueur envers le coupable, et consent à ce qu'il ne soit condamné qu'à des peines de police correctionnelle, si, avant le dixième jour accompli, il a rendu libre celui qu'il avait arrêté; alors la loi commue la peine en faveur de son repentir, et veut bien supposer que sa faute a été plutôt le résultat de l'irréflexion du moment que d'une préméditation tenant à des combinaisons criminelles; mais, passé le dixième jour, elle ne doute plus de la perversité de l'intention, et devient inflexible. Si même la détention ou séquestration a duré plus d'un mois, elle ne voit plus dans le coupable qu'un méchant tellement obstiné, tellement endurci, qu'il serait un fléau pour la société, s'il pouvait jamais rentrer dans son sein : elle l'en exclut pour toujours en le condamnant aux travaux forcés à perpétuité.

Enfin il est des circonstances particulières qui peuvent accompagner l'arrestation illégale, et qui lui donnent un tel caractère de gravité, que la loi considère alors le coupable comme atteint de brigandage et d'assassinat, et qu'elle prononce contre lui la peine de mort, peine destinée aux brigands et aux assassins.

Ces circonstances, dont la définition ne doit point dépendre de l'arbitraire du juge, sont spécifiées dans le Code : « Les coupables seront punis de mort, dit-il, « si l'arrestation a été exécutée avec le faux costume, « sous un faux nom, ou sur un faux ordre de l'autorité « publique;

« Si l'individu arrêté, détenu ou séquestré, a été me-
« nacé de la mort;

« S'il a été soumis à des tortures corporelles. »

Des attentats qui blessent l'ordre public à un tel degré
ne peuvent être trop sévèrement réprimés; ils doivent
être mis au même rang que les plus grands crimes contre
la paix publique.

Les dispositions que nous allons examiner maintenant
concernent les crimes ou délits qui tendent à empêcher ou
à détruire la preuve de l'état civil d'un enfant, ou à compromettre son existence.

Attentats contre l'état civil d'une personne.

Le Code pénal de 1791 ne contient qu'une seule disposition sur cette matière : il prononce douze ans de
fers contre celui qui a détruit la preuve de l'état civil
d'une personne.

L'expérience a fait reconnaître que cette disposition
était trop vague, et qu'il convenait de spécifier les différents cas, tels que le recélé ou la suppression d'un enfant,
la substitution d'un enfant à un autre, et la supposition
d'un enfant à une femme qui n'est point accouchée.

Nous ne parlerons point des édits et déclarations qui
furent rendus sous la dernière dynastie, relativement aux
recélés de grossesse : l'humanité eut long-temps à gémir
de lois si atroces.

L'Assemblée constituante fit disparaître cette législation si contraire aux mœurs d'un peuple civilisé, et particulièrement de la nation française.

Mais, pour éviter les détails auxquels s'étaient livrées
les anciennes lois, elle tomba dans l'excès opposé, et ne
détermina point du tout ce qui, en matière pénale, ne
peut être déterminé avec trop de soin. Les expressions du

nouveau Code ne laisseront point de doute que ceux-là seront condamnés à la peine de la réclusion, qui, par de fausses déclarations, donneront à un enfant une famille à laquelle il n'appartient point, et le priveront de celle à laquelle il appartient; ou qui, par un moyen quelconque, lui feront perdre l'état que la loi lui garantissait; ou enfin qui, étant chargés d'un enfant, ne le représenteront pas aux personnes qui ont droit de le réclamer.

Le Code Napoléon, pour assurer cet état aux enfants, exige que les naissances soient déclarées à l'officier de l'état civil, et désigne les personnes qu'il charge de faire ces déclarations. Depuis ce Code, on a remarqué que, faute d'une loi pénale, quelques personnes s'en étaient abstenues. Cette conduite est d'autant plus blâmable, qu'elles contreviennent à une loi sage dont le but est de veiller à l'intérêt d'enfants qui ne peuvent pas y veiller eux-mêmes; que la tendresse des parents eût dû être le garant de l'exécution de la loi; qu'enfin, s'il était possible de croire que le motif de ce délit fût l'espoir de soustraire un jour ces mêmes enfants aux lois sur la conscription, ils peuvent être assurés qu'ils les exposent, au contraire, à être appelés souvent plus tôt qu'ils ne le seraient, s'ils étaient en état de représenter leur acte de naissance. Le Code actuel punit ce délit.

Vous verrez, Messieurs, en parcourant les détails du projet, combien on a pris de précautions pour empêcher que l'intérêt personnel ou la négligence ne prive un enfant des moyens de reconnaître un jour la famille dont il est membre, et de réclamer les droits qui lui appartiennent comme membre de cette famille.

Parmi les délits que le Code prévoit, *je citerai l'exposition d'enfant.* Les peines de police correctionnelle auxquelles ce délit donnera lieu doivent être plus ou moins fortes, suivant le danger qu'on a fait courir à l'enfant; et ce danger est plus ou moins grand, suivant que le lieu de l'exposition est ou n'est pas solitaire. Il était impossible que la loi donnât une explication précise à cet égard; elle s'en rapporte aux juges; car le lieu le plus fréquenté peut quelquefois être solitaire, et le lieu le plus solitaire être très-fréquenté : cela dépend des circonstances.

Si l'enfant exposé dans un lieu solitaire a été mutilé ou estropié, ou si la mort est résultée de l'exposition, le coupable est puni comme s'il l'avait lui-même mutilé ou estropié, ou comme s'il lui avait lui-même donné la mort; car il ne pouvait se dissimuler que la privation absolue où il laissait l'enfant de toute espèce de secours l'exposait à cet événement, et il ne tenait qu'à lui de l'en préserver; dès qu'il ne l'a pas fait, la loi déclare qu'il en est la cause volontaire. et le soumet aux peines établies contre les auteurs de blessures ou d'homicides volontaires.

Il faut remarquer que, d'après le Code, l'exposition d'enfant n'est un délit que lorsque l'enfant exposé a moins de sept ans. Passé cet âge, la loi présume que l'enfant peut faire connaître les personnes entre les mains desquelles il se trouvait, et le lieu de leur demeure; qu'il peut, en un mot, fournir les renseignements nécessaires pour qu'il soit possible de retrouver la trace qu'on a voulu faire perdre.

C'est par les mêmes motifs que le Code, en prononçant des peines de police correctionnelle contre ceux qui porteraient à l'hospice un enfant dont ils se sont chargés gratuitement, ou pour lequel ils reçoivent une pension qui leur a été payée avec exactitude, ne parle que de l'enfant dont l'âge est au-dessous de sept ans accomplis.

Le législateur a craint que, tant qu'il n'aurait pas cet âge, il ne pût s'expliquer assez pour indiquer la maison où il a vécu jusqu'alors, et pour éclairer la justice de manière qu'elle puisse empêcher que son état civil ne soit perdu.

Tels sont les moyens par lesquels la loi tâche de mettre l'enfant à l'abri des atteintes directes et indirectes qu'on voudrait porter à ses droits.

Nous allons parler maintenant des précautions qu'elle prend contre l'enlèvement de mineurs.

Enlevement de mineurs.

Ce crime, enfanté par la cupidité ou par le déréglement des mœurs, souvent par l'une et par l'autre à la fois, présente un des plus dangereux attentats contre la faiblesse et l'inexpérience ; car l'enlèvement ne peut être fait que par violence ou par fraude, et en dérobant le mineur aux personnes qui le surveillaient. Le Code porte la réclusion contre celui qui se sera rendu coupable de ce crime : mais si la personne enlevée ou détournée est une fille au-dessous de seize ans accomplis, le crime étant plus grave, la peine est plus forte : c'est celle des travaux forcés à temps. Il est évident qu'un tel enlevement n'a pu avoir lieu que pour abuser de la personne, ou pour forcer les parents à consentir au mariage. L'homme n'est pas moins coupable quand la fille l'aurait suivi volontairement ; car c'est lui qui a été le corrupteur. Si cependant, lorsqu'il a commis l'enlèvement, il n'avait pas encore vingt-un ans, la loi se borne à prononcer contre lui des peines de police correctionnelle : elle le punit comme ayant commis une action très-répréhensible sans doute, et comme sachant très-bien que cette action était défendue par la loi ; mais elle ne veut

pas le punir aussi sévèrement que s'il était d'un âge qui
ne permît pas de douter qu'il a senti toutes les conséquences de son crime.

Si enfin le ravisseur a épousé la personne qu'il avait
enlevée, le sort du coupable dépendra du parti que
prendront ceux qui ont droit de demander la nullité du
mariage. S'ils ne la demandent point, la poursuite du
crime ne peut avoir lieu ; autrement, la peine qui serait
prononcée contre le coupable rejaillirait sur la personne
dont il a abusé, et qui, victime innocente de la faute de
son époux, serait réduite à partager sa honte. Il ne suffit
pas même, pour que l'époux puisse être poursuivi criminellement, que la nullité du mariage ait été demandée ;
il faut encore que le mariage soit en effet déclaré nul :
car il serait possible qu'à l'époque où l'action en nullité
serait intentée il existât une fin de non-recevoir contre
les parents, soit parce qu'ils auraient expressément ou ta-
citement approuvé le mariage, soit parce qu'il se serait
écoulé une année sans réclamation de leur part, depuis
qu'ils ont eu connaissance du mariage.

Ces fins de non-recevoir sont établies par l'article 183
du Code Napoléon. En ce cas, dès que le mariage ne
pourrait plus être attaqué, les considérations que je
viens d'exposer ne permettraient pas que la conduite de
l'époux fût recherchée ; et si l'intérêt de la société est
qu'aucun crime ne reste impuni, son plus grand intérêt,
en cette occasion, est de se montrer indulgente, et de ne
pas sacrifier à une vengeance tardive le bonheur d'une
famille entière.

La sollicitude du législateur s'est étendue jusqu'au
moment où l'homme vient de payer le dernier tribut à
la nature.

Infraction aux lois sur les inhumations.

Le Code Napoléon a fixé des règles pour constater les décès, et la loi pénale prononce des peines contre ceux qui ne font point les déclarations nécessaires pour que les décès soient constatés. Il importe que les déclarations soient faites non-seulement afin de connaître les changemens qui arrivent dans les familles, et de mettre les héritiers à portée de réclamer leurs droits, mais encore afin de ne pas laisser échapper la trace des crimes qui auraient pu occasionner la mort d'une personne. Ceux à qui la loi impose le devoir de faire ces déclarations ne doivent pas perdre de vue que, dans le cas où il s'éleverait quelques présomptions de mort violente, leur négligence les exposerait à être poursuivis comme recéleurs du cadavre d'une personne homicidée.

Le nouveau Code n'oublie pas non plus de punir ceux qui se rendent coupables de violations de tombeaux et de sépultures ; cet objet ne peut être indifférent ; les anciens ont toujours montré le respect le plus religieux pour les cendres des morts. Il suffit, pour s'en convaincre, de jeter un coup-d'œil sur leur législation, particulièrement sur celle des Grecs et des Romains. Les Gaulois étaient animés du même esprit que ceux dont ils envahirent le territoire. Une loi salique, dit Montesquieu, interdisait à celui qui avait dépouillé un cadavre le commerce des hommes, jusqu'à ce que les parents, acceptant la satisfaction, eussent demandé qu'il pût vivre parmi les hommes. Ce respect est si naturel, que le simple récit de telles violations inspire une horreur qu'on ne saurait contenir. Chez les sauvages même, le souvenir des morts enflamme leur imagination et produit en eux les émotions les plus vives.

Faux témoignage.

Le faux témoignage est un crime qui, dans tous les temps, a été puni des peines les plus sévères. L'édit de 1531, qui portait la peine de mort contre toute espèce de faux, comprenait en termes exprès le faux témoignage commis en justice. Cet édit fut modifié par celui de 1680, qui n'ordonna la peine de mort que pour les faux commis dans l'exercice d'une fonction publique, et autorisa les juges, pour les autres cas où il s'agirait de faux, à prononcer telle peine qu'ils jugeraient convenable, même celle de mort, suivant les circonstances. Les rédacteurs de la loi de 1791 ne voulurent pas abandonner à l'arbitraire la faculté de disposer ainsi de la vie des accusés.

Un des articles de cette loi porte que le faux témoin en matière criminelle sera puni de la peine de vingt ans de fers, et qu'il sera puni de mort s'il est intervenu condamnation à mort contre l'accusé dans le procès duquel aura été entendu le faux témoin.

Le nouveau Code s'est conformé à l'esprit qui a dicté cette disposition, et n'a fait d'autre changement que celui qui était nécessité par le nouvel ordre de peines; il ne distingue pas non plus si le faux témoin a été corrompu par argent; c'est un crime extrêmement grave, quel qu'en ait été le motif, que de faire perdre à un innocent l'honneur et la liberté, quelquefois même la vie, ou de faire rentrer dans la société un coupable qui, enhardi par l'impunité même, commettra bientôt de nouveaux forfaits : ainsi, en matière criminelle, la loi n'a nul égard aux ressorts qui ont pu faire mouvoir le faux témoin, Quant au faux témoignage dans toute autre matière, le nouveau Code prononce la réclusion; mais il punit plus

sévèrement le faux témoin qui s'est laissé corrompre par
argent, par une récompense quelconque ou par des promesses : il prononce contre lui le *minimum* de la peine
que doit subir le faux témoin en matière criminelle, c'est-
à-dire, celle des travaux forcés à temps.

Quant a la subornation de témoins en quelque matière que ce soit, les coupables seront condamnés à une
peine d'un degré supérieur à celle que subiront les faux
témoins dans la même affaire : les uns et les autres ne seront condamnés à la même peine que lorsque les faux
témoins devront être punis de mort. Cette subornation
est une espèce de provocation si dangereuse, qu'on a
pensé que le coupable devait être puni plus sévèrement
que la personne provoquée.

Enfin une disposition relative au faux serment, et qui
n'existait pas dans la loi de 1791, a été placée dans le
nouveau Code. Ce crime sera puni de la dégradation civique. Nulle peine ne convenait mieux au crime de faux
serment que celle qui consiste dans la destitution et l'exclusion du condamné de toutes fonctions ou emplois publics, et dans la privation de plusieurs droits civiques,
tels, par exemple, que celui d'être juré ou témoin. Le
coupable de faux serment s'est en effet rendu indigne de
jouir de ces avantages.

La poursuite de ce crime appartient surtout au
ministère public. Quant à la partie, ou le serment a
été déféré par elle, ou il l'a été d'office. Dans le premier
cas, la partie est repoussée par l'article 1363 du Code
Napoléon, qui porte que « lorsque le serment déféré ou
référé a été fait, l'adversaire n'est point recevable à en
« prouver la fausseté. » Cette disposition a pour but
d'empêcher que la partie qui est condamnée par l'effet
d'une déclaration à laquelle elle a consenti ne cherche à

recommencer le procès, sous prétexte que la déclaration est fausse, ce qui ne manquerait presque jamais d'arriver. Dans le second cas, qui est celui ou le serment a été déféré d'office par le juge, la partie intéressée peut être admise à prouver la fausseté de la déclaration : mais elle doit se conformer aux règles prescrites par le Code de procédure civile.

A l'égard du ministère public, la question de savoir si la partie est ou non recevable à prétendre que le serment est faux lui est étrangère. L'intérêt de la société demande que le crime de faux serment ne reste pas impuni; et quoique la partie ne puisse agir pour son intérêt privé, la peine due au crime ne doit pas moins être provoquée par le ministère public.

La dernière partie du chapitre relatif aux attentats contre les personnes concerne le délit de calomnie.

Attentats contre l'honneur.

Les anciennes lois ne prononçaient contre la calomnie que des peines arbitraires.

Les lois rendues depuis 1789 n'en ont point parlé; il est résulté de là que la calomnie n'a pas été suffisamment réprimée, et que l'envie ou la haine n'ont pas craint d'attaquer la réputation des hommes les plus recommandables. Depuis long-temps on désirait que le législateur mit un frein à de tels excès; car ou le fait qu'on s'est permis d'imputer à quelqu'un est défendu par la loi, ou il ne l'est pas. S'il est défendu, c'est aux juges qu'il appartient de verifier le fait et d'appliquer la peine. Tout bon citoyen doit le dénoncer; et si, au lieu de le déclarer à la justice, il le répand dans le public. soit par ses propos, soit par ses écrits, il est évident que cette conduite est dirigée par la méchanceté plutôt que par l'amour du bien. La malignité, qui saisit avidement ce qu'on lui présente comme ridicule ou odieux, convertit bientôt les allégations en preuves, et bientôt le poison de la calomnie a fait des ravages qui souvent ne s'arrêtent pas à la personne calomniée, mais portent la désolation dans toute sa famille. C'est surtout chez un peuple pour qui l'honneur est le plus grand des biens que la calomnie doit être sévèrement réprimée.

Le nouveau Code définit en ces termes le délit de calomnie :

« Sera coupable de délit de calomnie, celui qui, soit dans « des lieux ou réunions publiques, soit dans un acte au- « thentique et public, soit dans un écrit imprimé ou non, « qui aura été affiché, vendu ou distribué, aura imputé à « un individu quelconque des faits qui, s'ils existaient, ex- « poseraient celui contre lequel ils sont articulés à des pour- « suites criminelles ou correctionnelles, ou même l'expo- « seraient seulement au mépris ou à la haine des citoyens. »

On conçoit que cette disposition ne peut s'appliquer aux fonctionnaires ou autres qui, en donnant de la publicité à certains faits, ne font que remplir l'obligation où ils sont de les révéler ou de les réprimer.

A l'égard de ceux qui ne sont point dans le cas de l'exception, ils peuvent être poursuivis comme calomniateurs.

En vain prétendraient-ils que les faits sont notoires ; en vain demanderaient-ils qu'on les admette à la preuve ; ils ne seraient point écoutés ; de pareils débats ne serviraient qu'à donner plus d'éclat à cette publicité même qui constitue le délit. Si cependant l'auteur de l'imputation dénonce les faits, les juges doivent surseoir au jugement du délit de calomnie, jusqu'à ce qu'il soit décidé si la personne à qui ces faits sont imputés est réellement coupable; car, si elle était condamnée, on ne pourrait raisonnablement condamner le dénonciateur.

S'il est décidé que la personne dont l'honneur a été attaqué n'est pas coupable, soit parce que les faits ne sont point prouvés, soit parce qu'ils ne sont point défendus par la loi, l'auteur de l'imputation doit être déclaré convaincu de délit de calomnie, et puni des peines portées par la loi contre les calomniateurs. Ces peines sont un emprisonnement et une amende proportionnés à la gravité du fait déclaré calomnieux.

Le Code prononce une peine moindre contre celui qui, sans avoir donné auparavant de la publicité aux faits, s'est contenté de les dénoncer, et a depuis été reconnu les avoir dénoncés faussement. Le mal n'étant pas aussi considérable que dans le premier cas, la peine ne peut être aussi forte : elle ne doit pas cependant être trop faible, parce que c'est toujours un acte de méchanceté très-répréhensible.

Il est à remarquer cependant qu'il y a des faits qu'on peut répandre, quoique très-graves, sans être déclaré calomniateur; ce sont ceux dont on est en état de rapporter la preuve légale. Cette preuve légale résulte d'un jugement ou de tout autre acte authentique. Alors c'est au jugement, c'est à l'acte authentique que les faits doivent leur première publicité; ils ne pouvaient plus ensuite qu'être rappelés : or, la loi ne peut imputer à délit ce qui, par sa nature, doit être connu.

Le Code prononce une amende de 16 à 500 francs à l'égard des injures ou des expressions outrageantes qui ne renfermeraient l'imputation d'aucun fait précis, mais celle d'un vice déterminé, lorsqu'elles auront été proférées dans des lieux ou réunions publiques, ou insérées dans des écrits imprimés ou non, qui auraient été répandus et distribués.

Reprocher, par exemple, publiquement à quelqu'un un vice tel que l'ivrognerie ou la débauche, est un outrage qui ne doit pas être laissé impuni, si la personne offensée en demande réparation : mais l'injure n'est pas aussi grande que si quelques faits étaient précisés : le vague de l'injure en atténue la force, et l'amende est une peine suffisante.

Enfin, quelle que soit la quotité de l'amende qui sera prononcée, comme peine de la calomnie ou de l'injure, elle ne nuira jamais au paiement des dommages-intérêts que la partie offensée aura pu obtenir ; il suffit de se rappeler qu'aux termes de l'article 54 du Code, qui s'applique à tous les crimes et délits, lorsque les biens des condamnés seront insuffisants pour acquitter la totalité des condamnations, les restitutions et dommages-intérêts seront préférés à l'amende et à la confiscation.

Nous observerons, d'un autre côté, que l'auteur de l'imputation d'un vice n'a nul moyen de s'affranchir de la peine. Demanderait-il qu'on l'admît à la preuve ? la loi ne le permet pas. Voudrait-il dénoncer ? on ne dénonce que des faits précis et qualifiés crimes, délits ou contraventions. Cela ne peut s'appliquer à l'imputation d'un vice en général.

Nous n'avons point à nous occuper ici des autres injures que la loi punit, quoiqu'elles n'aient aucun caractère de publicité ; elles ne donnent lieu qu'à des peines de simple police, et ce sera l'objet du quatrième Livre.

Il nous reste à dire un mot sur les révélations de secrets.

A l'exception des révélations que la loi exige, parce qu'elles importent au salut public, tout dépositaire, par état ou profession, des secrets qu'on lui confie, ne peut les révéler sans encourir des peines de police correctionnelle : ne doit-on pas en effet considérer comme un délit grave des révélations qui souvent ne tendent à rien moins qu'à compromettre la réputation de la personne dont le secret est trahi, à détruire en elle une confiance devenue plus nuisible qu'utile, à déterminer ceux qui se trouvent dans la même situation à mieux aimer être victimes de leur silence que de l'indiscrétion d'autrui; enfin à ne montrer que des traîtres dans ceux dont l'état semble ne devoir offrir que des êtres bienfaisants et de vrais consolateurs? La nécessité de la peine, en pareille matière, est encore mieux sentie qu'elle ne pourrait être développée.

Telle est, Messieurs, l'analyse des principales dispositions de la partie du nouveau Code relative aux attentats contre les personnes. Vous avez remarqué les différences essentielles qu'offre la comparaison de ces dispositions avec le Code pénal et le Code correctionnel de 1791. Les lacunes que l'expérience a fait connaître ont été remplies; les distinctions qu'elle a recommandées ont été faites. S'il s'est présenté quelques difficultés, les regards de Sa Majesté, à qui rien n'échappe de tout ce qui peut être utile, les ont aperçues, et son génie les a fait disparaître. Nous espérons, Messieurs, que tant de soins réunis assureront à cet important ouvrage l'avantage glorieux d'être honoré de votre assentiment.

MOTIFS

Du Livre III, Titre II, Chapitre II,

PRÉSENTÉS

PAR MM. LE CHEVALIER FAURE, ET LES COMTES MARET
ET CORVETTO, Conseillers d'État.

Séance du 9 février 1810.

MESSIEURS,

Dans la dernière séance, nous avons eu l'honneur de vous soumettre un projet de loi destiné à faire partie du Code des délits et des peines, et relatif aux attentats contre les personnes.

Sa Majesté nous charge aujourd'hui de vous présenter un autre projet dépendant du même Code : il est relatif aux attentats contre les propriétés.

Les dispositions qu'il renferme doivent être également considérées comme la sanction de la loi civile. Tandis que le Code Napoléon règle les différentes manières dont on peut acquérir la propriété, le Code pénal détermine les différents cas où l'atteinte portée à la propriété constitue un crime ou délit. Ces cas sont très-variés. Ce qui appartient à autrui peut être soustrait par fraude ; il peut être enlevé par violence ; il peut être détruit par imprudence ou méchanceté. Chacun de ces actes est susceptible de nuances que le législateur doit saisir pour proportionner la peine au délit. Les motifs que nous allons donner des principales dispositions du projet vous feront connaître les grandes et nombreuses améliorations que promet le nouveau Code.

Nous parlerons d'abord des actes qualifiés vol.

Vol.

« Celui-là est coupable de vol, dit la loi, qui soustrait « frauduleusement une chose qui ne lui appartient pas. »

Le mot *frauduleusement* prouve qu'il faut aussi, pour qu'il y ait vol, que la chose soustraite appartienne à autrui. Si elle n'appartient à personne, il ne peut y avoir de fraude ; car l'expression est corrélative, et suppose que quelqu'un peut être trompé ou dépouillé.

La soustraction frauduleuse, étant un attentat à la propriété, doit être punie. Elle doit l'être plus ou moins, suivant qu'elle est précédée, accompagnée ou suivie de circonstances plus ou moins graves.

Avant de parler du degré d'influence que ces circonstances doivent avoir sur l'intensité de la peine, je ne puis me dispenser d'offrir à vos méditations un principe consacré par la nouvelle loi.

Ce principe consiste à rejeter l'action publique, et à n'admettre que l'action privée, c'est-à-dire, l'action en dommages-intérêts, à l'égard de toute espèce de fraude commise par les maris au préjudice de leurs femmes, par les femmes au préjudice de leurs maris, par un veuf ou ou une veuve, quant aux choses qui avaient appartenu à l'époux décédé ; enfin par les parents et alliés en ligne directe, ascendante ou descendante, les uns envers les autres.

Les rapports entre ces personnes sont trop intimes pour qu'il convienne, à l'occasion d'intérêts pécuniaires, de charger le ministère public de scruter des secrets de famille, qui peut-être ne devraient jamais être dévoilés, pour qu'il ne soit pas extrêmement dangereux qu'une accusation puisse être poursuivie dans des affaires où la ligne qui sépare le manque de délicatesse du véritable délit est souvent très-difficile à saisir; enfin pour que le ministère public puisse provoquer des peines dont l'effet ne se bornerait pas à répandre la consternation parmi tous les membres de la famille, mais qui pourraient encore être une source éternelle de divisions et de haines.

Loin que le silence du ministère public préjudicie à la partie privée, il ne pourra que lui être utile, puisque son action en réparations civiles lui est réservée, et qu'elle n'aura point à craindre, en la formant, que ses répétitions ne soient absorbées par les frais privilégiés d'une procédure criminelle.

Ces considérations puissantes ont nécessité la disposition spéciale dont nous venons de rendre compte. Mais comme une telle exception doit être renfermée dans le cercle auquel elle appartient, il en résulte que toute autre personne qui aurait recélé ou appliqué à son profit des objets provenant d'un vol dont le principal auteur serait compris dans l'exception, subirait la même peine que si elle-même eût commis le vol.

Souvent ces sortes de vols n'auraient pas lieu, si quelques étrangers ne les conseillaient ou ne les facilitaient.

La peine, au surplus, ne s'appliquera point à ceux qui auraient reçu les objets volés ou qui en auraient profité sans savoir qu'ils fussent volés.

Vous vous rappelez, Messieurs, qu'il résulte des articles 60 et 62 du Code, qu'on ne peut être puni pour avoir aidé, assisté ou facilité une action défendue par la loi, ou recélé une chose volée, que lorsqu'on l'a fait avec connaissance,

Après avoir parlé d'un cas particulier d'exception, nous allons faire connaître les peines établies par le nouveau Code en matière de vol.

Si le vol n'est accompagné d'aucune circonstance aggravante, il sera puni de peines de police correctionnelle, comme il l'a été jusqu'à ce jour.

Mais, si une ou plusieurs de ces circonstances existent, la rigueur de la peine devant être proportionnée à la gravité du crime, voici les bases sur lesquelles repose l'échelle proportionnelle.

La circonstance qui aggrave le plus le vol est la violence, parce qu'alors le crime offre tout à la fois un attentat contre la personne et un attentat contre la propriété.

Aussi le vol fait avec violence, quoique nulle autre circonstance n'existe, et qu'il n'ait laissé aucune trace de blessure, sera puni de la peine des travaux forcés à temps, ainsi qu'il l'était par la loi de 1791.

Mais si le vol, outre la violence, a été accompagné de plusieurs autres circonstances aggravantes ; par exemple, s'il a été commis la nuit et avec armes, ou si seulement la violence a laissé quelques traces de blessures ou de contusion, ce n'est plus la peine des travaux forcés à temps, mais celle des travaux forcés à perpétuité qui sera prononcée.

En effet, lorsque le vol porte un tel caractère, il est d'une nature si grave, que toute peine moins sévère ne serait pas assez répressive.

La loi du 26 floréal an 5 prononce la peine de mort à l'égard de tout vol commis dans une maison à l'aide de violences exercées sur les personnes qui s'y trouvaient, et lorsque ces violences auront laissé des traces, cette même loi veut aussi que la peine de mort ait lieu, si ceux qui ont commis le vol avec violence se sont introduits dans la maison par la force des armes.

Suivant le nouveau Code, le vol avec violence n'emportera la peine de mort que lorsqu'il aura été commis avec une réunion de circonstances dont l'ensemble présente un caractère si alarmant, que le crime doive être mis au même rang que l'assassinat.

Il faudra donc que le vol avec violence ait été en même temps commis la nuit par deux ou plusieurs personnes, avec armes apparentes ou cachées, et, de plus, à l'aide d'effraction extérieure, ou d'escalade, ou de fausses clefs, ou en prenant un faux titre, ou un faux costume, ou en alléguant un faux ordre.

Toutes ces circonstances réunies forment un corps de délit si grave, que la loi punit les coupables de la même peine que celui qui a commis un assassinat.

Il n'est pas même nécessaire, lorsque ce concours de circonstances existe, que les coupables aient commencé à exercer des violences; il suffit qu'ils aient menacé de faire usage de leurs armes.

A l'égard des vols commis dans les chemins publics, ces sortes de crimes qui portent toujours un caractère de violence, et qui menacent la sûreté individuelle, seront punis de la peine des travaux forcés à perpétuité : ici nous supposons qu'il n'y a eu de la part du coupable aucune attaque à dessein de tuer; autrement il subirait la peine due aux assassins.

Si le vol n'a été commis ni dans un chemin public, ni avec violence, mais avec une ou plusieurs des circonstances dont nous venons de parler, la peine sera plus ou moins forte, suivant que ces circonstances, soit par leur réunion, soit par leur nature particulière, influeront sur la gravité du délit.

Nous ajouterons que le vol, quoique dénué de toutes ces circonstances, sera puni plus rigoureusement que le vol simple, à raison de la qualité de l'auteur du vol et de la confiance nécessaire qu'a dû avoir en lui la personne volée; si, par exemple, le vol a été commis par un domestique envers son maître, ou par un aubergiste envers la personne qu'il aura logée, ou enfin, si c'est cette dernière qui a volé l'aubergiste.

Tous ces crimes seront punis de la réclusion : une peine plus forte empêcherait souvent qu'ils ne fussent dénoncés. C'est ce dont l'expérience n'a fourni que trop d'exemples.

Quant aux vols d'objets exposés à la foi publique, la loi de 1791 les punissait tous indistinctement d'une peine afflictive. Beaucoup de ces crimes restèrent impunis, parce que la peine était trouvée trop forte, et que l'on aimait mieux acquitter les coupables que de leur faire subir un châtiment qui excédait celui qu'ils paraissaient avoir mérité. La loi du 25 frimaire an 8 parut, et la connaissance de tous ces délits indistinctement fut attribuée aux tribunaux de police correctionnelle. Alors un nouvel inconvénient se fit apercevoir. La peine était insuffisante en plusieurs cas; et l'insuffisance de la peine produisit le même effet que l'impunité. Dès-lors ces sortes de délits se renouvelèrent fréquemment, et les tribunaux ont élevé de justes plaintes à cet égard.

La distinction que le nouveau Code établit apportera un remède efficace au mal.

Ou le vol aura été commis à l'égard d'objets qu'on ne pouvait se dispenser de confier à la foi publique, tels que les vols de bestiaux, d'instruments d'agriculture, de récoltes, ou de partie de récoltes qui se trouvaient dans les champs; en un mot, de choses qu'il est impossible de surveiller soi-même ou de faire surveiller. En ce cas, les coupables seront punis d'une peine afflictive.

Ou les objets volés pouvaient être gardés, de sorte que c'est volontairement qu'on les aura confiés à la foi publique. Dans ce dernier cas, ce n'est plus qu'un vol simple, qui dès-lors sera puni de peines de police correctionnelle.

Jusqu'à présent on avait regretté que des circonstances qui influaient sur la gravité du délit ne fussent pas définies; des interprétations arbitraires suppléaient à l'absence des définitions, ce qui était un grand mal, surtout en matière criminelle.

Le remède se trouvera dans le nouveau Code. Ainsi, par exemple, on s'est demandé sans cesse si l'effraction, pour être qualifiée extérieure, devait nécessairement être faite à l'entrée de la porte principale de la maison, ou si cette qualification appartenait également à l'effraction à l'aide de laquelle on s'était introduit dans les appartemens ou logemens particuliers. Le Code répond que l'effraction extérieure existe aussi dans ce dernier cas, parce que l'appartement particulier qu'on occupe dans une maison est, pour celui qui l'habite, sa maison même, et que beaucoup de maisons sont trop considérables, surtout dans les grandes villes, pour que la porte principale de l'édifice puisse rester fermée constamment, et que l'édifice entier puisse être habité par la même famille.

Une autre difficulté s'était présentée dans les cours criminelles. Elles n'étaient pas d'accord sur la question de savoir s'il fallait considérer comme vol fait à l'aide de fausses clefs celui qu'on aurait commis avec des clefs non imitées, ni contrefaites, ni altérées, mais qui n'avaient pas été destinées aux fermetures auxquelles elles ont été employées.

Le Code décide cette question et prononce l'affirmative. En effet, détourner une clef de sa destination pour l'employer à commettre un crime, n'est autre chose que convertir une clef véritable en une fausse clef. En un mot, toute clef n'est véritable que relativement à sa destination.

La seule différence que la loi admet entre cette clef dont il y a eu abus, et une clef contrefaite ou altérée, est que celle-ci est toujours fausse clef, et que la première ne le devient qu'au moment qu'on l'emploie comme on aurait fait d'une clef contrefaite.

A l'egard des fausses clefs proprement dites, la loi condamne celui qui les fabrique à des peines de police correctionnelle; elle veut même que, si c'est un serrurier, il subisse la peine de la réclusion. La faute doit être punie plus rigoureusement à raison de la facilité qu'on a eue de la commettre, et la confiance nécessairement attachée à cet état exige d'autant plus de précautions.

Nous terminerons cette partie en observant que la tentative de vol sera punie comme le vol même, quoique le vol n'eût donné lieu qu'à des peines de police correctionnelle. Une disposition spéciale était nécessaire sur ce point, vu que l'article 3 du Code en exige une à l'égard des tentatives de délits.

Nous allons examiner une autre espèce d'attentats à la propriété : ce sont ceux qui ont lieu par suite d'opérations de commerce, ou à l'aide d'entreprises réelles ou simulées; ce sont, d'une part, les banqueroutes, et de l'autre les escroqueries.

L'escroquerie est à la vérité comprise dans la banqueroute frauduleuse; mais ce dernier crime est beaucoup plus grave par sa cause et par ses effets.

Banqueroutes et escroqueries.

Le Code de commerce distingue deux espèces de banqueroute; la banqueroute simple et la banqueroute frauduleuse.

Les articles 586 et 587 de ce Code déterminent les divers cas qui constituent la banqueroute simple; ils consistent tous dans des imprudences ou négligences graves.

L'article 593 détermine ceux qui constituent la banqueroute frauduleuse.

Le mot *frauduleux* indique assez en quoi il consiste : nous nous abstiendrons de rapporter ses dispositions, à cause des nombreux détails qu'elles renferment.

Le nouveau Code prononce, comme a fait la loi de 1791, la peine des travaux forcés à temps contre les banqueroutiers frauduleux : on sent combien il est nécessaire d'établir une peine rigoureuse contre un crime destructif de cette confiance qui est l'âme du commerce, crime dont le contre-coup se fait souvent ressentir sur tant de familles réduites à leur tour à l'impossibilité de remplir leurs engagements.

Le nouveau Code porte contre le banqueroutier simple un emprisonnement d'un mois au moins, et de deux ans au plus. Il s'est conformé littéralement à la disposition de l'article 592 du Code de commerce.

On conçoit que l'amende ne pouvait, pour ce délit, être ajoutée à l'emprisonnement; car comment serait-il possible d'obtenir le paiement d'une amende de celui qui n'est pas en état de s'acquitter envers ses créanciers?

Une autre disposition relative à la faillite des agents de change ou courtiers est une conséquence nécessaire des dispositions du Code de commerce. Vous vous rappelez, Messieurs, qu'il est expressément établi par les articles 85 et 86 de ce Code, qu'un agent de change ou courtier ne peut, dans aucun cas, ni sous aucun prétexte, faire des opérations de commerce ou de banque pour son compte ; qu'il ne peut s'intéresser directement ni indirectement, sous son nom ou sous un nom interposé, dans aucune entreprise commerciale ; qu'il ne peut recevoir ni payer pour le compte de ses commettants ; qu'enfin il ne peut se rendre garant de l'exécution des marchés où il s'entremet.

S'il est absolument défendu à l'agent de change ou courtier de faire le commerce, il ne peut donc faire faillite qu'en prévariquant.

Passons ensuite à l'article 89 du même Code : il porte « qu'en cas de faillite, tout agent de change ou courtier « est poursuivi comme banqueroutier. » L'article n'avait pas besoin d'ajouter le mot *frauduleux* ; car la disposition relative à la banqueroute simple ne peut évidemment s'appliquer à un cas de prévarication dans l'exercice de fonctions si importantes et si délicates, à un cas de prévarication dont les effets peuvent être si désastreux pour les maisons de commerce. Il résulte de là que l'agent de change ou courtier, s'il est en état de faillite, doit être puni comme le banqueroutier frauduleux ; et que, s'il est en état de banqueroute frauduleuse, il doit être puni d'une peine plus forte que celle établie pour les cas ordinaires.

Ainsi, d'après le nouveau Code, la simple faillite de la part de l'agent de change ou courtier emportera la peine des travaux forcés à temps, et la banqueroute frauduleuse emportera celle des travaux forcés à perpétuité.

A l'égard de l'escroquerie, on a tâché, dans la nouvelle définition de ce qui constitue ce délit, d'éviter les inconvénients qui étaient résultés des rédactions précédentes.

Celle de la loi du 22 juillet 1791 était conçue de manière qu'on en a souvent abusé, tantôt pour convertir les procès civils en procès correctionnels, et par-là procurer à la partie poursuivante la preuve testimoniale et la contrainte par corps au mépris de la loi générale ; tantôt pour éluder la poursuite de faux en présentant l'affaire comme une simple escroquerie, et par-là procurer au coupable une espèce d'impunité, au grand préjudice de l'ordre public.

La loi du 2 frimaire an 2 ne remédia qu'à un seul de ces inconvénients. Elle put bien empêcher la confusion du faux avec l'escroquerie ; mais elle n'empêcha pas que la loi générale ne fût encore éludée.

Cet abus cessera sans doute d'après la rédaction du nouveau Code. La suppression du mot *dol* qui se trouvait dans les deux premières rédactions ôtera tout prétexte de supposer qu'un délit d'escroquerie existe par la seule intention de tromper. En approfondissant les termes de la définition, on verra que la loi ne veut pas que la poursuite en escroquerie puisse avoir lieu sans un concours de circonstances et d'actes antécédents qui excluent toute idée d'une affaire purement civile.

A la suite de cette définition on trouvera la réserve de peines plus graves, s'il y a crime de faux ; et les caractères auxquels ce crime peut être reconnu sont indiqués dans le chapitre concernant le faux, de manière à faire disparaître jusqu'à la plus légère incertitude.

Abus de confiance.

Le Code renferme plusieurs dispositions nouvelles sur les abus de confiance.

L'une atteint ceux qui auront abusé des besoins, des faiblesses ou des passions d'un mineur pour lui faire souscrire des actes préjudiciables à ses intérêts.

Depuis long-temps on gémissait de voir que cette espèce de corrupteurs de la jeunesse pouvait impunément ruiner les fils de famille. En vain le Code Napoléon déclare que la simple lésion donne lieu à la rescision en faveur du mineur émancipé contre toutes sortes de conventions; ces hommes sans pudeur se font payer plus cher leurs avances, à raison des risques qu'ils courent; ils prennent toutes leurs précautions pour éluder l'application de la loi civile; mais la crainte d'une peine correctionnelle pourra-les retenir, et les jeunes gens ne trouveront plus autant de facilité à se procurer des ressources désastreuses pour leur fortune, et quelquefois plus funestes encore sous le rapport des mœurs.

Une autre disposition, quoique applicable à un fait plus rare, était également sollicitée par l'expérience. Elle contient deux décisions à la fois. Voici l'exemple.

Un blanc-seing est destiné à être rempli d'un mandat, si le besoin l'exige : il se trouve entre les mains d'un tiers. Celui-ci le remplit d'une obligation. Le signataire réclame : il prouve la fraude. Comment ce délit sera-t-il qualifié? Ce sera, répond le Code, un abus de confiance, si le blanc-seing a été confié au tiers par le signataire qui l'a chargé d'écrire au-dessus de sa signature, non pas une obligation, mais un mandat. Dans ce cas, l'écriture est celle qui devait se trouver sur l'acte; seulement le tiers a fait ce qu'il ne lui était pas permis de faire.

Cette fraude est une véritable escroquerie. Mais c'est un faux, si le tiers n'a pas été chargé de remplir le blanc. Il n'y a point abus de confiance, puisque rien n'a été confié. Il y a faux, parce que la main qui a tracé l'écriture n'est point celle par qui le blanc devait être rempli, et qu'ainsi le blanc contient un corps d'écriture qu'il ne devait pas contenir.

Nous ne parlerons point ici de la peine que le coupable subira, s'il a commis un faux; cette peine est déterminée dans un autre titre.

S'il a commis seulement un abus de confiance, il sera condamné à des peines de police correctionnelle.

Nous passerons sous silence les modifications faites à loi du 9 germinal an 6, sur les loteries étrangères, et à celle du 16 pluviose an 12, sur les maisons de prêt.

Les dispositions principales de ces lois ont été placées dans le nouveau Code.

Nous nous abstiendrons également de parler de la disposition relative à ceux qui, dans les adjudications, auront entravé ou troublé la liberté des enchères. Le fond de cet article a été puisé dans la loi correctionnelle de 1791, et dans la loi particulière du 24 avril 1793. La nouvelle rédaction est beaucoup plus complète, et remplit plusieurs lacunes.

Préjudice porté aux manufactures, au commerce et aux arts.

Le Code s'occupe ensuite de divers délits qui portent un préjudice notable, non pas seulement aux intérêts de quelques personnes en particulier, mais encore à ceux du commerce en général. Plus les gouvernements ont senti combien la prospérité de l'État était intimement liée à celle du commerce, plus ils ont pris de précautions pour prévenir les fraudes qui pouvaient y porter atteinte. Sans doute ces fraudes rejaillissent tôt ou tard sur leurs auteurs, parce qu'elles leur font perdre le crédit nécessaire au succès de leurs opérations ; mais lorsqu'elles ont pour but de tromper sur la qualité, les dimensions ou la nature de la fabrication, à l'égard des produits de nos manufactures qui s'exportent à l'étranger, un si grand mal ne doit point rester impuni. C'est pour cette raison, et pour plusieurs autres dont nous parlerons dans un instant, que la loi du 22 germinal an 11 fut rendue. Les abus qu'elle prit soin de réprimer avaient été l'objet de vives réclamations, et il ne fallait rien moins que la crainte d'une juste peine pour en arrêter le cours.

Plusieurs dispositions de cette loi salutaire ont été rapportées dans le nouveau Code ; d'autres, que le besoin a sollicitées, y ont également trouvé place.

Le nouveau Code défend, comme l'a fait la loi de 1791, les coalitions entre les maîtres contre les ouvriers, et entre les ouvriers contre les maîtres.

Les maîtres se coalisent pour faire baisser le salaire des ouvriers, et les ouvriers pour faire augmenter leur paye.

Si cependant le salaire des ouvriers est trop modique, et qu'ils ne puissent subsister en France, ils iront chercher leurs moyens de subsistance en pays étranger. Si les maîtres sont obligés de donner aux ouvriers une paye trop forte, ils seront réduits à la triste nécessité ou de se ruiner, s'ils veulent soutenir la concurrence avec les autres établissements du même genre à qui les ouvriers ne font point la loi, ou de fermer leurs ateliers, au grand préjudice des ouvriers eux-mêmes.

Tel est l'effet que produisent aussi ces sortes de défenses ou d'interdictions que les ouvriers prononcent contre les directeurs d'ateliers et entrepreneurs d'ouvrages, et qu'ils prononcent même quelquefois les uns contre les autres. Ils croient par-là servir leur intérêt aux dépens de leur maître, et ils ne nuisent pas moins à leur propre intérêt.

Le Code prononce contre tous ces abus des peines de police correctionnelle, graduées suivant la nature du délit.

La loi regarde comme coupable de délit celui qui, dans la vue de nuire à l'industrie française, fait passer en pays étranger des directeurs, des ouvriers ou commis d'un établissement. Si chacun doit être libre de faire valoir son industrie et ses talents partout où il croit pouvoir en retirer le plus d'avantage, il convient de punir celui qui débauche des hommes nécessaires à un établissement, non pour procurer à ces hommes un plus grand bien, souvent incertain, mais pour causer la ruine de l'établissement même. Ces actes de méchanceté sont punis de peines de police correctionnelle.

La loi punit aussi correctionnellement celui qui communique à des Français résidant en France les secrets de la fabrique où il est employé : celui-ci ne fait point tort aux fabriques nationales en général, mais il préjudicie en particulier à la fabrique à laquelle ce secret appartient; il enlève à l'un le fruit de son invention, pour enrichir un autre à qui cette invention est étrangère; il décourage l'industrie par la crainte d'être frustrée de sa légitime récompense.

Mais la peine de la réclusion, c'est-à-dire une peine afflictive et infamante, attend quiconque aura communiqué de tels secrets à des étrangers ou à des Français résidant en pays étrangers. Ce n'est plus à un ou plusieurs particuliers qu'il fait tort : il nuit à la nation entière, qu'il prive d'une source de richesses; il contribue à diminuer la prospérité nationale, en contribuant à faire pencher la balance du commerce en faveur du pays étranger auquel il a sacrifié l'intérêt de la France.

Elles n'ont pas non plus échappé à la prévoyance du Code, ces manœuvres coupables qu'emploient des spéculateurs avides et de mauvaise foi pour opérer la hausse ou la baisse du prix des denrées ou des marchandises, ou des papiers et effets publics au-dessus, ou au dessous des prix qu'aurait déterminés la concurrence naturelle et libre du commerce. Le Code cite pour exemples de ces manœuvres les bruits faux ou calomnieux semés à dessein dans le public, les coalitions entre les principaux détenteurs de la marchandise ou denrée : il ajoute toute espèce de voie ou moyens frauduleux, parce qu'en effet ils sont si multipliés, qu'il ne serait guère plus facile de les détailler que de les prévoir.

La disposition ne peut s'appliquer à ces spéculations franches et loyales qui distinguent le vrai commerçant. Celles-ci, fondées sur des réalités, sont utiles à la société. Loin de créer tour à tour les baisses excessives et les hausses exagérées, elles tendent à les contenir dans les limites que comporte la nature des circonstances, et par-là servent le commerce en le préservant des secousses qui lui sont toujours funestes.

Une disposition du Code punit aussi de peines de police correctionnelle les paris qui auront été faits sur le hausse ou la baisse des effets publics.

La disposition suivante contient une explication essentielle. Voici les termes : « Sera réputée pari de ce genre « toute convention de vendre ou de livrer des effets « publics qui ne seront pas prouvés par le vendeur avoir

« existé à sa disposition au temps de la convention, ou
« avoir dû s'y trouver au temps de la livraison. »

Il résulte de cette définition, que le but de la loi est
de réprimer une foule de spéculateurs qui, sans avoir
aucune espèce de solvabilité, se livrent à ces jeux, et ne
craignent point de tromper ceux avec lesquels ils traitent.
La loi soumet le vendeur seul à la preuve qu'elle exige,
parce que c'est lui qui promet de livrer la chose ; mais si
la promesse de livrer existe de la part des deux contrac-
tants, la preuve est nécessaire pour l'un et pour l'autre ;
car tous deux sont respectivement vendeurs et acheteurs.

Ce moyen de répression, loin de nuire en aucune manière aux opérations des spéculateurs honnêtes et déli-
cats, les rendra moins périlleuses en les délivrant du
concours de ceux qui, n'ayant rien à perdre, osent tout
risquer.

Le Code contient aussi des dispositions non-seulement
contre ceux qui font usage de faux poids ou de fausses
mesures, mais encore contre ceux qui se servent d'autres
poids ou d'autres mesures que ceux qui ont été établis
par les lois de l'État. Ces deux actes n'étant pas susceptibles d'une assimilation parfaite, il a dû être établi
quelque différence dans les peines : un mot suffira pour
en faire sentir la nécessité.

En effet, l'usage de faux poids ou de fausses mesures
comprend nécessairement une fraude. Il n'en est pas de
même de l'usage des poids ou mesures anciennes : celui-
ci peut n'être pas accompagné de fraude ; et si la fraude
n'existe pas, ce n'est point un délit, c'est une contravention. Sans doute cette contravention doit être réprimée ; car la loi sur l'uniformité des poids et mesures est
d'une utilité qui ne peut être méconnue que par l'ignorance et les préjugés ; et ceux qui ne s'empressent pas de

se conformer à cette loi s'étonneront un jour d'avoir pu douter de sa sagesse. Au reste, lorsqu'ils sont trompés, ils ne peuvent pas prétendre que la loi doit venir à leur secours, comme s'ils l'avaient été par l'usage de faux poids ou de fausses mesures ayant la forme légale. Dans ce dernier cas, la loi les considérerait comme victimes d'une fraude dont ils n'ont pas dû se défier. Mais lorsqu'ils consentent à ce qu'on emploie à leur égard des poids ou mesures que la loi prohibe, ils se rendent complices d'une contravention : ils ont dû prévoir les risques auxquels ils se sont exposés, et la loi leur refuse toute action pour en obtenir la réparation. Ainsi le vendeur et même l'acheteur, quoique trompé, seront punis : le premier, pour avoir commis une fraude et une contravention, et on lui appliquera la peine relative à l'usage des faux poids et des fausses mesures; quant au second, c'est-à-dire à l'acheteur, il sera condamné, pour sa contravention, à une peine de simple police.

Je passe au délit de contrefaçon : il est évident que ce délit offre un attentat à la propriété. On peut contrefaire des ouvrages gravés ou peints, comme des ouvrages imprimés. Les règles d'après lesquelles la propriété d'un auteur est légalement reconnue, celles qui déterminent l'étendue et les bornes de cette propriété, ne sont point l'objet du Code pénal. Il ne s'agit ici que des peines qui doivent être subies par les contrefacteurs. Ces peines sont une amende et la confiscation de la chose contrefaite. Nous avons déjà dit, dans une autre occasion, que la confiscation et l'amende ne tournent jamais au profit de l'État qu'après que la partie lésée a été entièrement indemnisée.

Il est à considérer que le délit de contrefaçon exige une surveillance d'autant plus sévère, que son effet ne se borne pas à porter préjudice au propriétaire légitime; l'impunité d'un tel délit nuirait tout à la fois aux arts et au commerce, par le découragement qu'il apporterait parmi les auteurs et les éditeurs, puisqu'il n'en est aucun qui ne dût craindre pour lui le même sort. Disons plus, cette fraude rejaillirait sur l'Etat lui-même, qui tire son plus grand lustre de la prospérité des arts et du commerce

Délits des fournisseurs.

Le Code a prévu aussi une espèce de fraude dont la poursuite est réservée au Gouvernement seul, parce que l'intérêt de l'Etat est le seul qui en souffre. Je parle de l'inexécution des engagements contractés par les fournisseurs envers le Gouvernement. Si cette inexécution fait manquer le service, et qu'ils ne prouvent pas qu'elle est l'effet d'une force majeure, la loi les punit très-séverement; car il peut résulter les conséquences les plus fâcheuses de ce que le service n'a pas été fait au jour marqué. Le succès d'une bataille dépend quelquefois de l'exactitude la plus scrupuleuse à cet égard. Un moment perdu est souvent irréparable, ou ne peut se réparer que par de grands sacrifices. En un mot, il est impossible de calculer les suites d'une faute de cette espèce, et la peine que la loi porte contre les coupables est celle de la réclusion : elle ajoute une amende. Cet accessoire tient à la nature du délit, vu que les retards proviennent presque toujours de l'espoir d'augmenter les profits Nous avons dit que les fournisseurs ne sont pas punis, lorsqu'il est évident qu'une force majeure seule a causé ces retards. Ils ne le sont pas non plus, s'ils prouvent que la faute ne doit être imputée qu'à leurs agents : alors ce sont ces derniers qui doivent subir la peine. Mais la peine est plus forte si le crime a été facilité par des fonctionnaires publics ou des agents du Gouvernement. C'est un bien plus grand crime de participer au mal, lorsque par état on devait l'empêcher. La peine portée contre ces derniers est celle des travaux forcés à temps.

Nous n'avons pas besoin d'observer ici que ces dispositions relatives aux fournisseurs ne concernent que les fautes qu'ils peuvent avoir commises. S'ils avaient été d'intelligence avec l'ennemi, il faudrait se reporter au chapitre des crimes contre la sûreté de l'Etat.

Destructions et dommages.

Le Code, après s'être occupé des attentats à la propriété qui ont pour objet de s'enrichir aux dépens d'autrui, soit par fraude, soit par violence, s'occupe de ceux qui n'ont pour but que de satisfaire la vengeance ou la haine, et qui dès-lors dérivent uniquement de la méchanceté. Dans cette dernière espèce de crimes ou délits, le coupable ne prend point une chose qui appartient à autrui, afin d'en jouir lui-même; mais il détruit cette chose pour qu'un autre n'en jouisse pas. Au premier rang de ces attentats est le crime d'incendie. Ce crime, comme celui de l'empoisonnement, est l'acte qui caractérise la plus atroce lâcheté. Il n'en est point de plus effrayant, soit par la facilité des moyens, soit à cause de la rapidité des progrès, soit enfin par l'impossibilité de se tenir continuellement en garde contre le monstre capable d'un si grand forfait. L'empoisonnement même, sous certains rapports, semble n'être pas tout-à-fait aussi grave; car il n'offense que la personne qui doit en être la victime, tandis que l'autre crime s'étend jusqu'aux propriétés de ceux à qui l'on n'a voulu faire aucun mal, et tend à envelopper plusieurs familles dans une ruine commune. Il expose même la vie des personnes qui se trouvent dans le lieu incendié, et qui peuvent n'avoir pas le temps d'échapper aux flammes; ou si ce sont des récoltes qu'il incendie, ce feu peut se communiquer d'un champ à l'autre, et plonger un canton tout entier dans un état de détresse absolu. Un crime aussi exécrable mérite la mort; et telle est en effet la peine prononcée par le Code.

Si le crime d'incendie doit à juste titre être mis au même rang que l'assassinat, les menaces d'incendie doivent, par le même motif, être punies des mêmes peines que les menaces d'assassinat. Je ne répéterai point les observations que j'ai présentées dans la précédente séance, au sujet des menaces d'attentats contre les personnes.

On peut détruire des propriétés autrement que par le feu; comme les conséquences que ce crime entraîne ne sont pas en général aussi désastreuses que celles qui résultent du crime d'incendie, il emporte seulement la peine de la réclusion. Si cependant il en est résulté un homicide ou des blessures, celui par le fait duquel cet homicide ou ces blessures ont eu lieu, est considéré par la loi comme les ayant faits avec préméditation; car, en détruisant ou renversant un édifice, il savait que ces accidents pouvaient arriver; et l'acte de méchanceté dont il s'est rendu coupable ayant en effet produit ces accidents, ils doivent lui être imputés comme s'il les avait occasionnés à dessein.

Le Code défend aussi, sous des peines de police correctionnelle, de s'opposer par des voies de fait à l'exécution d'ouvrages que le Gouvernement a autorisés. Si le Gouvernement a été induit en erreur, il faut recourir aux autorités compétentes. Les retards occasionnés par les voies de fait doivent d'autant moins rester impunis, qu'ils peuvent causer un grand préjudice à l'intérêt public.

Si les propriétés qui ont été détruites sont des actes ou titres, la loi punit plus sévèrement la destruction des actes authentiques ou des effets de commerce ou de banque, que celle de toute autre pièce, parce que ces actes ou effets sont bien plus précieux, à raison des priviléges particuliers que la loi leur attache, et que dès-lors leur perte produit un bien plus grand mal. Aussi leur destruction est-elle punie d'une peine afflictive, tandis que celle des autres pièces ne donne lieu qu'à des peines de police correctionnelle.

Mais lorsqu'il s'agit de propriétés qu'on a non pas détruites, mais pillées ou dévastées, ce qui, relativement au propriétaire, produit souvent le même effet, si le pillage ou le dégât ont été commis à force ouverte, ce cas présente deux crimes à la fois : 1° l'action de piller ou dévaster ; 2° une sorte de rebellion qui a été employée pour en faciliter l'exécution. Cette complication demande une peine plus rigoureuse ; et en conséquence, le Code prononce la peine des travaux forcés à temps. La loi se relâche un peu de sa rigueur en faveur de ceux qui prouveront avoir été entraînés par des provocations ou sollicitations à prendre part à ces sortes de pillage : elle autorise les juges à ne condamner les coupables qu'à la peine de la réclusion : je dis *autorise*, car elle ne leur en impose pas la nécessité ; ils se détermineront suivant les circonstances, qui sont variées à l'infini. Enfin, si les choses pillées sont des objets de première nécessité, les coupables sont condamnés à une peine perpétuelle, et cette peine est la déportation. Ces crimes peuvent en effet avoir les suites les plus désastreuses : ils peuvent amener la guerre civile ; et il convient d'exclure à jamais de la société des hommes qui, par leurs excès, commettent le double crime de porter atteinte à la propriété individuelle, et d'exposer l'État aux plus grands dangers.

Je ne m'arrêterai point aux dispositions qui prononcent des peines de police correctionnelle contre ceux qui détruisent des productions de la terre nécessaires aux besoins de la vie, ou des instruments utiles à l'agriculture, ou qui font périr des animaux dont ils privent, sans aucune nécessité, le maître auquel ils appartiennent. La plupart de ces délits étaient prévus par les anciennes lois, mais plusieurs n'étaient pas assez punis : par exemple, l'ordonnance de 1669 ne prononçait point l'emprisonnement dans le cas d'arbres abattus ou mutilés de manière à les faire périr ; l'amende qu'elle prononçait était insuffisante : de là tant d'abus auxquels le nouveau Code remédiera.

A l'égard du délit qui se commet en inondant les propriétés d'autrui, faute d'avoir observé les réglements de l'autorité compétente sur la hauteur à laquelle on peut élever le déversoir, la loi n'avait jusqu'à présent parlé que de moulins et usines ; le nouveau Code parle aussi des étangs : la raison est la même, et de nombreuses réclamations se sont élevées pour leur rendre commune la disposition de la loi.

Quant aux droits de l'administration à cet égard, le Code pénal n'avait point à s'en occuper : des lois et décrets particuliers en déterminent l'étendue et les limites.

Je dois ajouter une observation.

La loi du 6 octobre 1791 ne distingue point lorsque l'inondation a causé des dégradations ou lorsqu'elle n'en a point occasionné : ces deux cas sont trop différents pour que la peine doive être la même. Le nouveau Code établit la distinction. Si aucune dégradation n'a eu lieu ; si, par exemple, il n'est résulté de l'inondation d'autre mal que d'avoir interrompu pendant quelque temps la communication par un chemin ou passage, une amende seule sera prononcée, ainsi que le veut la loi du 6 octobre.

Mais s'il y a eu des dégradations, le mal étant plus considérable, la désobéissance à l'autorité doit être plus sévèrement punie. Le Code porte un emprisonnement, outre l'amende. Cet emprisonnement, quoique de courte durée, suffira pour l'efficacité de l'exemple.

Il ne me reste plus qu'à dire un mot sur quelques délits qu'on ne peut attribuer à la méchanceté, mais qui sont l'effet de l'imprudence ou du défaut de précaution.

De tout temps il a existé des ordonnances et des règlements qui ont prescrit l'observation de différentes règles pour prévenir les incendies. Si l'une de ces règles avait été négligée, et qu'un incendie eût eu lieu, les contrevenants étaient condamnés à l'amende. Telle était entre autres l'ordonnance de police du 15 novembre 1781, concernant les incendies, règlement fait pour la ville de Paris. La loi du 6 octobre 1791 a depuis généralisé une partie de ses sages dispositions, et elles se retrouveront dans le nouveau Code.

Le Code s'est enfin occupé des précautions qui ont pour objet de prévenir les maladies épizootiques. Les lois et réglements qui concernent ces maladies sont une branche particulière de législation à laquelle le Code n'a point entendu porter atteinte. Il se borne à quelques mesures générales, applicables à tous les temps et à tous les lieux. Une personne a-t-elle en sa possession des animaux ou bestiaux infectés de maladie contagieuse, ou soupçonnés de l'être, elle doit en avertir sur-le-champ le maire de la commune où ils se trouvent, et, sans attendre que le maire ait répondu, les tenir renfermés : autrement, dans l'intervalle qui s'écoulerait entre l'avertissement et la réponse, la communication libre qu'on leur laisserait pourrait occasionner une contagion parmi les autres animaux. Première précaution ordonnée sous peine d'un emprisonnement et d'une amende.

Si l'administration trouve que ces animaux ne sont infectés d'aucune maladie contagieuse, et que dès-lors nul danger ne s'oppose à ce qu'on les laisse communiquer avec d'autres, le possesseur peut, d'après la décision administrative, leur rendre la liberté.

Il doit, au contraire, se l'interdire strictement lorsque la décision est prohibitive. Deuxième précaution dont on ne peut s'écarter sans encourir un emprisonnement plus long, et une amende plus forte que dans le premier cas.

Si même, pour n'avoir pas respecté la prohibition, une contagion était survenue, le Code veut que l'emprisonnement soit de deux ans au moins, et cinq ans au plus, et que l'amende puisse être prononcée dans une proportion qui ne pourra être moindre de cent francs, ni excéder mille.

Le Code ne pourrait s'étendre davantage en cette partie sans se livrer à une multitude de détails extrêmement fastidieux, et qui appartiennent à la classe des dispositions réglementaires.

Telle est, Messieurs, l'analyse des principales dispositions du chapitre relatif aux attentats contre la propriété. A cet égard il est beaucoup de délits emportant des peines de police correctionnelle qui seront prévenus, si les gardes champêtres, les gardes forestiers et autres officiers de police exercent avec une sévère exactitude la surveillance qui leur est confiée. Ils seront donc plus coupables que les autres, lorsqu'eux-mêmes commettront ces délits. Aussi une disposition particulière rend plus forte à leur égard la peine de police correctionnelle. Cette disposition ne s'applique qu'aux attentats contre la propriété.

Je terminerai par quelques observations sur une disposition générale qui s'applique à toutes les parties du Code.

Observations générales.

Au milieu d'un si grand nombre de délits de police correctionnelle que le Code a prévus, il est facile de concevoir que plus d'une fois des actes qualifiés délits seront accompagnés de circonstances particulières, qui, loin de les aggraver, les atténueront sensiblement. La justice reconnaîtra peut-être en même temps que le dommage éprouvé par la personne lésée est extrêmement modique; il pourrait dès-lors en résulter que le *minimum* de la peine déterminée par la loi pour le cas général serait trop fort, et que les juges se trouveraient placés dans l'alternative fâcheuse d'user envers le coupable d'une rigueur dont l'excès leur paraîtrait injuste, ou de le renvoyer absous, en sacrifiant le devoir du magistrat à un sentiment inspiré par l'humanité.

Une disposition qui termine la partie du Code dont nous nous occupons en ce moment porte que, si le préjudice n'excède pas vingt-cinq francs, et que les circonstances paraissent atténuantes, les juges sont autorisés à réduire l'emprisonnement, et l'amende même, jusqu'au *minimum* des peines de police. Au moyen de cette précaution, la conscience du juge sera rassurée, et la peine sera proportionnée au délit.

Il n'était pas possible d'établir une règle semblable à l'égard des crimes. Tout crime emporte peine afflictive ou infamante, mais tout crime n'emporte pas la même espèce de peine; tandis qu'en matière de délits de police correctionnelle, la peine est toujours soit l'emprisonnement, soit l'amende, soit l'un et l'autre ensemble.

Cela posé, la réduction des peines de police correctionnelle ne frappe que sur la quotité de l'amende et sur la durée de l'emprisonnement.

Au contraire, les peines établies pour les crimes étant de différentes espèces, il faudrait, lorsqu'un crime serait atténué par quelque circonstance qui porterait le juge à considérer la peine comme trop rigoureuse, quant à son espèce; il faudrait, disons-nous, que le juge fût autorisé à changer l'espèce de peine, et à descendre du degré fixé par la loi à un degré inférieur, par exemple, à prononcer la réclusion au lieu des travaux forcés à temps, ou bien à substituer le carcan à la réclusion. Ce changement, cette substitution ne serait pas une réduction de peine proprement dite, elle serait une véritable commutation de peine. Or, le droit de commutation de peine est placé par la constitution dans les attributions du souverain; il fait partie du droit de faire grâce : c'est au souverain seul qu'il appartient de décider, en matière de crimes, si telle circonstance vérifiée au procès est assez atténuante pour justifier une commutation. La seule exception laissée au pouvoir judiciaire est dans les cas d'excuse; encore faut-il que le fait allégué pour excuse soit admis comme tel par la loi avant qu'on puisse descendre, en cas de preuves, à une peine inférieure;

Il résulte de ces observations qu'en fait de peine afflic-

tive ou infamante, le juge doit se renfermer dans les limites que la loi lui a tracées; qu'il ne peut dire que la
faute est excusable que lorsque la loi a prévu formellement les circonstances sur lesquelles l'excuse est fondée,
et que toute application d'une peine inférieure à celle
fixée par la loi est un acte de clémence qui ne peut
émaner que du prince, unique source de toutes les
grâces

Vous venez d'entendre, Messieurs, les motifs des
principales dispositions du projet de loi qui vous est soumis : en examinant ses détails, vous serez convaincus,
nous osons l'espérer, que dans cette partie, comme dans
toutes les autres de la législation pénale, on a tâché d'atteindre le plus haut degré de perfection possible. Nos
efforts pour perfectionner le Code ont été secondés par
les sages observations de votre commission. Si ce monument, fruit de longues et profondes méditations, est ré-
commandé par vos suffrages, il réunira tous les titres à
la confiance publique.

MOTIFS DU LIVRE IV,

PRÉSENTÉS

PAR MM. LE COMTE RÉAL, LES CHEVALIERS FAURE ET GIUNTI, Conseillers d'État.

Séance du 10 février 1810.

Messieurs,

Nous avons l'honneur de vous présenter le quatrième et dernier Livre du *Code des* DÉLITS ET DES PEINES, celui qui établit les *peines de police simple*, et qui définit et classe les diverses *contraventions* auxquelles ces peines seront appliquées.

Ceux qui m'ont précédé à cette tribune vous ont parlé de *crimes*, de *délits*; et, au moment où ils ont déroulé sous vos yeux cette épouvantable série d'attentats qu'il faut prévoir, chacun de vous, jetant un regard sur le passé, a vu dans ce tableau de crimes possibles, et presque prophétisés, la véritable et sanglante histoire des passions, des fureurs et de la dépravation de l'homme.

Je viens mettre sous vos yeux des tableaux moins sévères, rappeler des souvenirs moins tristes; et, dans cette série de fautes que la morale réprouve encore, et que la loi punit, du moins vous ne verrez plus de *crimes*, plus de *délits*, mais de simples *contraventions*. Dans l'énumération des peines, vous ne m'entendrez point parler de mort, de sang versé; plus de fers, plus de travaux forcés; un *emprisonnement* de quelques jours, une légère *amende*, suffiront pour proportionner ici la *peine* à la *contravention*.

Les dispositions contenues dans les trois premiers Livres, les *peines* qui y sont déterminées, établissent le Code de *Police de sûreté*; elles ont pour objet et auront pour résultat de s'assurer de la personne de tous les malfaiteurs qui, de temps en temps, et sur diverses parties du territoire, signalent leur funeste existence par des attentats à la vie ou à la propriété des citoyens.

Les dispositions renfermées dans le quatrième Livre que nous vous présentons ont pour objet, auront aussi pour résultat nécessaire le maintien habituel de l'ordre et de la tranquillité dans toutes les parties de l'Empire.

Cette quatrième partie, concourant par des moyens différents au même résultat, était le complément nécessaire et indispensable des trois premières.

Ainsi, par exemple, effrayés ou atteints par les dispositions précédentes, les brigands ne peuvent infester les grandes routes, et le voyageur peut les fréquenter avec sécurité. La partie du Code que nous vous présentons va plus loin; et sur ces routes, devenues sûres par le bienfait des précédentes dispositions, elle maintient l'ordre qui en procure l'usage, qui en écarte les accidents; et si les précédentes dispositions mettent le voyageur à l'abri des attentats du voleur, celles que nous présentons le défendent contre l'insolence et la tyrannie du roulier.

Ainsi lorsque les dispositions précédentes garantissent les propriétés des ravages de l'incendie, en punissant de mort l'incendiaire volontaire, la loi de *police* donne à la propriété une garantie nouvelle, en éveillant l'attention, en punissant des imprudences qui causent des incendies accidentels.

Au Code qui poursuit et supplicie la méchanceté qui commet les crimes il a donc fallu joindre celui qui châtie l'imprudence, cause de tant d'accidents et de malheurs.

Et pendant que les dispositions précédentes assurent le repos de la cité par le supplice du criminel consommé qui lui fait la guerre, les dispositions du Code de *police simple* arrivent au même but en faisant la guerre aux petites passions, à ces *contraventions* légères dont l'habitude ne conduit que trop souvent aux plus grands crimes.

Plusieurs des dispositions contenues dans ce Code ne seraient point déplacées dans un cours de morale ; et c'est ainsi que le Code sévère des *délits et des peines,* ce Code vengeur des crimes, arrive par degrés au Code du bon voisinage et de l'urbanité.

Avant l'Assemblée constituante, les dispositions qui forment aujourd'hui le *Code de police simple* étaient disséminées et perdues dans un grand nombre de volumes, dans une infinité de règlements et d'ordonnances de police, dont plusieurs, de date très-ancienne, n'étaient plus en harmonie ni avec les mœurs, ni avec les habitudes nationales.

Chaque province, chaque ville, chaque quartier avait ses lois, ses usages locaux, sa jurisprudence particulière ; et dans cette partie de la législation qui touche de plus près le peuple, et surtout dans la partie pénale de cette législation, l'arbitraire et le caprice classaient le délit, infligeaient, graduaient, et quelquefois créaient la peine.

Après s'être occupée du grand ouvrage de la *police de sûreté*, l'Assemblée constituante tira du chaos la législation relative à *la police simple*, et, par la loi du 19 juillet 1791, en créa le Code sous le nom de *police municipale*.

Le Code des *délits et des peines* du 3 brumaire an 4

(articles 595 et 596) rapporta les dispositions de la loi du 19 juillet 1791, relatives à la forme de procéder, et aux règles d'instruction à observer par les tribunaux de *police municipale* et *correctionnelle*, et interdit en conséquence aux municipalités tout exercice du pouvoir judiciaire que la loi de 1791 leur avait attribué.

Le même Code de brumaire, après avoir (art. 600) spécifié les peines de police simple, ne consacra qu'un seul article (l'article 605) à la classification des délits qui en seraient passibles ; et il admit au nombre de ces délits, les délits mentionnés dans le titre II de la loi du 28 septembre 1791, sur la *police rurale*, et qui, suivant les dispositions de cette loi, *étaient dans le cas d'être jugés par voie de police municipale*.

Un second article (l'art. 606) laissait au tribunal de police le pouvoir de graduer selon les circonstances, et le plus ou le moins de gravité du délit, les *peines* qu'il était chargé de prononcer, sans néanmoins qu'elles pussent, en aucun cas, être au-dessous d'une amende de la valeur d'une journée de travail, ou d'un jour d'emprisonnement, ni s'élever au-dessus de la valeur de trois journées de travail, ou de trois jours d'emprisonnement.

Un troisième article (l'art. 607) prononçait sur la récidive ; et dans ce cas, les peines devant suivre la proportion réglée par les lois des 19 juillet et 28 septembre 1791, et ces peines alors excédant la compétence du tribunal de police, ne pouvaient être prononcées que par le tribunal de *police correctionnelle*.

Enfin un quatrième et dernier article (l'art. 608) définissait la *récidive*.

Cette législation ainsi réduite présentait des lacunes à remplir.

13

La dernière disposition de l'article 605, comparée à quelques dispositions des articles empruntés à la loi du 28 septembre, faisait naitre sur la compétence quelques incertitudes.

Quelques délits soumis à la *police simple* paraissaient assez graves pour être réclamés par la *police correctionnelle*; et réciproquement quelques contraventions attribuées à celle-ci appartenaient évidemment à la *police simple*.

Presque la totalité des dispositions empruntées à la loi du 28 septembre 1791 paraissent étrangères à la *police simple*, et sont réclamées par le *Code rural*.

La peine prononcée contre la *récidive*, et surtout le changement de juridiction, qui donne les juges, et qui applique *les peines du délit* à ce qui n'est qu'une *contravention*, ont paru répugner aux principes.

Enfin cette latitude accordée au juge, par une heureuse innovation, pour l'application de la *peine*, cette latitude, dis-je, resserrait l'équité du juge dans un espace encore trop étroit, et la même *peine* pesait trop également sur des *délits* de force inégale.

Dans le projet soumis à votre sanction, vous trouverez les dispositions que désirait le dernier état des choses, et les lacunes seront remplies.

Les limites de la compétence ont été indiquées par des lignes très-prononcées.

On a restitué tous les *délits* à la *police correctionnelle*, qui a rendu à la *police simple* toutes les *contraventions*.

On a renvoyé au *Code rural* toutes les dispositions qui lui appartenaient franchement; quelques *contraventions* mixtes sont restées seules dans le domaine de la *police simple*.

La *récidive* jugée *par les mêmes juges* trouve une punition plus proportionnée à la *contravention*, et plus conforme aux principes.

Enfin, dans ce projet dont je vais, en très-peu de lignes, vous tracer l'économie, vous verrez que, par le moyen d'une simple classification, combinée avec une plus grande latitude donnée au juge, nous avons évité ce que l'arbitraire du juge, ce que l'arbitraire de la loi pouvaient avoir de dangereux, pour obtenir de l'équité du juge et de la sevérité de la loi une punition bien juste, bien proportionnée à la *contravention.*

Le Livre IV est distribué en deux chapitres.

Le premier traite *des peines.*

Le second traite des *contraventions* et *peines.*

Le chapitre premier spécifie les *peines,* en détermine l'étendue, la durée.

Ces peines sont l'*emprisonnement,* l'*amende,* et la *confiscation* de certains objets saisis.

L'*emprisonnement* ne peut être moindre d'un jour, ni en excéder cinq.

Les *amendes* peuvent être prononcées depuis un franc jusqu'à quinze francs.

Le projet conserve et renouvelle la disposition qui se trouvait dans le Code de l'Assemblée constituante, et qui applique l'*amende* au profit de la commune où la *con-travention* a été commise.

On a cru devoir répéter dans ce chapitre une disposi-tion déjà consacrée dans un des précédents, et qui statue qu'en cas d'insuffisance des biens, les restitutions et les indemnités dues à la partie lésée sont préférées à l'amende.

Le paiement de l'amende, les restitutions, indemnités et frais entraîneront la *contrainte par corps,* mais avec ces différences, que, pour le *paiement de l'amende*, le condamné ne pourra être détenu plus de quinze jours, s'il justifie de son insolvabilité; au lieu que pour le paiement des restitutions, etc., le condamné doit garder prison jusqu'à parfait paiement, à moins que ces dernières condamnations ne soient prononcées au profit de l'Etat.

Le chapitre II se subdivise en trois sections; et chaque section comprend une classe de *contraventions* qui est punie par une *peine* proportionnée à la gravité de la *contravention*.

Les *contraventions* de la première classe sont punies d'une *amende*, depuis un franc jusqu'à cinq francs.

De toutes les *contraventions* classées dans cette première section, il n'y en a que deux qui soient passibles de l'*emprisonnement*; encore le juge n'est-il point forcé de le prononcer, mais il le peut suivant les circonstances.

Dans ce cas, l'*emprisonnement* sera de trois jours au plus.

L'*emprisonnement* pendant trois jours au plus sera toujours prononcé en cas de récidive.

Les *contraventions* de la deuxième classe sont punies d'une *amende* qui ne peut être moindre de six francs, et qui ne peut en excéder dix.

L'*amende* de cinq jours au plus est toujours appliquée en cas de *récidive*.

Les contraventions de la troisième classe sont punies d'une *amende* de onze à quinze francs inclusivement.

Suivant les circonstances, l'*emprisonnement* pendant cinq jours au plus *pourra* être prononcé contre quelques-unes des contraventions classées dans cette troisième section;

Et l'*emprisonnement* pendant cinq jours aura toujours lieu en cas de *récidive*.

C'est en établissant cette classification, c'est en accordant en même temps au juge le droit d'élever, dans la proportion autorisée par la classification, la quotité de l'*amende*, ou d'augmenter, dans les cas prévus, la durée de l'*emprisonnement*, que nous avons pu nous assurer que le texte de la loi ne serait ni éludé ni forcé, et que le juge jouirait cependant de l'indépendance raisonnable et suffisante dont il a besoin pour faire bonne justice : indépendance réclamée par Montesquieu, qui prononce que, *dans l'exercice de la police, c'est plutôt le magistrat qui punit que la loi.*

À la suite du chapitre IV se trouve, dans l'art. 484 et dernier, une disposition générale qui s'applique au Code entier, et qui mérite toute votre attention. Cet article dit :

« En tout ce qui n'est pas réglé par le présent Code, « en matière de *crimes, délits* et *contraventions,* les « cours et tribunaux continueront d'observer et de faire « exécuter les dispositions des lois et des règlements actuellement en vigueur. »

Cette disposition était d'absolue nécessité. Elle maintient les dispositions pénales, sans lesquelles quelques lois, des Codes entiers, des règlements généraux d'une utilité reconnue, resteraient sans exécution.

Ainsi cette dernière disposition maintient les lois et règlements actuellement en vigueur, relatifs,

Aux dispositions du Code rural, qui ne sont point entrées dans ce Code ;

Aux taxes, contributions directes ou indirectes, droits réunis, de douanes et d'octrois ;

Aux tarifs pour le prix de certaines denrées ou de certains salaires;

Aux calamités publiques, comme épidémies, épizooties, contagions, disettes, inondations;

Aux entreprises de services publics, comme coches, messageries, voitures publiques de terre et d'eau, voitures de place, numéros ou indications de noms sur voitures, postes aux lettres et postes aux chevaux;

A la formation, entretien et conservation des rues, chemins, voies publiques, ponts et canaux;

A la mer, à ses rades, rivages et ports, et aux pêcheries maritimes;

A la navigation intérieure, à la police des eaux et aux pêcheries;

A la chasse, aux bois, aux forêts;

Aux matières générales de commerce, affaires et expéditions maritimes, bourses ou rassemblements commerciaux, police des foires et marchés;

Aux commerces particuliers d'orfévrerie, bijouterie, joaillerie, de serrurerie et de gens de marteau; de pharmacie et apothicairerie; de poudres et salpêtres; des arquebusiers et artificiers; des cafetiers, restaurateurs, marchands et débitants de boissons; de cabaretiers et aubergistes;

A la garantie des matières d'or et d'argent;

A la police des maisons de débauche et de jeu;

A la police des fêtes, cérémonies et spectacles;

A la construction, entretien, solidité, alignement des édifices et aux matières de voiries;

Aux lieux d'inhumation et sépulture;

A l'administration, police et discipline des hospices, maisons sanitaires et lazarets; aux écoles, aux maisons de dépôt, d'arrêt, de justice et de peine, de détention correctionnelle et de police ; aux maisons ou lieux de fabrique, manufactures ou ateliers ; à l'exploitation des mines et des usines ;

Au port d'armes ;

Au service des gardes nationales ;

A l'état civil, etc. , etc.

Vous connaissez maintenant, Messieurs, dans son ensemble et dans ses détails, ce nouveau Code qui doit donner le mouvement au Code d'Instruction criminelle que vous avez sanctionné dans votre avant-dernière session.

Vous pouvez maintenant apprécier ce bel ouvrage, et reconnaître quelle immense supériorité lui donnent sur celui de l'Assemblée constituante les nombreuses améliorations qu'il a reçues.

Le Code des *Délits et des Peines* de 1791 était déjà sans doute un monument magnifique élevé à l'humanité, à la raison sur les ruines d'institutions barbares ; mais on ne peut pas se dissimuler que ses auteurs travaillaient sur un volcan, et qu'ils n'ont pas toujours pu écouter la voix de la raison.

Vingt ans d'ailleurs se sont écoulés depuis que cette immense machine a été mise à exécution ; et pendant ces vingt ans, au nombre desquels se trouvent les longues et instructives années qui ont précédé Brumaire, pendant ces vingt ans, une expérience de tous les jours en a signalé les défauts, les parties faibles, les lacunes.

Soumises à un examen sévère, toutes les parties de ce grand ouvrage ont été l'objet d'une longue méditation ; d'innombrables lacunes ont été remplies ; tous les articles conservés ont été refondus ; toutes les définitions, rendues plus complètes, ont gagné de clarté et de précision ; des parties entières toutes nouvelles ont été ajou-

tées. Les juges cesseront enfin d'être les aveugles appli-
cateurs d'un texte qui produisait, par son inflexibilité
même, tous les maux d'un atroce arbitraire. L'immense
bienfait de la latitude accordée aux juges débarrassera
enfin leur raison de ces entraves d'acier qui la tenaient
dans un homicide esclavage ; tous les crimes seront atteints, tous les criminels seront punis, parce que cette
latitude permettra enfin au juge d'appliquer une peine
qui, pouvant être toujours proportionnée au délit, ne
sera jamais cruelle, ne sera jamais dérisoire. L'impunité
de beaucoup de criminels est due à l'aveugle inflexibilité
de la loi ancienne, autant peut-être qu'à la faiblesse des
jurés et à la mauvaise composition du jury.

Ce Code présente à la société une sécurité plus grande,
en plaçant les hommes repris de justice, les vagabonds
et les mendiants sous la surveillance légale de la haute
police.

En insérant dans son Code ce moyen puissant d'ordre
et de sûreté publics, le législateur ne hasarde point une
théorie nouvelle dont les résultats soient incertains. Ce
moyen, la force des choses l'avait créé ; et, en l'adoptant,
en lui donnant enfin une existence légale, le législateur
n'a fait autre chose que consacrer une mesure dont une
longue expérience avait proclamé l'efficacité. En la léga-
lisant, il lui imprime une nouvelle force ; il la dépouille
de tout ce qu'elle pouvait offrir d'inquiétant et d'irrégu-
lier, en intéressant les tribunaux à son maintien, en les
associant à son exécution.

Vous n'hésiterez donc pas, Messieurs, à revêtir de
votre sanction ce nouveau Code, digne de prendre place
dans cette grande et majestueuse collection de Codes ho-
norés du nom de leur illustre auteur. Ce Code portera
aussi le nom de NAPOLÉON, non pas seulement parse

qu'il aura été promulgué sous son règne, facile honneur dont pouvaient se contenter les monarques dont on a dit, légèrement sans doute, qu'ils étaient seulement les rois d'un grand règne; il portera le nom de NAPOLÉON parce qu'il est aussi son ouvrage, parce que ce guerrier législateur en a éclairé la discussion, parce qu'il l'a enrichi de ses inspirations, parce que ce Code porte l'empreinte de sa sagesse et de son génie.

Heureux, Messieurs, d'associer vos travaux à ses travaux !

Heureux d'assister à cette époque où sa main puissante, sa main créatrice lance ainsi dans l'espace des siècles ses lois immortalisées par son nom !

Epoque miraculeuse, époque héroïque où chaque année de son règne est signalée

Par la conquête d'un empire ;

Par une paix toujours glorieuse, toujours généreuse, parce que toujours la force et la modération l'ont dictée;

Par la confection de travaux immenses;

Par des projets nouveaux dont la conception seule aurait suffi pour immortaliser un autre monarque.

S'il combat, s'il triomphe, s'il pardonne comme César, il consolide et pacifie comme Auguste;

Économe et magnifique, il change aussi la vieille cité en une cité de marbre;

Et au moment où il rétablit et agrandit encore l'Empire de Charlemagne, au moment où il restitue à l'Italie régénérée la Rome des Césars, il donne à la grande nation des Codes qui font oublier ceux qui portent le nom de Justinien.

Ainsi, couvert de tous les genres de gloire, de tous les faits glorieux qui, pris séparément, ont illustré tant de héros, tant de siècles, le héros du dix-neuvième continue de marquer, par d'impérissables monuments, chacun des pas qu'il fait dans sa marche triomphale qui le conduit à l'immortalité.

FIN DES MOTIFS.

CODE PÉNAL.

(Décrété le 12 février 1810. Promulgué le 22 du même mois.)

DISPOSITIONS PRÉLIMINAIRES.

ARTICLE PREMIER.

L'INFRACTION que les lois punissent des peines de police est une *contravention*.

L'infraction que les lois punissent de peines correctionnelles est un *délit*.

L'infraction que les lois punissent d'une peine afflictive ou infamante est un *crime*.

2. Toute tentative de *crime* qui aura été manifestée par des actes extérieurs et suivie d'un commencement d'exécution, si elle n'a été suspendue, ou n'a manqué son effet que par des circonstances fortuites ou indépendantes de la volonté de l'auteur, est considérée comme le *crime* même.

3. Les tentatives de *délits* ne sont considérées comme *délits* que dans les cas déterminés par une disposition spéciale de la loi.

4. Nulle contravention, nul délit, nul crime, ne peuvent être punis de peines qui n'étaient pas prononcées par la loi avant qu'ils fussent commis.

5. Les dispositions du présent Code ne s'appliquent pas aux contraventions, délits et crimes *militaires*.

LIVRE PREMIER.

6. Les peines en matière criminelle sont ou afflictives et infamantes, ou seulement infamantes.

7. Les peines afflictives et infamantes sont,

1° La mort;

2° Les travaux forcés à perpétuité;

3° La déportation;

4° Les travaux forcés à temps;

5° La réclusion.

La marque et la confiscation générale peuvent être prononcées concurremment avec une peine afflictive, dans les cas déterminés par la loi.

8. Les peines infamantes sont,

1° Le carcan;

2° Le bannissement;

3° La dégradation civique.

9. Les peines en matière correctionnelle sont,

1° L'emprisonnement à temps dans un lieu de correction;

2° L'interdiction à temps de certains droits civiques, civils ou de famille;

3° L'amende;

10. La condamnation aux peines établies par la loi est toujours prononcée sans préjudice des restitutions et dommages et intérêts qui peuvent être dus aux parties.

11. Le renvoi sous la surveillance spéciale de la haute police, l'amende et la confiscation spéciale, soit du corps du délit, quand la propriété en appartient au

condamné, soit des choses produites par le délit, soit de celles qui ont servi ou qui ont été destinées à le commettre, sont des peines communes aux matières criminelle et correctionnelle.

CHAPITRE PREMIER.

Des peines en matière criminelle.

12. Tout condamné à mort aura la tête tranchée.

13. Le coupable condamné à mort pour parricide sera conduit sur le lieu de l'exécution, en chemise, nu-pieds, et la tête couverte d'un voile noir.

Il sera exposé sur l'échafaud pendant qu'un huissier fera au peuple lecture de l'arrêt de condamnation ; il aura ensuite le poing droit coupé, et sera immédiatement exécuté à mort.

14. Les corps des suppliciés seront délivrés à leurs familles, si elles les réclament, à la charge par elles de les faire inhumer sans aucun appareil.

15. Les hommes condamnés aux travaux forcés seront employés aux travaux les plus pénibles ; ils traîneront à leurs pieds un boulet, ou seront attachés deux à deux avec une chaîne, lorsque la nature du travail auquel ils seront employés le permettra.

16. Les femmes et les filles condamnées aux travaux forcés n'y seront employées que dans l'intérieur d'une maison de force.

17. La peine de la déportation consistera à être transporté et à demeurer à perpétuité dans un lieu déterminé par le Gouvernement, hors du territoire continental de l'Empire.

Si le déporté rentre sur le territoire de l'Empire, il sera, sur la seule preuve de son identité, condamné aux travaux forcés à perpétuité.

Le déporté qui ne sera pas rentré sur le territoire de l'Empire, mais qui sera saisi dans des pays occupés par les armées françaises, sera reconduit dans le lieu de sa déportation.

18. Les condamnations aux travaux forcés à perpétuité et à la déportation, emporteront mort civile.

Néanmoins le Gouvernement pourra accorder au déporté, dans le lieu de la déportation, l'exercice des droits civils, ou de quelques-uns de ces droits.

19. La condamnation à la peine des travaux forcés à temps sera prononcée pour cinq ans au moins, et vingt ans au plus.

20. Quiconque aura été condamné à la peine des travaux forcés à perpétuité, sera flétri, sur la place publique, par l'application d'une empreinte avec un fer brûlant, sur l'épaule droite.

Les condamnés à d'autres peines ne subiront la flétrissure que dans les cas où la loi l'aurait attachée à la peine qui leur est infligée.

Cette empreinte sera des lettres T. P. pour les coupables condamnés aux travaux forcés à perpétuité; de la lettre T., pour les coupables condamnés aux travaux forcés à temps, lorsqu'ils devront être flétris.

La lettre F. sera ajoutée dans l'empreinte, si le coupable est un faussaire.

21. Tout individu de l'un ou de l'autre sexe, condamné à la peine de la réclusion, sera renfermé dans une maison de force, et employé à des travaux dont le produit pourra être en partie appliqué à son profit, ainsi qu'il sera réglé par le Gouvernement.

La durée de cette peine sera au moins de cinq années, et de dix ans au plus.

22. Quiconque aura été condamné à l'une des peines des travaux forcés à perpétuité, des travaux forcés à temps, ou de la réclusion, avant de subir sa peine, sera attaché au carcan sur la place publique : il y demeurera exposé aux regards du peuple durant une heure; au-dessus de sa tête sera placé un écriteau portant, en caractères gros et lisibles, ses noms, sa profession, son domicile, sa peine, et la cause de sa condamnation.

23. La durée de la peine des travaux forcés à temps, et de la peine de la réclusion, se comptera du jour de l'exposition.

24. La condamnation à la peine du carcan sera exécutée de la manière prescrite par l'article 22.

25. Aucune condamnation ne pourra être exécutée les jours de fêtes nationales ou religieuses, ni les dimanches.

26. L'exécution se fera sur l'une des places publiques du lieu qui sera indiqué par l'arrêt de condamnation.

27. Si une femme condamnée à mort se déclare et s'il est vérifié qu'elle est enceinte, elle ne subira la peine qu'après sa délivrance.

28. Quiconque aura été condamné à la peine des travaux forcés à temps, du bannissement, de la réclusion ou du carcan, ne pourra jamais être juré, ni expert, ni être employé comme témoin dans les actes, ni déposer en justice, autrement que pour y donner de simples renseignements.

Il sera incapable de tutelle et de curatelle, si ce n'est de ses enfants et sur l'avis seulement de sa famille.

Il sera déchu du droit de port d'armes et du droit de servir dans les armées de l'Empire.

29. Quiconque aura été condamné à la peine des travaux forcés à temps ou de la réclusion, sera de plus, pendant la durée de sa peine, en état d'interdiction légale ; il lui sera nommé un curateur pour gérer et administrer ses biens, dans les formes prescrites pour la nomination des curateurs aux interdits.

30. Les biens du condamné lui seront remis après qu'il aura subi sa peine, et le curateur lui rendra compte de son administration.

31. Pendant la durée de la peine, il ne pourra lui être remis aucune somme, aucune provision, aucune portion de ses revenus.

32. Quiconque aura été condamné au bannissement, sera transporté, par ordre du Gouvernement, hors du territoire de l'Empire.

La durée du bannissement sera au moins de cinq années, et de dix ans au plus.

33. Si le banni, durant le temps de son bannissement, rentre sur le territoire de l'Empire, il sera, sur la seule preuve de son identité, condamné à la peine de la déportation.

34. La dégradation civique consiste dans la destitution et l'exclusion du condamné de toutes fonctions ou emplois publics, et dans la privation de tous les droits énoncés en l'article 28.

35. La durée du bannissement se comptera du jour où l'arrêt sera devenu irrévocable.

36. Tous arrêts qui porteront la peine de mort, des travaux forcés à perpétuité ou à temps, la déportation, la réclusion, la peine du carcan, le bannissement et la dégradation civique, seront imprimés par extrait.

Ils seront affichés dans la ville centrale du département, dans celle où l'arrêt aura été rendu, dans la commune du lieu où le délit aura été commis, dans celle où se fera l'exécution et dans celle du domicile du condamné.

37. La confiscation générale est l'attribution des biens d'un condamné au domaine de l'État.

Elle ne sera la suite nécessaire d'aucune condamnation; elle n'aura lieu que dans les cas où la loi la prononce expressément.

38. La confiscation générale demeure grevée de toutes les dettes légitimes, jusqu'à concurrence de la valeur des biens confisqués, de l'obligation de fournir aux enfants ou autres descendants une moitié de la portion dont le père n'aurait pu les priver.

De plus, la confiscation générale demeure grevée de la prestation des aliments à qui il en est dû de droit.

39. L'Empereur pourra disposer des biens confisqués, en faveur, soit des père, mère ou autres ascendants, soit de la veuve, soit des enfants, ou autres descendants légitimes, naturels ou adoptifs, soit des autres parents du condamné.

CHAPITRE II.

Des peines en matière correctionnelle.

40. Quiconque aura été condamné à la peine d'emprisonnement, sera renfermé dans une maison de correction; il y sera employé à l'un des travaux établis dans cette maison, selon son choix.

La durée de cette peine sera au moins de six jours, et de cinq années au plus; sauf les cas de récidive ou autres où la loi aura déterminé d'autres limites.

14.

La peine à un jour d'emprisonnement est de vingt-quatre heures;

Celle à un mois est de trente jours.

41. Les produits du travail de chaque détenu pour délit correctionnel seront appliqués, partie aux dépenses communes de la maison, partie à lui procurer quelques adoucissements, s'il les mérite, partie à former pour lui, au temps de sa sortie, un fonds de réserve; le tout ainsi qu'il sera ordonné par des règlements d'administration publique.

42. Les tribunaux, jugeant correctionnellement, pourront, dans certains cas, interdire, en tout ou en partie, l'exercice des droits civiques, civils et de famille suivants :

1º De vote et d'élection;

2º D'éligibilité;

3º D'être appelé ou nommé aux fonctions de juré ou autres fonctions publiques, ou aux emplois de l'administration, ou d'exercer ces fonctions ou emplois;

4º De port d'armes;

5º De vote et de suffrage dans les délibérations de famille;

6º D'être tuteur, curateur, si ce n'est de ses enfants, et sur l'avis seulement de la famille;

7º D'être expert ou employé comme témoin dans les actes;

8º De témoignage en justice, autrement que pour y faire des simples déclarations.

43. Les tribunaux ne prononceront l'interdiction mentionnée dans l'article précédent, que lorsqu'elle aura été autorisée ou ordonnée par une disposition particulière de la loi.

CHAPITRE III.

Des peines et des autres condamnations qui peuvent être prononcées pour crimes ou délits.

44. L'effet du renvoi sous la surveillance de la haute police de l'État, sera de donner au Gouvernement, ainsi qu'à la partie intéressée, le droit d'exiger, soit de l'individu placé dans cet état, après qu'il aura subi sa peine, soit de ses père et mère, tuteur ou curateur, s'il est en âge de minorité, une caution solvable de bonne conduite, jusqu'à la somme qui sera fixée par l'arrêt ou le jugement : toute personne pourra être admise à fournir cette caution.

Faute de fournir ce cautionnement, le condamné demeure à la disposition du Gouvernement, qui a le droit d'ordonner, soit l'éloignement de l'individu d'un certain lieu, soit sa résidence continue dans un lieu déterminé de l'un des départements de l'Empire.

45. En cas de désobéissance à cet ordre, le Gouvernement aura le droit de faire arrêter et détenir le condamné, durant un intervalle de temps qui pourra s'étendre jusqu'à l'expiration du temps fixé pour l'état de la surveillance spéciale.

46. Lorsque la personne mise sous la surveillance spéciale du Gouvernement, et ayant obtenu sa liberté sous caution, aura été condamnée par un arrêt ou jugement devenu irrévocable, pour un ou plusieurs crimes, ou pour un ou plusieurs délits commis dans l'intervalle déterminé par l'acte de cautionnement, les cautions seront contraintes, même par corps, au paiement des sommes portées dans cet acte.

Les sommes recouvrées seront affectées de préférence

aux restitutions, aux dommages-intérêts et frais adjugés aux parties lésées par ces crimes ou ces délits.

47. Les coupables condamnés aux travaux forcés à temps et à la réclusion seront de plein droit, après qu'ils auront subi leur peine, et pendant toute la vie, sous la surveillance de la haute police de l'État.

48. Les coupables condamnés au bannissement seront de plein droit sous la même surveillance pendant un temps égal à la durée de la peine qu'ils auront subie.

49. Devront être renvoyés sous la même surveillance, ceux qui auront été condamnés pour crimes ou délits qui intéressent la sûreté intérieure ou extérieure de l'État.

50. Hors les cas déterminés par les articles précédents, les condamnés ne seront placés sous la surveillance de la haute police de l'État, que dans le cas où une disposition particulière de la loi l'aura permis.

51. Quand il y aura lieu à restitution, le coupable sera condamné en outre, envers la partie, à des indemnités dont la détermination est laissée à la justice de la cour ou du tribunal, lorsque la loi ne les aura pas réglées, sans qu'elles puissent jamais être au-dessous du quart des restitutions, et sans que la cour ou le tribunal puisse, du consentement même de la partie, en prononcer l'application à une œuvre quelconque.

52. L'exécution des condamnations à l'amende, aux restitutions, aux dommages-intérêts et aux frais, pourra être poursuivie par la voie de la contrainte par corps.

53. Lorsque des amendes et des frais seront prononcés au profit de l'État, si, après l'expiration de la peine afflictive ou infamante, l'emprisonnement du condamné, pour l'acquit de ces condamnations pécuniaires, a duré

une année complète, il pourra, sur la preuve acquise par les voies de droit, de son absolue insolvabilité, obtenir sa liberté provisoire.

La durée de l'emprisonnement sera réduite à six mois, s'il s'agit d'un délit; sauf, dans tous les cas, à reprendre la contrainte par corps, s'il survient au condamné quelque moyen de solvabilité.

54. En cas de concurrence de l'amende ou de la confiscation avec les restitutions et les dommages-intérêts, sur les biens insuffisants du condamné, ces dernières condamnations obtiendront la préférence.

55. Tous les individus condamnés pour un même crime, ou pour un même délit, sont tenus solidairement des amendes, des restitutions, des dommages - intérêts et des frais.

CHAPITRE IV.

Des peines de la récidive pour crimes et délits.

56. Quiconque, ayant été condamné pour crime, aura commis un second crime emportant la dégradation civique, sera condamné à la peine du carcan.

Si le second crime emporte la peine du carcan ou le bannissement, il sera condamné à la peine de la réclusion.

Si le second crime entraîne la peine de la réclusion, il sera condamné à la peine des travaux forcés à temps et à la marque ;

Si le second crime entraîne la peine des travaux forcés à temps, ou la déportation, il sera condamné à la peine des travaux forcés à perpétuité ;

Si le second crime entraîne la peine des travaux forcés à perpétuité, il sera condamné à la peine de mort.

57. Quiconque, ayant été condamné pour un crime, aura commis un délit de nature à être puni correctionnellement, sera condamné au *maximum* de la peine portée par la loi, et cette peine pourra être élevée jusqu'au double.

58. Les coupables condamnés correctionnellement à un emprisonnement de plus d'une année, seront aussi, en cas de nouveau délit, condamnés au *maximum* de la peine portée par la loi, et cette peine pourra être élevée jusqu'au double : ils seront de plus mis sous la surveillance spéciale du Gouvernement, pendant au moins cinq années, et dix ans au plus.

FIN DU LIVRE PREMIER.

LIVRE II.

DES PERSONNES PUNISSABLES, EXCUSABLES OU RESPON-
SABLES POUR CRIMES OU POUR DÉLITS.

(Décrété le 13 février 1810. Promulgué le 23 du même mois.)

CHAPITRE UNIQUE.

59. Les complices d'un crime ou d'un délit seront punis de la même peine que les auteurs mêmes de ce crime ou de ce délit, sauf les cas où la loi en aurait disposé autrement.

60. Seront punis comme complices d'une action qualifiée crime ou délit, ceux qui, par dons, promesses, menaces, abus d'autorité ou de pouvoir, machinations ou artifices coupables, auront provoqué à cette action, ou donné des instructions pour la commettre;

Ceux qui auront procuré des armes, des instruments, ou tout autre moyen qui aura servi à l'action, sachant qu'ils devaient y servir.

Ceux qui auront, avec connaissance, aidé ou assisté l'auteur ou les auteurs de l'action, dans les faits qui l'auront préparée ou facilitée, ou dans ceux qui l'auront consommée; sans préjudice des peines qui seront spécialement portées par le présent Code contre les auteurs de complots ou de provocations attentatoires à la sûreté intérieure ou extérieure de l'État, même dans le cas où le crime qui était l'objet des conspirateurs ou des provocateurs n'aurait pas été commis.

61. Ceux qui, connaissant la conduite criminelle des malfaiteurs exerçant des brigandages ou des violences contre la sûreté de l'Etat, la paix publique, les personnes ou les propriétés, leur fournissent habituellement logement, lieu de retraite ou de réunion, seront punis comme leurs complices.

62. Ceux qui sciemment auront recélé, en tout ou en partie, des choses enlevées, détournées ou obtenues à l'aide d'un crime ou d'un délit, seront aussi punis comme complices de ce crime ou délit.

63. Néanmoins, et à l'égard des recéleurs désignés dans l'article précédent, la peine de mort, des travaux forcés à perpétuité, ou de la déportation, lorsqu'il y aura lieu, ne leur sera appliquée qu'autant qu'ils seront convaincus d'avoir eu, au temps du recélé, connaissance des circonstances auxquelles la loi attache les peines de ces trois genres : sinon, ils ne subiront que la peine des travaux forcés à temps.

64. Il n'y a ni crime ni délit, lorsque le prévenu était en état de démence au temps de l'action, ou lorsqu'il a été contraint par une force à laquelle il n'a pu résister.

65. Nul crime ou délit ne peut être excusé, ni la peine mitigée, que dans les cas et dans les circonstances où la loi déclare le fait excusable, ou permet de lui appliquer une peine moins rigoureuse.

66. Lorsque l'accusé aura moins de seize ans, s'il est décidé qu'il a agi *sans discernement*, il sera acquitté ; mais il sera, selon les circonstances, remis à ses parents, ou conduit dans une maison de correction, pour y être élevé et détenu pendant tel nombre d'années que le jugement déterminera, et qui toutefois ne pourra excéder l'époque où il aura accompli sa vingtième année.

67. S'il est décidé qu'il a agi avec *discernement*, les peines seront prononcées ainsi qu'il suit :

S'il a encouru la peine de mort, des travaux forcés à perpétuité, ou de la déportation, il sera condamné à la peine de dix à vingt ans d'emprisonnement dans une maison de correction;

S'il a encouru la peine des travaux forcés à temps, ou de la réclusion, il sera condamné à être renfermé dans une maison de correction pour un temps égal au tiers au moins et à la moitié au plus de celui auquel il aurait pu être condamné à l'une de ces peines.

Dans tous ces cas, il pourra être mis, par l'arrêt ou le jugement, sous la surveillance de la haute police, pendant cinq ans au moins et dix ans au plus.

S'il a encouru la peine du carcan ou du bannissement, il sera condamné à être enfermé, d'un an à cinq ans, dans une maison de correction.

68. Dans aucun des cas prévus par l'article précédent, le condamné ne subira l'exposition publique.

69. Si le coupable n'a encouru qu'une peine correctionnelle, il pourra être condamné à telle peine correctionnelle qui sera jugée convenable, pourvu qu'elle soit au-dessous de la moitié de celle qu'il aurait subie s'il avait eu seize ans.

70. Les peines des travaux forcés à perpétuité, de la déportation et des travaux forcés à temps, ne seront prononcées contre aucun individu âgé de soixante-dix ans accomplis au moment du jugement.

71. Ces peines seront remplacées, à leur égard, par celle de la réclusion, soit à perpétuité, soit à temps, et selon la durée de la peine qu'elle remplacera.

72. Tout condamné à la peine des travaux forcés à perpétuité ou à temps, dès qu'il aura atteint l'âge de

soixante-dix ans accomplis, en sera relevé, et sera renfermé dans la maison de force pour tout le temps à expirer de sa peine, comme s'il n'eût été condamné qu'à la réclusion.

73. Les aubergistes et hôteliers convaincus d'avoir logé, plus de vingt-quatre heures, quelqu'un qui, pendant son séjour, aurait commis un crime ou un délit, seront civilement responsables des restitutions, des indemnités et des frais adjugés à ceux à qui ce crime ou ce délit aurait causé quelque dommage, faute par eux d'avoir inscrit sur leur registre le nom, la profession et le domicile du coupable; sans préjudice de leur responsabilité, dans le cas des articles 1952 et 1953 du Code Napoléon.

74. Dans les autres cas de responsabilité civile qui pourront se présenter dans les affaires criminelles, correctionnelles ou de police, les cours et tribunaux devant qui ces affaires seront portées, se conformeront aux dispositions du Code Napoléon, livre III, titre IV, chapitre II.

—

FIN DU LIVRE DEUXIÈME.

LIVRE III.

(Décrété le 15 février 1810. Promulgué le 25 du même mois.)

TITRE PREMIER.

Des Crimes et des Délits contre la chose publique.

CHAPITRE PREMIER.

Des Crimes et Délits contre la sûreté de l'Etat.

SECTION PREMIÈRE.

Des Crimes et Délits contre la sûreté extérieure de l'Etat.

75. Tout Français qui aura porté les armes contre la France, sera puni de mort.

Ses biens seront confisqués.

76. Quiconque aura pratiqué des machinations ou entretenu des intelligences avec les puissances étrangères ou leurs agents, pour les engager à commettre des hostilités ou entreprendre la guerre contre la France, ou pour leur en procurer les moyens, sera puni de mort, et ses biens seront confisqués.

Cette disposition aura lieu dans le cas même où lesdites machinations ou intelligences n'auraient pas été suivies d'hostilités.

77. Sera également puni de mort et de la confiscation de ses biens quiconque aura pratiqué des manœuvres ou entretenu des intelligences avec les ennemis de l'État, à l'effet de faciliter leur entrée sur le territoire et dépendances de l'Empire français, ou de leur livrer des villes, forteresses, places, postes, ports, magasins, arsenaux, vaisseaux ou bâtiments appartenant à la France, ou de fournir aux ennemis des secours en soldats, hommes, argent, vivres, armes ou munitions, ou de seconder les progrès de leurs armes sur les possessions ou contre les forces françaises de terre ou de mer, soit en ébranlant la fidélité des officiers, soldats, matelots ou autres, envers l'Empereur et l'État, soit de toute autre manière.

78. Si la correspondance avec les sujets d'une puissance ennemie, sans avoir pour objet l'un des crimes énoncés en l'article précédent, a néanmoins eu pour résultat de fournir aux ennemis des instructions nuisibles à la situation militaire ou politique de la France ou de ses alliés, ceux qui auront entretenu cette correspondance seront punis du bannissement, sans préjudice de plus fortes peines, dans le cas où ces instructions auraient été la suite d'un concert constituant un fait d'espionnage.

79. Les peines exprimées aux articles 76 et 77 seront les mêmes, soit que les machinations ou manœuvres énoncées en ces articles aient été commises envers la France, soit qu'elles l'aient été envers les alliés de la France, agissant contre l'ennemi commun.

80. Sera puni des peines exprimées en l'art. 76, tout fonctionnaire public, tout agent du Gouvernement, ou toute autre personne qui, chargée ou instruite officiellement ou à raison de son état, du secret d'une négociation ou d'une expédition, l'aura livré aux agents d'une puissance étrangère ou de l'ennemi.

81. Tout fonctionnaire public, tout agent, tout préposé du Gouvernement, chargé, à raison de ses fonctions, du dépôt des plans de fortifications, arsenaux, ports ou rades, qui aura livré ces plans ou l'un de ces plans à l'ennemi ou aux agents de l'ennemi, sera puni de mort, et ses biens seront confisqués.

Il sera puni du bannissement, s'il a livré ces plans aux agents d'une puissance étrangère, neutre ou alliée.

82. Toute autre personne qui, étant parvenue, par corruption, fraude ou violence, à soustraire lesdits plans, les aura livrés ou à l'ennemi ou aux agents d'une puissance étrangère, sera punie comme le fonctionnaire ou agent mentionné dans l'article précédent, et selon les distinctions qui y sont établies.

Si lesdits plans se trouvaient, sans le préalable emploi de mauvaises voies, entre les mains de la personne qui les a livrés, la peine sera, au premier cas mentionné dans l'article 81, la déportation ;

Et au second cas du même article, un emprisonnement de deux à cinq ans.

83. Quiconque aura recélé, ou aura fait recéler les espions ou les soldats ennemis envoyés à la découverte, et qu'il aura connus pour tels, sera condamné à la peine de mort.

84. Quiconque aura, par des actions hostiles non approuvées par le Gouvernement, exposé l'État à une déclaration de guerre, sera puni du bannissement ; et, si la guerre s'en est suivie, de la déportation.

85. Quiconque aura, par des actes non approuvés par le Gouvernement, exposé des Français à éprouver des représailles, sera puni du bannissement.

SECTION II.

Des Crimes contre la sûreté intérieure de l'État.

§. Ier.

Des Attentats et Complots dirigés contre l'Empereur et sa famille.

86. L'attentat ou complot contre la vie ou contre la personne de l'Empereur, est crime de lèse-majesté; ce crime est puni comme parricide, et emporte de plus la confiscation des biens.

87. L'attentat ou le complot contre la vie ou la personne des membres de la famille impériale;

L'attentat ou le complot dont le but sera,

Soit de détruire ou de changer le Gouvernement ou l'ordre de successibilité au trône,

Soit d'exciter les citoyens ou habitants à s'armer contre l'autorité impériale,

Seront punis de la peine de mort et de la confiscation des biens.

88. Il y a attentat dès qu'un acte est commis ou commencé pour parvenir à l'exécution de ces crimes, quoiqu'ils n'aient pas été consommés.

89. Il y a complot dès que la résolution d'agir est concertée et arrêtée entre deux conspirateurs ou un plus grand nombre, quoiqu'il n'y ait pas eu d'attentat.

90. S'il n'y a pas eu de complot arrêté, mais une proposition faite et non agréée d'en former un pour arriver au crime mentionné dans l'article 86, celui qui aura fait une telle proposition sera puni de la réclusion.

L'auteur de toute proposition non agréée tendante à l'un des crimes énoncés dans l'article 87, sera puni du bannissement.

§. II.

Des crimes tendant à troubler l'Etat par la guerre civile, l'illégal emploi de la force armée, la dévastation et le pillage publics.

91. L'attentat ou le complot dont le but sera, soit d'exciter la guerre civile en armant ou en portant les citoyens ou habitants à s'armer les uns contre les autres,

Soit de porter la dévastation, le massacre et le pillage dans une ou plusieurs communes,

Seront punis de la peine de mort, et les biens des coupables seront confisqués.

92. Seront punis de mort et de la confiscation de leurs biens, ceux qui auront levé ou fait lever des troupes armées, engagé ou enrôlé, fait engager ou enrôler des soldats, ou leur auront fourni ou procuré des armes ou munitions, sans ordre ou autorisation du pouvoir légitime.

93. Ceux qui, sans droit ou motif légitime, auront pris le commandement d'un corps d'armée, d'une troupe, d'une flotte, d'une escadre, d'un bâtiment de guerre, d'une place forte, d'un poste, d'un port, d'une ville,

Ceux qui auront retenu, contre l'ordre du Gouvernement, un commandement militaire quelconque;

Les commandants qui auront tenu leur armée ou troupe rassemblée, après que le licenciement ou la séparation en auront été ordonnés,

Seront punis de la peine de mort, et leurs biens seront confisqués.

94. Toute personne qui, pouvant disposer de la force publique, en aura requis ou ordonné, fait requérir ou ordonner l'action ou l'emploi contre la levée des

gens de guerre légalement établie, sera punie de la déportation.

Si cette réquisition ou cet ordre ont été suivis de leur effet, le coupable sera puni de mort, et ses biens seront confisqués.

95. Tout individu qui aura incendié ou détruit par l'explosion d'une mine, des édifices, magasins, arsenaux, vaisseaux, ou autres propriétés appartenant à l'État, sera puni de mort, et ses biens seront confisqués.

96. Quiconque, soit pour envahir des domaines, propriétés ou deniers publics, places, villes, forteresses, postes, magasins, arsenaux, ports, vaisseaux ou bâtimens appartenant à l'État, soit pour piller ou partager des propriétés publiques ou nationales, ou celles d'une généralité de citoyens, soit enfin pour faire attaque ou résistance envers la force publique agissant contre les auteurs de ces crimes, se sera mis a la tête de bandes armées, ou y aura exercé une fonction ou commandement quelconque, sera puni de mort, et ses biens seront confisqués.

Les mêmes peines seront appliquées à ceux qui auront dirigé l'association, levé ou fait lever, organisé ou fait organiser les bandes, ou leur auront, sciemment et volontairement, fourni ou procuré des armes, munitions et instrumens de crime, ou envoyé des convois de subsistances, ou qui auront de toute autre manière pratiqué des intelligences avec les directeurs ou commandants des bandes.

97. Dans le cas où l'un ou plusieurs des crimes mentionnés aux articles 86, 87 et 91 auront été exécutés ou simplement tentés par une bande, la peine de mort avec confiscation des biens sera appliquée, sans distinction de grades, à tous les individus faisant partie de la

bande et qui auront été saisis sur le lieu de la réunion séditieuse.

Sera puni des mêmes peines, quoique non saisi sur le lieu, quiconque aura dirigé la sédition, ou aura exercé dans la bande un emploi ou commandement quelconque.

98. Hors le cas où la réunion séditieuse aurait eu pour objet ou résultat l'un ou plusieurs des crimes énoncés aux articles 86, 87 et 91, les individus faisant partie des bandes, dont il est parlé ci-dessus, sans y exercer aucun commandement ni emploi, et qui auront été saisis sur les lieux, seront punis de la déportation.

99. Ceux qui, connaissant le but et le caractère desdites bandes, leur auront, sans contrainte, fourni des logements, lieux de retraite ou de réunion, seront condamnés à la peine des travaux forcés à temps.

100. Il ne sera prononcé aucune peine, pour le fait de sédition, contre ceux qui, ayant fait partie de ces bandes sans y exercer aucun commandement, et sans y remplir aucun emploi ni fonction, se seront retirés au premier avertissement des autorités civiles ou militaires, ou même depuis, lorsqu'ils n'auront été saisis que hors des lieux de la réunion séditieuse, sans opposer de résistance et sans armes.

Ils ne seront punis, dans ces cas, que des crimes particuliers qu'ils auraient personnellement commis ; et néanmoins, ils pourront être renvoyés, pour cinq ans ou au plus jusqu'à dix, sous la surveillance spéciale de la haute police.

101. Sont compris dans le mot *armes*, toutes machines, tous instruments ou ustensiles tranchants, perçants ou contondants.

Les couteaux et ciseaux de poche, les cannes simples,

ne seront réputés armes qu'autant qu'il en aura été fait usage pour tuer, blesser ou frapper.

Disposition commune aux deux paragraphes de la présente Section.

102. Seront punis comme coupables des crimes et complots mentionnés dans la présent· section, tous ceux qui, soit par discours tenus dans des lieux ou réunions publics, soit par placards affichés, soit par des écrits imprimés, auront excité directement les citoyens ou habitants à les commettre.

Néanmoins, dans le cas où lesdites provocations n'auraient été suivies d'aucun effet, leurs auteurs seront simplement punis du bannissement.

SECTION III.

De la révélation et de la non-révélation des crimes qui compromettent la sûreté intérieure ou extérieure de l'État.

103. Toutes personnes qui, ayant eu connaissance de complots formés ou de crimes projetés contre la sûreté intérieure ou extérieure de l'État, n'auront pas fait la déclaration de ces complots ou crimes, et n'auront pas révélé au Gouvernement, ou aux autorités administratives ou de police judiciaire, les circonstances qui en seront venues à leur connaissance, le tout dans les vingt-quatre heures qui auront suivi ladite connaissance, seront, lors même qu'elles seraient reconnues exemptes de toute complicité, punies, pour le seul fait de non-révélation, de la manière et selon les distinctions qui suivent.

104. S'il s'agit du crime de lèse-majesté, tout individu

qui, au cas de l'article précédent, n'aura point fait les déclarations qui y sont prescrites, sera puni de la réclusion.

105. A l'égard des autres crimes ou complots mentionnés au présent chapitre, toute personne qui, en étant instruite, n'aura pas fait les déclarations prescrites par l'article 103, sera punie d'un emprisonnement de deux à cinq ans, et d'une amende de 500 francs à 2000 francs.

106. Celui qui aura eu connaissance desdits crimes ou complots non révélés, ne sera point admis à excuse sur le fondement qu'il ne les aurait point approuvés, ou même qu'il s'y serait opposé, et aurait cherché à en dissuader leurs auteurs.

107. Néanmoins, si l'auteur du complot ou crime est époux, même divorcé, ascendant ou descendant, frère ou sœur, ou allié aux mêmes degrés, de la personne prévenue de réticence, celle-ci ne sera point sujette aux peines portées par les articles précédents; mais elle pourra être mise, par l'arrêt ou jugement, sous la surveillance spéciale de la haute police pendant un temps qui n'excédera point dix ans.

108. Seront exemptés des peines prononcées contre les auteurs de complots ou d'autres crimes attentatoires à la sûreté intérieure ou extérieure de l'État, ceux des coupables qui, avant toute exécution ou tentative de ces complots ou de ces crimes, et avant toutes poursuites commencées, auront les premiers donné aux autorités mentionnées en l'article 103, connaissance de ces complots ou crimes et de leurs auteurs ou complices, ou qui, même depuis le commencement des poursuites, auront procuré l'arrestation desdits auteurs ou complices.

Les coupables qui auront donné ces connaissances ou procuré ces arrestations, pourront néanmoins être condamnés à rester pour la vie ou à temps sous la surveillance spéciale de la haute police.

CHAPITRE II.

Des Crimes et Délits contre les Constitutions de l'Empire.

SECTION PREMIÈRE.

Crimes et Délits relatifs à l'exercice des Droits civiques.

109. Lorsque, par attroupement, voies de fait ou menaces, on aura empêché un ou plusieurs citoyens d'exercer leurs droits civiques, chacun des coupables sera puni d'un emprisonnement de six mois au moins et de deux ans au plus, et de l'interdiction du droit de voter et d'être éligible, pendant cinq ans au moins et dix ans au plus.

110. Si ce crime a été commis par suite d'un plan concerté pour être exécuté soit dans tout l'Empire, soit dans un ou plusieurs départements, soit dans un ou plusieurs arrondissements communaux, la peine sera le bannissement.

111. Tout citoyen qui, étant chargé, dans un scrutin, du dépouillement des billets contenant les suffrages des citoyens, sera surpris falsifiant ces billets ou en soustrayant de la masse, ou y en ajoutant, ou inscrivant sur les billets des votants non lettrés des noms autres que ceux qui lui auraient été déclarés, sera puni de la peine du carcan.

112. Toutes autres personnes coupables des faits énoncés dans l'article précédent seront punies d'un emprisonnement de six mois au moins et de deux ans au plus,

et de l'interdiction du droit de voter et d'être éligibles pendant cinq ans au moins et dix ans au plus.

113. Tout citoyen qui aura, dans les élections, acheté ou vendu un suffrage à un prix quelconque, sera puni d'interdiction des droits de citoyen et de toute fonction ou emploi public, pendant cinq ans au moins et dix ans au plus.

Seront, en outre, le vendeur et l'acheteur du suffrage, condamnés chacun à une amende double de la valeur des choses reçues ou promises.

SECTION II.

Attentats à la Liberté.

114. Lorsqu'un fonctionnaire public, un agent ou un préposé du Gouvernement, aura ordonné ou fait quelque acte arbitraire, et attentatoire à la liberté individuelle, soit aux droits civiques d'un ou de plusieurs citoyens, soit aux Constitutions de l'Empire, il sera condamné à la peine de la dégradation civique.

Si néanmoins il justifie qu'il a agi par ordre de ses supérieurs, pour des objets du ressort de ceux-ci, et sur lesquels il leur était dû obéissance hiérarchique, il sera exempt de la peine, laquelle sera, dans ce cas, appliquée seulement aux supérieurs qui auront donné l'ordre.

115. Si c'est un ministre qui a ordonné ou fait les actes ou l'un des actes mentionnés en l'article précédent, et si, après les invitations mentionnées dans les art. 63 et 67 du sénatus-consulte du 28 floréal an 12, il a refusé ou négligé de faire réparer ces actes dans les délais fixés par ledit sénatus-consulte, il sera puni du bannissement.

116. Si les ministres prévenus d'avoir ordonné ou

autorisé l'acte contraire aux Constitutions, prétendent
que la signature à eux imputée leur a été surprise, ils
seront tenus, en faisant cesser l'acte, de dénoncer celui
qu'ils déclareront auteur de la surprise ; sinon ils seront
poursuivis personnellement.

117. Les dommages-intérêts qui pourraient être prononcés à raison des attentats exprimés dans l'article 114,
seront demandés, soit sur la poursuite criminelle, soit
par la voie civile, et seront réglés, eu égard aux personnes, aux circonstances et au préjudice souffert, sans
qu'en aucun cas, et quel que soit l'individu lésé, lesdits
dommages-intérêts puissent être au-dessous de 25 fr.
pour chaque jour de détention illégale et arbitraire et
pour chaque individu.

118. Si l'acte contraire aux Constitutions a été fait
d'après une fausse signature du nom d'un ministre ou
d'un fonctionnaire public, les auteurs du faux et ceux
qui en auront sciemment fait usage seront punis des travaux forcés à temps, dont le *maximum* sera toujours
appliqué dans ce cas.

119. Les fonctionnaires publics chargés de la police
administrative ou judiciaire, qui auront refusé ou négligé
de déférer à une réclamation légale tendant à constater
les détentions illégales et arbitraires, soit dans les maisons destinées à la garde des détenus, soit partout ailleurs, et qui ne justifieront pas les avoir dénoncées à
l'autorité supérieure, seront punis de la dégradation
civique, et tenus des dommages-intérêts, lesquels seront
réglés comme il est dit dans l'article 117.

120. Les gardiens et concierges des maisons de dépôt,
d'arrêt, de justice ou de peine, qui auront reçu un prisonnier sans mandat ou jugement, ou sans ordre provisoire du Gouvernement ; ceux qui l'auront retenu ou au-

ront refusé de le représenter à l'officier de police ou au porteur de ses ordres, sans justifier de la défense du procureur impérial ou du juge; ceux qui auront refusé d'exhiber leurs registres à l'officier de police, seront, comme coupables de détention arbitraire, punis de six mois à deux ans d'emprisonnement, et d'une amende de seize fr. à deux cents fr.

121. Seront, comme coupables de forfaiture, punis de la dégradation civique, tout officier de police judiciaire, tous procureurs généraux ou impériaux, tous substituts, tous juges, qui auront provoqué, donné ou signé un jugement, une ordonnance ou un mandat, tendant à la poursuite personnelle ou accusation, soit d'un ministre, soit d'un membre du Sénat, du Conseil d'État ou du Corps législatif, sans les autorisations prescrites par les Constitutions; ou qui, hors les cas de flagrant délit ou de clameur publique, auront, sans les mêmes autorisations, donné ou signé l'ordre ou le mandat de saisir ou arrêter un ou plusieurs ministres, ou membres du Sénat, du Conseil d'État ou du Corps législatif.

122. Seront aussi punis de la dégradation civique les procureurs généraux ou impériaux, leurs substituts, les juges ou les officiers publics qui auront retenu ou fait retenir un individu hors des lieux déterminés par le Gouvernement ou par l'administration publique, ou qui auront traduit un citoyen devant une cour d'assises ou une cour spéciale, sans qu'il ait été préalablement mis légalement en accusation.

SECTION III.

Coalitions des Fonctionnaires.

123. Tout concert de mesures contraires aux lois, pratiqué soit par la réunion d'individus ou de corps dé-

positaires de quelque partie de l'autorité publique, soit par députation ou correspondance entre eux, sera puni d'un emprisonnement de deux mois au moins et de six mois au plus, contre chaque coupable, qui pourra de plus être condamné à l'interdiction des droits civiques, et de tout emploi public, pendant dix ans au plus.

124. Si, par l'un des moyens exprimés ci-dessus, il a été concerté des mesures contre l'exécution des lois ou contre les ordres du Gouvernement, la peine sera le bannissement.

Si ce concert a eu lieu entre les autorités civiles et les corps militaires ou leurs chefs, ceux qui en seront les auteurs ou provocateurs seront punis de la déportation; les autres coupables seront bannis.

125. Dans le cas où ce concert aurait eu pour objet ou résultat un complot attentatoire à la sûreté intérieure de l'Etat, les coupables seront punis de mort, et leurs biens seront confisqués.

126. Seront coupables de forfaiture, et punis de la dégradation civique,

Les fonctionnaires publics qui auront, par délibération, arrêté de donner des démissions dont l'objet ou l'effet serait d'empêcher ou de suspendre soit l'administration de la justice, soit l'accomplissement d'un service quelconque.

SECTION IV.

Empiétements des autorités administratives et judiciaires.

127. Seront coupables de forfaiture, et punis de la dégradation civique,

1.° Les juges, les procureurs généraux ou impériaux,

ou leurs substituts, les officiers de police, qui se seront immiscés dans l'exercice du pouvoir législatif, soit par des règlements contenant des dispositions législatives, soit en arrêtant ou en suspendant l'exécution d'une ou de plusieurs lois, soit en délibérant sur le point de savoir si les lois seront publiées ou exécutées.

2° Les juges, les procureurs généraux ou impériaux ou leurs substituts, les officiers de police judiciaire, qui auraient excédé leur pouvoir, en s'immisçant dans les matières attribuées aux autorités administratives, soit en faisant des règlements sur ces matières, soit en défendant d'exécuter les ordres émanés de l'administration, ou qui, ayant permis ou ordonné de citer des administrateurs pour raison de l'exercice de leurs fonctions, auraient persisté dans l'exécution de leurs jugements ou ordonnances, nonobstant l'annulation qui en aurait été prononcée, ou le conflit qui leur aurait été notifié.

128. Les juges qui, sur la revendication formellement faite par l'autorité administrative d'une affaire portée devant eux, auront néanmoins procédé au jugement avant la décision de l'autor.té supérieure, seront punis chacun d'une amende de 16 francs au moins. et de 150 francs au plus.

Les officiers du ministère public qui auront fait des réquisitions ou donné des conclusions pour ledit jugement, seront punis de la même peine.

129. La peine sera d'une amende de 100 francs au moins, et de 500 francs au plus contre chacun des juges qui, après une réclamation légale des parties intéressées ou de l'autorité administrative, auront, sans autorisation du Gouvernement, rendu des ordonnances ou décerné des mandats contre ses agents ou préposés prévenus de crimes ou délits commis dans l'exercice de leurs fonctions.

La même peine sera appliquée aux officiers du ministère public ou de police, qui auront requis lesdites ordonnances ou mandats.

130. Les préfets, sous-préfets, maires et autres administrateurs qui se seront immiscés dans l'exercice du pouvoir législatif, comme il est dit au n° 1er de l'art. 127, ou qui se seront ingérés de prendre des arrêtés généraux tendant à intimer des ordres ou des défenses quelconques à des cours ou tribunaux, seront punis de la dégradation civique.

131. Lorsque ces administrateurs entreprendront sur les fonctions judiciaires en s'ingérant de connoître de droits et intérêts privés du ressort des tribunaux, et qu'après la réclamation des parties ou de l'une d'elles, ils auront néanmoins décidé l'affaire avant que l'autorité supérieure ait prononcé, ils seront punis d'une amende de 16 francs au moins, et de 150 francs au plus.

CHAPITRE III.

(Décrété le 16 février 1810. Promulgué le 26 du même mois.)

Crimes et Délits contre la paix publique.

SECTION PREMIÈRE.

Du Faux.

§. 1er.

Fausse monnaie.

132. Quiconque aura contrefait ou altéré les monnaies d'or ou d'argent ayant cours légal en France, ou participé à l'émission ou exposition desdites monnaies contrefaites ou altérées, ou à leur introduction sur le territoire français, sera puni de mort, et ses biens seront confisqués.

133. Celui qui aura contrefait ou altéré des monnaies de billon ou de cuivre ayant cours légal en France, ou participé à l'émission ou exposition desdites monnaies contrefaites ou altérées, ou à leur introduction sur le territoire français, sera puni des travaux forcés à perpétuité.

134. Tout individu qui aura, en France, contrefait ou altéré des monnaies étrangères, ou participé à l'émission, exposition ou introduction en France de monnaies étrangères contrefaites ou altérées, sera puni des travaux forcés à temps.

135. La participation énoncée aux précédents articles ne s'applique point à ceux qui, ayant reçu pour bonnes des pièces de monnaie contrefaites ou altérées, les ont remises en circulation.

Toutefois celui qui aura fait usage desdites pièces après en avoir vérifié ou fait vérifier les vices, sera puni d'une amende triple au moins et sextuple au plus de la somme représentée par les pièces qu'il aura rendues à la circulation, sans que cette amende puisse, en aucun cas, être inférieure à 16 francs.

136. Ceux qui auront eu connaissance d'une fabrique ou d'un dépôt de monnaies d'or, d'argent, billon ou cuivre ayant cours légal en France, contrefaites ou altérées, et qui n'auront pas, dans les vingt-quatre heures, révélé ce qu'ils savent aux autorités administratives ou de police judiciaire, seront, pour le seul fait de non-révélation, et lors même qu'ils seraient reconnus exempts de toute complicité, punis d'un emprisonnement d'un mois à deux ans.

137. Sont néanmoins exceptés de la disposition précédente les ascendants et descendants, époux même di-

vorcés, et les frères et sœurs des coupables, ou les alliés de ceux-ci aux mêmes degrés.

138. Les personnes coupables des crimes mentionnés aux articles 132 et 133, seront exemptes de peines, si, avant la consommation de ces crimes et avant toutes poursuites, elles en ont donné connaissance et révélé les auteurs aux autorités constituées, ou si, même après les poursuites commencées, elles ont procuré l'arrestation des autres coupables.

Elles pourront néanmoins être mises pour la vie, ou à temps, sous la surveillance spéciale de la haute police.

§. II.

Contrefaction des Sceaux de l'Etat, des Billets de banque, des Effets publics, et des Poinçons, Timbres et Marques.

139. Ceux qui auront contrefait le sceau de l'Etat ou fait usage du sceau contrefait;

Ceux qui auront contrefait ou falsifié, soit des effets émis par le trésor public avec son timbre, soit des billets de banques autorisées par la loi, ou qui auront fait usage de ces effets et billets contrefaits ou falsifiés, ou qui les auront introduits dans l'enceinte du territoire français,

Seront punis de mort et leurs biens seront confisqués.

140. Ceux qui auront contrefait ou falsifié, soit un ou plusieurs timbres nationaux, soit les marteaux de l'Etat servant aux marques forestières, soit le poinçon ou les poinçons servant à marquer les matières d'or ou d'argent, ou qui auront fait usage des papiers, effets, timbres, marteaux ou poinçons falsifiés ou contrefaits, seront punis des travaux forcés à temps, dont le *maximum* sera toujours appliqué dans ce cas.

141. Sera puni de la réclusion, quiconque s'étant indûment procuré les vrais timbres, marteaux ou poinçons ayant l'une des destinations exprimées en l'article 140, en aura fait une application ou usage préjudiciable aux droits ou intérêts de l'Etat.

142. Ceux qui auront contrefait les marques destinées à être apposées au nom du Gouvernement sur les diverses espèces de denrées ou de marchandises, ou qui auront fait usage de ces fausses marques;

Ceux qui auront contrefait le sceau, timbre ou marque d'une autorité quelconque, ou d'un établissement particulier de banque ou de commerce, ou qui auront fait usage des sceaux, timbres ou marques contrefaits,

Seront punis de la réclusion.

143. Sera puni du carcan, quiconque, s'étant indûment procuré les vrais sceaux, timbres ou marques ayant l'une des destinations exprimées en l'article 142, en aura fait une application ou usage préjudiciable aux droits ou intérêts de l'Etat, d'une autorité quelconque, ou même d'un établissement particulier.

144. Les dispositions des articles 136, 137 et 138, sont applicables aux crimes mentionnés dans l'article 139.

§. III.

Des Faux en écritures publiques ou authentiques, et de commerce ou de banque.

145. Tout fonctionnaire ou officier public qui, dans l'exercice de ses fonctions, aura commis un faux,

Soit par fausses signatures,

Soit par altération des actes, écritures ou signatures,

Soit par supposition de personnes,

Soit par des écritures faites ou intercalées sur des registres ou d'autres actes publics, depuis leur confection ou clôture,

Sera puni des travaux forcés à perpétuité.

146. Sera aussi puni des travaux forcés à perpétuité, tout fonctionnaire ou officier public qui, en rédigeant des actes de son ministère, en aura frauduleusement dénaturé la substance ou les circonstances, soit en écrivant des conventions autres que celles qui auraient été tracées ou dictées par les parties, soit en constatant comme vrais des faits faux, ou comme avoués des fa.ts qui ne l'étaient pas.

147. Seront punis des travaux forcés à temps toutes autres personnes qui auront commis un faux en écriture authentique et publique, ou en écriture de commerce ou de banque.

Soit par contrefaçon ou altération d'écritures ou de signatures,

Soit par fabrication de conventions, dispositions, obligations ou décharges, ou par leur insertion après coup dans ces actes,

Soit par addition ou altération de clauses, de déclarations ou de faits que ces actes avaient pour objet de recevoir et de constater.

148. Dans tous les cas exprimés au présent paragraphe, celui qui aura fait usage des actes faux sera puni des travaux forcés à temps.

149. Sont exceptés des dispositions ci-dessus, les faux commis dans les passe-ports et feuilles de route, sur lesquels il sera particulièrement statué ci-après.

§. IV.

Du Faux en écriture privée.

150. Tout individu qui aura, de l'une des manières exprimées en l'article 147, commis un faux en écriture privée, sera puni de la réclusion.

151. Sera puni de la même peine celui qui aura fait usage de la pièce fausse.

152. Sont exceptés des dispositions ci-dessus, les faux certificats de l'espèce dont il sera ci-après parlé.

§. V.

Des Faux commis dans les Passe-ports, Feuilles de route et Certificats.

153. Quiconque fabriquera un faux passe-port ou falsifiera un passe-port originairement véritable, ou fera usage d'un passe-port fabriqué ou falsifié, sera puni d'un emprisonnement d'une année au moins, et de cinq ans au plus.

154. Quiconque prendra, dans un passe-port, un nom supposé, ou aura concouru comme témoin à faire délivrer le passe-port sous le nom supposé, sera puni d'un emprisonnement de trois mois à un an.

Les logeurs et aubergistes qui sciemment inscriront sur leurs registres, sous des noms faux ou supposés, les personnes logées chez eux, seront punis d'un emprisonnement de six jours au moins, et d'un mois au plus.

155. Les officiers publics qui délivreront un passe-port à une personne qu'ils ne connaîtront pas personnellement, sans avoir fait attester ses noms et qualités par deux citoyens à eux connus, seront punis d'un emprisonnement d'un mois à six mois.

Si l'officier public, instruit de la supposition du nom, a néanmoins délivré le passe-port sous le nom supposé, il sera puni du bannissement.

156. Quiconque fabriquera une fausse feuille de route, ou falsifiera une feuille de route originairement véritable, ou fera usage d'une feuille de route fabriquée ou falsifiée, sera puni, savoir :

D'un emprisonnement d'une année au moins, et de cinq ans au plus, si la fausse feuille de route n'a eu pour objet que de tromper la surveillance de l'autorité publique ;

Du bannissement, si le trésor public a payé au porteur de la fausse feuille des frais de route qui ne lui étaient pas dus ou qui excédaient ceux auxquels il pouvait avoir droit, le tout néanmoins au-dessous de cent francs ;

Et de la réclusion, si les sommes indûment reçues par le porteur de la feuille s'élèvent à cent francs ou au-delà.

157. Les peines portées en l'article précédent seront appliquées, selon les distinctions qui y sont posées, à toute personne qui se sera fait délivrer, par l'officier public, une feuille de route sous un nom supposé.

158. Si l'officier public était instruit de la supposition de nom lorsqu'il a délivré la feuille, il sera puni, savoir :

Dans le premier cas posé par l'article 156, du bannissement ;

Dans le second cas du même article, de la réclusion ;

Et dans le troisième cas, des travaux forcés à temps.

159. Toute personne qui, pour se rédimer elle-même ou en affranchir une autre d'un service public quel-

conque, fabriquera, sous le nom d'un médecin, chirurgien, ou autre officier de santé, un certificat de maladie ou d'infirmité, sera punie d'un emprisonnement de deux à cinq ans.

160. Tout médecin, chirurgien, ou autre officier de santé qui, pour favoriser quelqu'un, certifiera faussement des maladies ou infirmités propres à dispenser d'un service public, sera puni d'un emprisonnement de deux à cinq ans.

S'il y a été mu par dons ou promesses, il sera puni du bannissement : les corrupteurs seront, en ce cas, punis de la même peine.

161. Quiconque fabriquera, sous le nom d'un fonctionnaire ou officier public, un certificat de bonne conduite, indigence ou autres circonstances propres à appeler la bienveillance du Gouvernement ou des particuliers sur la personne y désignée, et à lui procurer places, crédit ou secours, sera puni d'un emprisonnement de six mois à deux ans.

La même peine sera appliquée, 1° à celui qui falsifiera un certificat de cette espèce, originairement véritable, pour l'approprier à une personne autre que celle à laquelle il a été primitivement délivré; 2° à tout individu qui se sera servi du certificat ainsi fabriqué ou falsifié.

162. Les faux certificats de toute autre nature, et d'où il pourrait résulter, soit lésion envers des tiers, soit préjudice envers le trésor public, seront punis, selon qu'il y aura lieu, d'après les dispositions des paragraphes 3 et 4 de la présente section.

Dispositions communes.

163. L'application des peines portées contre ceux qui ont fait usage de monnaies, billets, sceaux, timbres, marteaux, poinçons, marques et écrits faux, contrefaits, fabriqués ou falsifiés, cessera toutes les fois que le faux n'aura pas été connu de la personne qui aura fait usage de la chose fausse.

164. Dans tous les cas où la peine du faux n'est point accompagnée de la confiscation des biens, il sera prononcé contre les coupables une amende dont le *maximum* pourra être porté jusqu'au quart du bénéfice illégitime que le faux aura procuré ou était destiné à procurer aux auteurs du crime, à leurs complices ou à ceux qui ont fait usage de la pièce fausse. Le *minimum* de cette amende ne pourra être inférieur à 100 francs.

165. La marque sera infligée à tout faussaire condamné soit aux travaux forcés à temps, soit même à la réclusion.

SECTION II.

De la Forfaiture et des Crimes et Délits des Fonctionnaires publics dans l'exercice de leurs fonctions.

166. Tout crime commis par un fonctionnaire public dans ses fonctions est une forfaiture.

167. Toute forfaiture pour laquelle la loi ne prononce pas de peines plus graves est punie de la dégradation civique.

168. Les simples délits ne constituent pas les fonctionnaires en forfaiture.

§. 1er.

Des Soustractions commises par les Dépositaires publics.

169. Tout percepteur, tout commis à une perception, dépositaire ou comptable public, qui aura détourné ou soustrait des deniers publics ou privés, ou effets actifs en tenant lieu, ou des pièces, titres, actes, effets mobiliers qui étaient entre ses mains en vertu de ses fonctions, sera puni des travaux forcés à temps, si les choses détournées ou soustraites sont d'une valeur au-dessus de trois mille francs.

170. La peine des travaux forcés à temps aura lieu également, quelle que soit la valeur des deniers ou des effets détournés ou soustraits, si cette valeur égale ou excède soit le tiers de la recette ou du dépôt, s'il s'agit de deniers ou effets une fois reçus ou déposés, soit le cautionnement, s'il s'agit d'une recette ou d'un dépôt attaché à une place sujette à cautionnement, soit enfin le tiers du produit commun de la recette pendant un mois, s'il s'agit d'une recette composée de rentrées successives et non sujette à cautionnement.

171. Si les valeurs détournées ou soustraites sont au-dessous de trois mille francs, et en outre inférieures aux mesures exprimées en l'article précédent, la peine sera un emprisonnement de deux ans au moins, et de cinq ans au plus, et le condamné sera de plus déclaré à jamais incapable d'exercer aucune fonction publique.

172. Dans les cas exprimés aux trois articles précédents, il sera toujours prononcé contre le condamné une amende dont le *maximum* sera le quart des restitutions et indemnités, et le *minimum* le douzième.

173. Tout juge, administrateur, fonctionnaire ou

officier public qui aura détruit, supprimé, soustrait ou détourné les actes et titres dont il était dépositaire en cette qualité, ou qui lui auront été remis ou communiqués à raison de ses fonctions, sera puni des travaux forcés à temps.

Tous agents, préposés ou commis, soit du Gouvernement, soit des dépositaires publics, qui se seront rendus coupables des mêmes soustractions, seront soumis à la même peine.

§. II.

Des Concussions commises par des Fonctionnaires publics

174. Tous fonctionnaires, tous officiers publics, leurs commis ou préposés, tous percepteurs des droits, taxes, contributions, deniers, revenus publics ou communaux, et leurs commis ou préposés, qui se seront rendus coupables du crime de concussion, en ordonnant de percevoir ou en exigeant ou recevant ce qu'ils savaient n'être pas dû, ou excéder ce qui était dû pour droits, taxes, contributions, deniers ou revenus, ou pour salaires ou traitements, seront punis, savoir, les fonctionnaires ou les officiers publics, de la peine de la réclusion ; et leurs commis ou préposés, d'un emprisonnement de deux ans au moins et de cinq ans au plus.

Les coupables seront de plus condamnés à une amende dont le *maximum* sera le quart des restitutions et des dommages-intérêts, et le *minimum* le douzième.

§. III.

Des Délits de Fonctionnaires qui se seront ingérés dans des Affaires ou Commerces incompatibles avec leur qualité.

175. Tout fonctionnaire, tout officier public, tout agent du Gouvernement, qui, soit ouvertement, soit par actes simulés, soit par interposition de personnes, aura pris ou reçu quelque intérêt que ce soit, dans les actes, adjudications, entreprises ou régies dont il a ou avait, au temps de l'acte, en tout ou en partie, l'administration ou la surveillance, sera puni d'un emprisonnement de six mois au moins et de deux ans au plus, et sera condamné à une amende qui ne pourra excéder le quart des restitutions et des indemnités, ni être au-dessous du douzième.

Il sera de plus déclaré à jamais incapable d'exercer aucune fonction publique.

La présente disposition est applicable à tout fonctionnaire ou agent du Gouvernement qui aura pris un intérêt quelconque dans une affaire dont il était chargé d'ordonnancer le paiement ou de faire la liquidation.

176. Tout commandant des divisions militaires, des départements ou des places et villes, tout préfet ou sous-préfet qui aura, dans l'étendue des lieux où il a droit d'exercer son autorité, fait ouvertement, ou par des actes simulés, ou par interposition de personnes, le commerce des grains, grenailles, farines, substances farineuses, vins ou boissons, autres que ceux provenant de ses propriétés, sera puni d'une amende de cinq cents francs au moins, de dix mille francs au plus, et de la confiscation des denrées appartenant à ce commerce.

§. IV.

De la Corruption des Fonctionnaires publics.

177. Tout fonctionnaire public de l'ordre administratif ou judiciaire, tout agent ou préposé d'une administration publique qui aura agréé des offres ou promesses, ou reçu des dons ou présents pour faire un acte de sa fonction ou de son emploi, même juste, mais non sujet à salaire, sera puni du carcan, et condamné à une amende double de la valeur des promesses agréées ou des choses reçues, sans que ladite amende puisse être inférieure à deux cents francs.

La présente disposition est applicable à tout fonctionnaire, agent ou préposé de la qualité ci-dessus exprimée, qui, par offres ou promesses agréées, dons ou présents reçus, se sera abstenu de faire un acte qui entrait dans l'ordre de ses devoirs.

178. Dans le cas où la corruption aurait pour objet un fait criminel emportant une peine plus forte que celle du carcan, cette peine plus forte sera appliquée aux coupables.

179. Quiconque aura contraint ou tenté de contraindre par voies de fait ou menaces, corrompu ou tenté de corrompre par promesses, offres, dons ou présents, un fonctionnaire, agent ou préposé, de la qualité exprimée en l'article 177, pour obtenir, soit une opinion favorable, soit des procès verbaux, états, certificats ou estimations contraires à la vérité, soit des places, emplois, adjudications, entreprises ou autres bénéfices quelconques, soit enfin tout autre acte du ministère du fonctionnaire, agent ou préposé, sera puni des mêmes peines que le fonctionnaire, agent ou préposé corrompu.

Toutefois, si les tentatives de contrainte ou corruption n'ont eu aucun effet, les auteurs de ces tentatives seront simplement punis d'un emprisonnement de trois mois au moins, et de six mois au plus, et d'une amende de 100 à 300 francs.

180. Il ne sera jamais fait au corrupteur restitution des choses par lui livrées, ni de leur valeur : elles seront confisquées au profit des hospices des lieux où la corruption aura été commise.

181. Si c'est un juge prononçant en matière criminelle, ou un juré qui s'est laissé corrompre, soit en faveur, soit au préjudice de l'accusé, il sera puni de la réclusion, outre l'amende ordonnée par l'article 177.

182. Si, par l'effet de la corruption, il y a eu condamnation à une peine supérieure à celle de la réclusion, cette peine, quelle qu'elle soit, sera appliquée au juge ou juré coupable de corruption.

183. Tout juge ou administrateur qui se sera décidé par faveur pour une partie, ou par inimitié contre elle, sera coupable de forfaiture et puni de la dégradation civique.

§. V.

Des Abus d'autorité.

Ire CLASSE.

Des Abus d'autorité contre les Particuliers.

184. Tout juge, tout procureur général ou impérial, tout substitut, tout administrateur ou tout autre officier de justice ou de police, qui se sera introduit dans le domicile d'un citoyen hors les cas prévus par la loi, et sans les formalités qu'elle a prescrites, sera puni d'une amende de seize francs au moins, et de deux cents francs au plus.

185. Tout juge ou tribunal, tout administrateur ou autorité administrative, qui, sous quelque prétexte que ce soit, même du silence ou de l'obscurité de la loi, aura dénié de rendre la justice qu'il doit aux parties, après en avoir été requis, et qui aura persévéré dans son déni, après avertissement ou injonction de ses supérieurs, pourra être poursuivi, et sera puni d'une amende de deux cents francs au moins, et de cinq cents francs au plus, et de l'interdiction de l'exercice des fonctions publiques depuis cinq ans jusqu'à vingt.

186. Lorsqu'un fonctionnaire ou un officier public, un administrateur, un agent ou un préposé du Gouvernement ou de la police, un exécuteur des mandats de justice ou jugements, un commandant en chef ou en sous-ordre de la force publique, aura, sans motif légitime, usé ou fait user de violence envers les personnes, dans l'exercice ou à l'occasion de l'exercice de ses fonctions, il sera puni selon la nature et la gravité de ses violences, et en élevant la peine suivant la règle posée par l'art. 198 ci-après.

187. Toute suppression, toute ouverture de lettres confiées à la poste, commise ou facilitée par un fonctionnaire ou un agent du Gouvernement ou de l'administration des postes, sera punie d'une amende de seize francs à trois cents francs. Le coupable sera, de plus, interdit de toute fonction ou emploi public pendant cinq ans au moins et dix ans au plus.

IIe CLASSE.

Des Abus d'autorité contre la chose publique.

188. Tout fonctionnaire public, agent ou préposé du Gouvernement, de quelque état et grade qu'il soit, qui

aura requis ou ordonné, fait requérir ou ordonner l'action ou l'emploi de la force publique contre l'exécution d'une loi ou contre la perception d'une contribution légale, ou contre l'exécution, soit d'une ordonnance ou mandat de justice, soit de tout autre ordre émané de l'autorité légitime, sera puni de la réclusion.

189. Si cette réquisition ou cet ordre ont été suivis de leur effet, la peine sera la déportation.

190. Les peines énoncées aux art. 188 et 189 ne cesseront d'être applicable aux fonctionnaires ou préposés qui auraient agi par ordre de leurs supérieurs, qu'autant que cet ordre aura été donné par ceux-ci pour des objets de leur ressort, et sur lesquels il leur était dû obéissance hiérarchique ; dans ce cas, les peines portées ci-dessus ne seront appliquées qu'aux supérieurs qui les premiers auront donné cet ordre.

191. Si, par suite desdits ordres ou réquisitions, il survient d'autres crimes punissables de peines plus fortes que celles exprimées aux articles 188 et 189, ces peines plus fortes seront appliquées aux fonctionnaires, agents ou préposés coupables d'avoir donné lesdits ordres, ou fait lesdites réquisitions.

§. VI.

De quelques Délits relatifs à la tenue des Actes de l'état civil.

192. Les officiers de l'état civil qui auront inscrit leurs actes sur de simples feuilles volantes, seront punis d'un emprisonnement d'un mois au moins et trois mois au plus, et d'une amende de 16 francs à 200 francs.

193. Lorsque, pour la validité d'un mariage, la loi prescrit le consentement des pères, mères ou autres personnes, et que l'officier de l'état civil ne se sera point as-

suré de l'existence de ce consentement, il sera puni d'une amende de 16 fr. à 300 fr., et d'une emprisonnement de six mois au moins, et d'un an au plus.

194. L'officier de l'état civil sera aussi puni de 16 fr. à 300 fr. d'amende, lorsqu'il aura reçu, avant le terme prescrit par l'article 228 du Code Napoléon, l'acte de mariage d'une femme ayant déjà été mariée.

195. Les peines portées aux articles précédents contre les officiers de l'état civil leur seront appliquées, lors même que la nullité de leurs actes n'aurait pas été demandée, ou aurait été couverte ; le tout sans préjudice des peines plus fortes, prononcées en cas de collusion, et sans préjudice aussi des autres dispositions pénales du titre V du Livre Ier du Code Napoléon.

§. VII.

De l'Exercice de l'autorité publique illégalement anticipé ou prolongé.

196. Tout fonctionnaire public qui sera entré en exercice de ses fonctions sans avoir prêté le serment, pourra être poursuivi, et sera puni d'une amende de 16 francs à 150 francs.

197. Tout fonctionnaire public révoqué, destitué, suspendu ou interdit légalement, qui, après en avoir eu la connaissance officielle, aura continué l'exercice de ses fonctions, ou qui, étant électif ou temporaire, les aura exercées après avoir été remplacé, sera puni d'un emprisonnement de six mois au moins, et de deux ans au plus, et d'une amende de cent francs à cinq cent francs. Il sera interdit de l'exercice de toute fonction publique pour cinq ans au moins, et dix ans au plus, à compter du jour où il aura subi la peine : le tout sans préjudice des plus fortes peines

portées contre les officiers ou les commandants militaires,
par l'article 93 du présent Code.

Disposition particulière.

198. Hors les cas où la loi règle spécialement les peines encourues pour crimes ou délits commis par les fonctionnaires ou officiers publics, ceux d'entre eux qui auront participé à d'autres crimes ou délits qu'ils étaient
chargés de surveiller ou de réprimer, seront punis comme
il suit :

S'il s'agit d'un délit de police correctionnelle, ils subiront toujours le *maximum* de la peine attachée à l'espèce de délit;

Et s'il s'agit de crimes emportant peine afflictive, ils
seront condamnés, savoir :

A la réclusion, si le crime emporte contre tout autre coupable la peine du bannissement ou du carcan;

Aux travaux forcés à temps, si le crime emporte contre tout autre coupable la peine de la réclusion;

Et aux travaux forcés à perpétuité, lorsque le crime
emportera contre tout autre coupable la peine de la déportation ou celle des travaux forcés à temps.

Au-delà des cas qui viennent d'être exprimés, la
peine commune sera appliquée sans aggravation.

SECTION III.

*Des troubles apportés à l'ordre public par les Mi-
nistres des Cultes dans l'exercice de leur ministère.*

§. Ier.

*Des Contraventions propres à compromettre l'Etat
civil des Personnes.*

199. Tout ministre d'un culte qui procédera aux cérémonies religieuses d'un mariage sans qu'il lui ait été

justifié d'un acte de mariage préalablement reçu par les officiers de l'état civil, sera, pour la première fois, puni d'une amende de 16 francs à 100 francs.

200. En cas de nouvelles contraventions de l'espèce exprimée en l'article précédent, le ministre de culte qui les aura commises, sera puni, savoir :

Pour la première récidive, d'un emprisonnement de deux à cinq ans;

Et pour la seconde, de la déportation.

§. II.

Des Critiques , Censures ou Provocations dirigées contre l'Autorité publique dans un discours pastoral prononcé publiquement.

201. Les ministres des cultes qui prononceront, dans l'exercice de leur ministère, et en assemblée publique, un discours contenant la critique ou censure du Gouvernement, d'une loi, d'un décret impérial ou de tout autre acte de l'autorité publique, seront punis d'un emprisonnement de trois mois à deux ans.

202. Si le discours contient une provocation directe à la désobéissance aux lois ou autres actes de l'autorité publique, ou s'il tend à soulever ou armer une partie des citoyens contre les autres, le ministre du culte qui l'aura prononcé sera puni d'un emprisonnement de deux à cinq ans, si la provocation n'a été suivie d'aucun effet ; et du bannissement, si elle a donné lieu à désobéissance, autre toutefois que celle qui aurait dégénéré en sédition ou révolte.

203. Lorsque la provocation aura été suivie d'une sédition ou révolte dont la nature donnera lieu contre l'un ou plusieurs des coupables à une peine plus forte que

celle du bannissement, cette peine, quelle qu'elle soit, sera appliquée au ministre coupable de la provocation

§. III.

Des Critiques, Censures ou Provocations dirigées contre l'Autorité publique dans un écrit pastoral.

204. Tout écrit contenant des instructions pastorales, en quelque forme que ce soit, et dans lequel un ministre de culte se sera ingéré de critiquer ou censurer, soit le Gouvernement, soit tout acte de l'autorité publique, emportera la peine du bannissement contre le ministre qui l'aura publié.

205. Si l'écrit mentionné en l'article précédent contient une provocation directe à la désobéissance, aux lois ou autres actes de l'autorité publique, ou s'il tend à soulever ou armer une partie des citoyens contre les autres, le ministre qui l'aura publié sera puni de la déportation.

206. Lorsque la provocation contenue dans l'écrit pastoral aura été suivie d'une sédition ou révolte dont la nature donnera lieu contre l'un ou plusieurs des coupables à une peine plus forte que celle de la déportation, cette peine, quelle qu'elle soit, sera appliquée au ministre coupable de la provocation.

§. IV.

De la Correspondance des Ministres des cultes avec des cours ou puissances étrangères, sur des matières de religion.

207. Tout ministre d'un culte qui aura, sur des questions ou matières religieuses, entretenu une correspondance avec une cour ou puissance étrangère, sans en avoir préalablement informé le ministre de l'Empereur

chargé de la surveillance des cultes, et sans avoir obtenu son autorisation, sera, pour ce seul fait, puni d'une amende de 100 fr. à 500 fr., et d'un emprisonnement d'un mois à deux ans.

208. Si la correspondance mentionnée en l'article précédent a été accompagnée ou suivie d'autres faits contraires aux dispositions formelles d'une loi ou d'un décret de l'Empereur, le coupable sera puni du bannissement, à moins que la peine résultant de la nature de ces faits ne soit plus forte, auquel cas cette peine plus forte sera seule appliquée.

SECTION IV.

Résistance, Désobéissance, et autres Manquements envers l'Autorité publique.

§. 1er.

Rebellion.

209. Toute attaque, toute résistance avec violence et voies de fait envers les officiers ministériels, les gardes champêtres ou forestiers, la force publique, les préposés à la perception des taxes et des contributions, leurs porteurs de contraintes, les préposés des douanes, les séquestres, les officiers ou agents de la police administrative ou judiciaire, agissant pour l'exécution des lois, des ordres ou ordonnances de l'autorité publique, des mandats de justice ou jugements, est qualifiée, selon les circonstances, crime ou délit de rebellion.

210. Si elle a été commise par plus de vingt personnes armées, les coupables seront punis des travaux forcés à temps; et s'il n'y a pas eu port d'armes, ils seront punis de la réclusion.

211. Si la rebellion a été commise par une réunion

armée de trois persounes ou plus, jusqu'à vingt inclusivement, la peine sera la réclusion ; s'il n'y a pas eu port
d'armes, la peine sera un emprisonnement de six mois au
moins et de deux ans au plus.

212. Si la rebellion n'a été commise que par une ou
deux personnes, avec armes, elle sera punie d'un emprisonnement de six mois à deux ans ; et si elle a lieu sans
armes, d'un emprisonnement de six jours à six mois.

213. En cas de rebellion avec bande ou attroupement,
l'article 100 du présent Code sera applicable aux rebelles
sans fonctions ni emplois dans la bande, qui se seront retirés au premier avertissement de l'autorité publique, ou même depuis, s'ils n'ont été saisis que hors
du lieu de la rebellion, et sans nouvelle résistance et sans
armes.

214. Toute réunion d'individus pour un crime ou un
délit, est réputée réunion armée, lorsque plus de deux
personnes portent des armes ostensibles.

215. Les personnes qui se trouveraient munies d'armes
cachées, et qui auraient fait partie d'une troupe ou réunion non réputée armée, seront individuellement punies
comme si elles avaient fait partie d'une troupe ou réu-
niou armée.

216. Les auteurs des crimes et délits commis pendant
le cours et à l'occasion d'une rebellion seront punis des
peines prononcées contre chacun de ces crimes, si elles
sont plus fortes que celles de la rebellion.

217. Sera puni comme coupable de la rebellion quiconque y aura provoqué, soit par des discours tenus
dans des lieux ou réunions publics, soit par placards affiché, soit par écrits imprimés.

Dans le cas où la rebellion n'aurait pas eu lieu, la

provocateur sera puni d'un emprisonnement de six jours au moins, et d'un an au plus.

218. Dans tous les cas où il sera prononcé, pour fait de rebellion, une simple peine d'emprisonnement, les coupables pourront être condamnés en outre à une amende de seize francs à deux cents francs.

219. Seront punies comme réunions de rebelles, celles qui auront été formées avec ou sans armes, et accompagnées de violences ou de menaces contre l'autorité administrative, les officiers et les agents de police, ou contre la force publique,

1.° Par les ouvriers ou journaliers, dans les ateliers publics ou manufactures;

2.° Par les individus admis dans les hospices;

3.° Par les prisonniers, prévenus, accusés ou condamnés.

220. La peine appliquée pour rebellion à des prisonniers prévenus, accusés ou condamnés relativement à d'autres crimes ou délits, sera par eux subie, savoir :

Par ceux qui, à raison des crimes ou délits qui ont causé leur détention, sont ou seraient condamnés à une peine non capitale ni perpétuelle, immédiatement après l'expiration de cette peine;

Et par les autres, immédiatement après l'arrêt ou jugement en dernier ressort, qui les aura acquittés ou renvoyés absous du fait pour lequel ils étaient détenus.

221. Les chefs d'une rebellion, et ceux qui l'auront provoquée, pourront être condamnés à rester, après l'expiration de leur peine, sous la surveillance spéciale de la haute police pendant cinq ans au moins, et dix ans au plus.

§. II.

Outrages et Violences envers les Dépositaires de l'autorité et de la force publique.

222. Lorsqu'un ou plusieurs magistrats de l'ordre administratif ou judiciaire auront reçu dans l'exercice de leurs fonctions, ou à l'occasion de cet exercice, quelque outrage par paroles tendant à inculper leur honneur ou leur délicatesse, celui qui les aura ainsi outragés sera puni d'un emprisonnement d'un mois à deux ans.

Si l'outrage a eu lieu à l'audience d'une cour ou d'un tribunal, l'emprisonnement sera de deux à cinq ans.

223. L'outrage fait par gestes ou menaces à un magistrat dans l'exercice ou à l'occasion de l'exercice de ses fonctions sera puni d'un mois à six mois d'emprisonnement; et si l'outrage a eu lieu à l'audience d'une cour ou d'un tribunal, il sera puni d'un emprisonnement d'un mois à deux ans.

224. L'outrage fait par paroles, gestes ou menaces à tout officier ministériel, ou agent dépositaire de la force publique, dans l'exercice ou à l'occasion de l'exercice de ses fonctions, sera puni d'une amende de 16 fr. à 200 fr.

225. La peine sera de six jours à un mois d'emprisonnement, si l'outrage mentionné en l'article précédent a été dirigé contre un commandant de la force publique.

226. Dans le cas des articles 222, 223 et 225, l'offenseur pourra être, outre l'emprisonnement, condamné à faire réparation, soit à la première audience, soit par écrit, et le temps de l'emprisonnement prononcé contre lui ne sera compté qu'à dater du jour où la réparation aura eu lieu.

227. Dans le cas de l'article 224, l'offenseur pourra de même, outre l'amende, être condamné à faire répara-

tion à l'offensé; et s'il retarde ou refuse, il y sera contraint par corps.

228. Tout individu qui, même sans armes, et sans qu'il en soit résulté de blessures, aura frappé un magistrat dans l'exercice de ses fonctions, ou à l'occasion de cet exercice, sera puni d'un emprisonnement de deux à cinq ans.

Si cette voie de fait a eu lieu à l'audience d'une cour ou d'un tribunal, le coupable sera puni du carcan.

229. Dans l'un et l'autre des cas exprimés en l'article précédent, le coupable pourra de plus être condamné à s'éloigner, pendant cinq à dix ans, du lieu où siège le magistrat, et d'un rayon de deux myriamètres.

Cette disposition aura son exécution à dater du jour où le condamné aura subi sa peine,

Si le condamné enfreint cet ordre avant l'expiration du temps fixé, il sera puni du bannissement.

230. Les violences de l'espèce exprimée en l'article 228, dirigées contre un officier ministériel, un agent de la force publique, ou un citoyen chargé d'un ministère de service public, si elles ont eu lieu pendant qu'ils exerçaient leur ministère ou à cette occasion, seront punies d'un emprisonnement d'un mois à six mois.

231. Si les violences exercées contre les fonctionnaires et agents désignés aux articles 228 et 230 ont été la cause d'effusion de sang, blessures ou maladie, la peine sera la réclusion; si la mort s'en est suivie dans les quarante jours, le coupable sera puni de mort.

232. Dans le cas même où ces violences n'auraient pas causé d'effusion de sang, blessures ou maladie, les coups seront punis de la réclusion, s'ils ont été portés avec préméditation ou guet-apens.

233. Si les blessures sont du nombre de celles qui

portent le caractère de meurtre, le coupable sera puni de mort.

§. III.

Refus d'un Service dû légalement.

234. Tout commandant, tout officier ou sous-officier de la force publique qui, après en avoir été légalement requis par l'autorité civile, aura refusé de faire agir la force à ses ordres, sera puni d'un emprisonnement d'un mois à trois mois, sans préjudice des réparations civiles qui pourraient être dues aux termes de l'article 11 du présent Code.

235. Les lois pénales et règlements relatifs à la conscription militaire continueront de recevoir leur execution.

236. Les témoins et jurés qui auront allégué une excuse reconnue fausse, seront condamnés, outre les amendes prononcées pour la non-comparution, à un emprisonnement de six jours à deux mois.

§. IV.

Evasion de détenus   Recèlement de criminels.

237. Toutes les fois qu'une évasion de détenus aura lieu, les huissiers, les commandants en chef ou en sous-ordre, soit de la gendarmerie, soit de la force armée servant d'escorte ou garnissant les postes, les concierges, gardiens, geôliers, et tous autres préposés à la conduite, au transport ou à la garde des détenus, seront punis ainsi qu'il suit.

238. Si l'évadé était prévenu de délits de police, ou de crimes simplement infamants, ou s'il était prisonnier de guerre, les préposés à sa garde ou conduite seront pu-

nis, en cas de négligence, d'un emprisonnement de six jours à deux mois;

Et en cas de connivence, d'un emprisonnement de six mois à deux ans.

Ceux qui, n'étant pas chargés de la garde ou de la conduite du détenu, auront procuré ou facilité son évasion, seront punis de six jours à trois mois d'emprisonnement.

239. Si les détenus évadés, ou l'un d'eux, étaient prévenus ou accusés d'un crime de nature à entraîner une peine afflictive à temps, ou condamnés pour l'un de ces crimes, la peine sera, contre les préposés à la garde ou conduite, en cas de négligence, un emprisonnement de deux à six mois;

En cas de connivence, la réclusion.

Les individus non chargés de la garde des détenus, qui auront procuré ou facilité l'évasion, seront punis d'un emprisonnement de trois mois à deux ans.

240. Si les évadés ou l'un d'eux sont prévenus ou accusés de crimes de nature à entraîner le peine de mort ou des peines perpétuelles, ou s'ils sont condamnés à l'une de ces peines, leurs conducteurs ou gardiens seront punis d'un an à deux ans d'emprisonnement, en cas de négligence, et des travaux forcés à temps, en cas de connivence.

Les individus non chargés de la conduite ou de la garde, qui auront facilité ou procuré l'évasion, seront punis d'un emprisonnement d'un an au moins, et de cinq ans au plus.

241. Si l'évasion a eu lieu ou a été tentée avec violence ou bris de prison, les peines contre ceux qui l'auront favorisée en fournissant des instruments propres à l'opérer, seront, au cas que l'évadé fût de la qualité

exprimée en l'article 238, trois mois à deux ans d'emprisonnement;

Au cas de l'article 239, deux à cinq ans d'emprisonnement; et au cas de l'article 240, la réclusion.

242. Dans tous les cas ci-dessus, lorsque les tiers qui auront procuré ou facilité l'évasion, y seront parvenus en corrompant les gardiens ou geôliers, ou de connivence avec eux, ils seront punis des mêmes peines que lesdits gardiens ou geôliers.

243. Si l'évasion avec bris ou violence a été favorisée par transmission d'armes, les gardiens et conducteurs qui y auront participé seront punis des travaux forcés à perpétuité; les autres personnes, des travaux forcés à temps.

244. Tous ceux qui auront connivé à l'évasion d'un détenu, seront solidairement condamnés, à titre de dommages-intérêts, à tout ce que la partie civile du détenu aurait eu droit d'obtenir contre lui.

245. A l'égard des détenus qui se seront évadés ou qui auront tenté de s'évader par bris de prison ou par violence, ils seront, pour ce seul fait, punis de six mois à un an d'emprisonnement, et subiront cette peine immédiatement après l'expiration de celle qu'ils auront encourue pour le crime ou délit à raison duquel ils étaient détenus, ou immédiatement après l'arrêt ou jugement qui les aura acquittés ou renvoyés absous dudit crime ou délit; le tout sans préjudice des plus fortes peines qu'ils auraient pu encourir pour d'autres crimes qu'ils auraient commis dans leurs violences.

246. Quiconque sera condamné pour avoir favorisé une évasion, ou des tentatives d'évasion, à un emprisonnement de plus de six mois, pourra, en outre, être mis

sous la surveillance spéciale de la haute police, pour un intervalle de cinq à dix ans.

247. Les peines d'emprisonnement ci-dessus établies contre les conducteurs ou les gardiens en cas de négligence seulement, cesseront lorsque les évadés seront repris ou représentés, pourvu que ce soit dans les quatre mois de l'évasion, et qu'ils ne soient pas arrêtés pour d'autres crimes ou délits commis postérieurement.

248. Ceux qui auront recélé ou fait recéler des personnes qu'ils savaient avoir commis des crimes emportant peine afflictive, seront punis de trois mois d'emprisonnement au moins, et de deux ans au plus.

Sont exceptés de la présente disposition les ascendants ou descendants, époux ou épouse même divorcés, frères ou sœurs des criminels recélés, ou leurs alliés au même degré.

§. V.

Bris de Scellés et Enlevement de pièces dans les Dépôts publics.

249. Lorsque des scellés apposés soit par ordre du Gouvernement, soit par suite d'une ordonnance de justice rendue en quelque matière que ce soit, auront été brisés, les gardiens seront punis, pour simple négligence, de six jours à six mois d'emprisonnement.

250. Si le bris de scellés s'applique à des papiers et effets d'un individu prévenu ou accusé d'un crime emportant la peine de mort, des travaux forcés à perpétuité, ou de la déportation, ou qui soit condamné à l'une de ces peines, le gardien négligent sera puni de six mois à deux ans d'emprisonnement.

251. Quiconque aura, à dessein, brisé des scellés apposés sur des papiers ou effets de la qualité énoncée en

l'art. precédent, ou participé au bris des scellés, sera puni de la réclusion ; et si c'est le gardien lui-même, il sera puni des travaux forcés à temps.

252. A l'égard de **tous** autres bris de scellés, les coupables seront punis de six mois à deux ans d'emprisonnement ; et si c'est le gardien lui-même, il sera puni de deux à cinq ans de la même peine.

253. Tout vol commis à l'aide d'un bris de scellés sera puni comme vol commis à l'aide d'effraction.

254. Quant aux soustractions, destructions et enlevements de pièces ou de procédures criminelles, ou d'autres papiers, registres, actes et effets, contenus dans des archives, greffes ou dépôts publics, ou remis à un dépositaire public en cette qualité, les peines seront, contre les greffiers, archivistes, notaires ou autres dépositaires négligents, de trois mois à un an d'emprisonnement, et d'une amende de cent francs à trois cents francs.

255. Quiconque se sera rendu coupable des soustractions, enlevements ou destructions mentionnés en l'article précédent, sera puni de la réclusion.

Si le crime est l'ouvrage du dépositaire lui-même, il sera puni des travaux forcés à temps.

256. Si le bris de scellés, les soustractions, enlevements ou destruction de pièces ont été commis avec violence envers les personnes, la peine sera, contre toute personne, celle des travaux forcés à temps, sans préjudice de peines plus fortes, s'il y a lieu, d'après la nature des violences et des autres crimes qui y seraient joints.

§. VI.

Dégradations de monuments.

257. Quiconque aura détruit, abattu, mutilé ou dégradé des monuments, statues et autres objets destinés à l'utilité ou à la décoration publique, et élevés par l'autorité publique ou avec son autorisation, sera puni d'un emprisonnement d'un mois à deux ans, et d'une amende de 100 francs à 500 fr.

§. VII.

Usurpation de titres ou fonctions.

258. Quiconque, sans titre, se sera immiscé dans des fonctions publiques, civiles ou militaires, ou aura fait les actes d'une de ces fonctions, sera puni d'un emprisonnement de deux à cinq ans, sans préjudice de la peine de faux, si l'acte porte le caractère de ce crime.

259. Toute personne qui aura publiquement porté un costume, un uniforme ou une décoration qui ne lui appartenaient pas, ou qui se sera attribué des titres impériaux qui ne lui auraient pas été légalement conférés, sera punie d'un emprisonnement de six mois à deux ans.

§. VIII.

Entraves au libre exercice des Cultes.

260. Tout particulier qui, par des voies de fait ou des menaces, aura contraint ou empéché une ou plusieurs personnes d'exercer l'un des cultes autorisés, d'assister à l'exercice de ce culte, de célébrer certaines fêtes, d'observer certains jours de repos, et, en conséquence, d'ouvrir ou de fermer leurs ateliers, boutiques ou maga-

sins, et de faire ou quitter certains travaux, sera puni, pour ce seul fait, d'une amende de 16 fr. à 200 fr., et d'un emprisonnement de six jours à deux mois.

261. Ceux qui auront empêché, retardé ou interrompu les exercices d'un culte par des troubles ou désordres causés dans le temple ou autre lieu destiné ou servant actuellement à ces exercices, seront punis d'une amende de 16 fr. à 300 fr., et d'un emprisonnement de six jours à trois mois.

262. Toute personne qui aura, par paroles ou gestes, outragé les objets d'un culte dans les lieux destinés ou servant actuellement à son exercice, ou les ministres de ce culte dans leurs fonctions, sera punie d'une amende de 16 fr. à 500 fr., et d'un emprisonnement de quinze jours à six mois.

263. Quiconque aura frappé le ministre d'un culte dans ses fonctions, sera puni du carcan.

264. Les dispositions du présent paragraphe ne s'appliquent qu'aux troubles, outrages ou voies de fait dont la nature ou les circonstances ne donneront pas lieu à de plus fortes peines, d'après les autres dispositions du présent Code.

SECTION V.

Association de malfaiteurs, Vagabondage et Mendicité.

§. Ier.

Association de malfaiteurs.

265. Toute association de malfaiteurs envers les personnes ou les propriétés, est un crime contre la paix publique.

266. Ce crime existe par le seul fait d'organisation de bandes ou de correspondance entre elles et leurs chefs ou commandants, ou de convention tendant à rendre compte ou à faire distribution ou partage du produit des méfaits.

267. Quand ce crime n'aurait été accompagné ni suivi d'aucun autre, les auteurs, directeurs de l'association, et les commandants en chef ou en sous-ordre de ces bandes seront punis des travaux forcés à temps.

268. Seront punis de la réclusion tous autres individus chargés d'un service quelconque dans ces bandes, et ceux qui auront sciemment et volontairement fourni aux bandes ou à leurs divisions, des armes, munitions, instruments de crime, logement, retraite ou lieu de réunion.

§. 11.

Vagabondage.

269. Le vagabondage est un délit.

270. Les vagabonds ou gens sans aveu sont ceux qui n'ont ni domicile certain, ni moyens de subsistance, et qui n'exercent habituellement ni métier ni profession.

271. Les vagabonds ou gens sans aveu qui auront été légalement déclarés tels, seront, pour ce seul fait, punis de trois à six mois d'emprisonnement, et demeureront, après avoir subi leur peine, à la disposition du Gouvernement pendant le temps qu'il déterminera, eu égard à leur conduite.

272. Les individus déclarés vagabonds par jugement pourront, s'ils sont étrangers, être conduits, par les ordres du Gouvernement, hors du territoire de l'Empire.

273. Les vagabonds nés en France pourront, après un jugement même passé en force de chose jugée, être réclamés par délibération du conseil municipal de la commune où ils sont nés, ou cautionnés par un citoyen solvable.

Si le Gouvernement accueille la réclamation ou agrée la caution, les individus ainsi réclamés ou cautionnés seront, par ses ordres, renvoyés ou conduits dans la commune qui les a réclamés, ou dans celle qui leur sera assignée pour résidence, sur la demande de la caution.

§. III.

Mendicité.

274. Toute personne qui aura été trouvée mendiant dans un lieu pour lequel il existera un établissement public organisé afin d'obvier à la mendicité, sera punie de trois à six mois d'emprisonnement, et sera, après l'expiration de sa peine, conduite au dépôt de mendicité.

275. Dans les lieux où il n'existe point encore de tels établissements, les mendiants d'habitude valides seront punis d'un mois à trois mois d'emprisonnement.

S'ils ont été arrêtés hors du canton de leur résidence, ils seront punis d'un emprisonnement de six mois à deux ans.

276. Tous mendiants, même invalides, qui auront usé de menaces, ou seront entrés sans permission du propriétaire ou des personnes de sa maison, soit dans une habitation, soit dans un enclos en dépendant,

Ou qui feindront des plaies ou infirmités,

Ou qui mendieront en réunion, à moins que ce ne soit le mari et la femme, le père ou la mère et leurs jeunes enfants, l'aveugle et son conducteur,

Seront punis d'un emprisonnement de six mois à deux ans.

Dispositions communes aux Vagabonds et Mendiants.

277. Tout mendiant ou vagabond qui aura été saisi travesti d'une manière quelconque,

Ou porteur d'armes, bien qu'il n'en ait usé ni menacé,

Ou muni de limes, crochets ou autres instruments propres, soit à commettre des vols ou d'autres délits, soit à lui procurer les moyens de pénétrer dans les maisons,

Sera puni de deux à cinq ans d'emprisonnement.

278. Tout mendiant ou vagabond qui sera trouvé porteur d'un ou de plusieurs effets d'une valeur supérieure à cent francs, et qui ne justifiera point d'où ils lui proviennent, sera puni de la peine portée en l'article 276.

279. Tout mendiant ou vagabond qui aura exercé quelque acte de violence que ce soit envers les personnes, sera puni de la réclusion, sans préjudice de peines plus fortes, s'il y a lieu, à raison du genre et des circonstances de la violence.

280. Tout vagabond ou mendiant qui aura commis un crime emportant la peine des travaux forcés à temps, sera en outre marqué.

281. Les peines établies par le présent Code contre les individus porteurs de faux certificats, faux passe-ports ou fausses feuilles de route, seront toujours, dans leur espèce, portées au *maximum* quand elles seront appliquées à des vagabonds ou mendiants.

282. Les vagabonds ou mendiants qui auront subi les peines portées par les articles précédents, demeure-

ront, à la fin de cos peines, à la disposition du Gouvernement.

SECTION VI.

Délits commis par la voie d'Ecrits, Images ou Gravures distribués sans nom d'Auteur, Imprimeur ou Graveur.

283. Toute publication ou distribution d'ouvrages, écrits, avis, bulletins, affiches, journaux, feuilles périodiques ou autres imprimés, dans lesquels ne se trouvera pas l'indication vraie des noms, profession et demeure de l'auteur ou de l'imprimeur, sera, pour ce seul fait, punie d'un emprisonnement de six jours à six mois, contre toute personne qui aura sciemment contribué à la publication ou distribution.

284. Cette disposition sera réduite à des peines de simple police,

1° A l'égard des crieurs, afficheurs, vendeurs ou distributeurs qui auront fait connaître la personne de laquelle ils tiennent l'écrit imprimé ;

2° A l'égard de quiconque aura fait connaître l'imprimeur ;

3° A l'égard même de l'imprimeur qui aura fait connaître l'auteur.

285. Si l'écrit imprimé contient quelques provocations à des crimes ou délits, les crieurs, afficheurs, vendeurs et distributeurs seront punis comme complices des provocateurs, à moins qu'ils n'aient fait connaître ceux dont ils tiennent l'écrit contenant la provocation.

En cas de révélation, ils n'encourront qu'un emprisonnement de six jours à trois mois, et la peine de complicité ne restera applicable qu'à ceux qui n'auront point fait connaître les personnes dont ils auront reçu l'écrit imprimé, et à l'imprimeur, s'il est connu.

286. Dans tous les cas ci-dessus, il y aura confiscation des exemplaires saisis.

287. Toute exposition ou distribution de chansons, pamphlets, figures ou images contraires aux bonnes mœurs, sera punie d'une amende de seize francs à cinq cents francs, d'un emprisonnement d'un mois à un an, et de la confiscation des planches et des exemplaires imprimés ou gravés de chansons, figures ou autres objets du délit.

288. La peine d'emprisonnement et l'amende prononcée par l'article précédent seront réduites à des peines de simple police,

1° A l'égard des crieurs, vendeurs ou distributeurs qui auront fait connaître la personne qui leur a remis l'objet du délit;

2° A l'égard de quiconque aura fait connaître l'imprimeur ou le graveur;

3° A l'égard même de l'imprimeur ou du graveur qui auront fait connaître l'auteur ou la personne qui les aura chargés de l'impression ou de la gravure.

289. Dans tous les cas exprimés en la présente section, et où l'auteur sera connu, il subira le *maximum* de la peine attachée à l'espèce du délit.

Disposition particulière.

290. Tout individu qui, sans y avoir été autorisé par la police, fera le métier de crieur ou afficheur d'écrits imprimés, dessins ou gravures, même munis des noms d'auteurs, imprimeurs, dessinateurs ou graveurs, sera puni d'un emprisonnement de six jours à deux mois.

SECTION VII.

Des Associations ou Réunions illicites.

291. Nulle association de plus de vingt personnes, dont le but sera de se réunir tous les jours ou à certains jours marqués pour s'occuper d'objets religieux, littéraires, politiques ou autres, ne pourra se former qu'avec l'agrément du Gouvernement, et sous les conditions qu'il plaira à l'autorité publique d'imposer à la société.

Dans le nombre de personnes indiqué par le présent article ne sont pas comprises celles domiciliées dans la maison où l'association se réunit.

292. Toute association de la nature ci-dessus exprimée qui se sera formée sans autorisation, ou qui, après l'avoir obtenue, aura enfreint les conditions à elle imposées, sera dissoute.

Les chefs, directeurs ou administrateurs de l'association seront en outre punis d'une amende de seize francs à deux cents francs.

293. Si, par discours, exhortations, invocations ou prières, en quelque langue que ce soit, ou par lecture, affiche, publication ou distribution d'écrits quelconques, il a été fait, dans ces assemblées, quelques provocations à des crimes ou à des délits, la peine sera de cent francs à trois cents francs d'amende, et de trois mois à deux ans d'emprisonnement, contre les chefs, directeurs et administrateurs de ces associations, sans préjudice des peines plus fortes qui seraient portées par la loi contre les individus personnellement coupables de la provocation, lesquels, en aucun cas, ne pourront être punis d'une peine moindre que celle infligée aux chefs, directeurs et administrateurs de l'association.

294. Tout individu qui, sans la permission de l'autorité municipale, aura accordé ou consenti l'usage de sa maison ou de son appartement, en tout ou en partie, pour la réunion des membres d'une association même autorisée, ou pour l'exercice d'un culte, sera puni d'une amende de seize francs à deux cents francs.

(Décrété le 17 février 1810. Promulgué le 27 du même mois.)

TITRE II.

CRIMES ET DÉLITS CONTRE DES PARTICULIERS.

CHAPITRE PREMIER.

Crimes et Délits contre les Personnes.

SECTION PREMIÈRE.

Meurtre et autres Crimes capitaux, Menaces d'attentats contre les Personnes.

§. 1er.

Meurtre, Assassinat, Parricide, Infanticide, Empoisonnement.

295. L'HOMICIDE commis volontairement est qualifié meurtre.

296. Tout meurtre commis avec préméditation ou de guet-apens, est qualifié assassinat.

297. La préméditation consiste dans le dessein formé, avant l'action, d'attenter à la personne d'un individu déterminé, ou même de celui qui sera trouvé ou rencontré, quand même ce dessein serait dépendant de quelque circonstance ou de quelque condition.

298. Le guet-apens consiste à attendre plus ou moins de temps, dans un ou divers lieux, un individu, soit pour

lui donner la mort, soit pour exercer sur lui des actes de violence.

299. Est qualifié parricide le meurtre des pères ou mères légitimes, naturels ou adoptifs, ou de tout autre ascendant légitime.

300. Est qualifié infanticide le meurtre d'un enfant nouveau-né.

301. Est qualifié empoisonnement tout attentat à la vie d'une personne, par l'effet de substances qui peuvent donner la mort plus ou moins promptement, de quelque manière que ces substances aient été employées ou administrées, et quelles qu'en aient été les suites.

302. Tout coupable d'assassinat, de parricide, d'infanticide et d'empoisonnement, sera puni de mort, sans préjudice de la disposition particulière contenue en l'article 13, relativement au parricide.

303. Seront punis comme coupables d'assassinat, tous malfaiteurs, quelle que soit leur dénomination, qui, pour l'exécution de leurs crimes, emploient des tortures ou commettent des actes de barbarie.

304. Le meurtre emportera la peine de mort, lorsqu'il aura précédé, accompagné ou suivi un autre crime ou délit.

En tout autre cas, le coupable de meurtre sera puni de la peine des travaux forcés à perpétuité.

§. II.

Menaces.

305. Quiconque aura menacé, par écrit anonyme ou signé, d'assassinat, d'empoisonnement, ou de tout autre attentat contre les personnes qui serait punissable de la peine de mort, des travaux forcés à perpétuité, ou de la

déportation, sera puni de la peine des travaux forcés à temps, dans le cas où la menace aurait été faite avec ordre de déposer une somme d'argent dans un lieu indiqué, ou de remplir toute autre condition.

306. Si cette menace n'a été accompagnée d'aucun ordre ou condition, la peine sera d'un emprisonnement de deux ans au moins, et de cinq ans au plus, et d'une amende de cent francs à six cents francs.

307. Si la menace faite avec ordre ou sous condition a été verbale, le coupable sera puni d'un emprisonnement de six mois à deux ans, et d'une amende de vingt-cinq francs à trois cents francs.

308. Dans les cas prévus par les deux précédents articles, le coupable pourra de plus être mis, par l'arrêt ou le jugement, sous la surveillance de la haute police pour cinq ans au moins et dix ans au plus.

SECTION II.

Blessures et Coups volontaires non qualifiés Meurtre, et autres Crimes et Délits volontaires.

309. Sera puni de la peine de la réclusion, tout individu qui aura fait des blessures ou porté des coups, s'il est résulté de ces actes de violence une maladie ou incapacité de travail personnel pendant plus de vingt jours.

310. Si le crime mentionné au précédent article a été commis avec préméditation ou guet-apens, la peine sera celle des travaux forcés à temps.

311. Lorsque les blessures ou les coups n'auront occasionné aucune maladie ni incapacité de travail personnel de l'espèce mentionnée en l'article 309, le coupable sera puni d'un emprisonnement d'un mois à deux ans, et d'une amende de seize francs à deux cents francs.

S'il y a eu préméditation ou guet-apens, l'emprisonnement sera de deux ans à cinq ans, et l'amende de cinquante francs à cinq cents francs.

312. Dans les cas prévus par les articles 309, 310 et 311, si le coupable a commis le crime envers ses père ou mère légitimes, naturels ou adoptifs, ou autres ascendants légitimes, il sera puni ainsi qu'il suit :

Si l'article auquel le cas se référera prononce l'emprinement et l'amende, le coupable subira la peine de la réclusion ;

Si l'article prononce la peine de la réclusion, il subira celle des travaux forcés à temps ;

Si l'article prononce la peine des travaux forcés à temps, il subira celle des travaux forcés à perpétuité.

313. Les crimes et les délits prévus dans la présente section et dans la section précédente, s'ils sont commis en réunion séditieuse, avec rebellion ou pillage, sont imputables aux chefs, auteurs, instigateurs et provocateurs de ces réunions, rebellions ou pillages, qui seront punis comme coupables de ces crimes ou de ces délits, et condamnés aux mêmes peines que ceux qui les auront personnellement commis.

314. Tout individu qui aura fabriqué ou débité des stilets, tromblons ou quelque espèce que ce soit d'armes prohibées par loi ou par des règlemens d'administration publique, sera puni d'un emprisonnement de six jours à six mois.

Celui qui sera porteur desdites armes sera puni d'une amende de seize francs à deux cents francs.

Dans l'un et l'autre cas, les armes seront confisquées.

Le tout sans préjudice de plus forte peine, s'il y échet, en cas de complicité de crime.

315. Outre les peines correctionnelles mentionnées dans les articles précédents, les tribunaux pourront pro-

noncer le renvoi sous la **surveillance** de la haute police depuis deux ans jusqu'à dix ans.

316. Toute personne coupable du crime de castration subira la peine des travaux forcés à perpétuité.

Si la mort en est résultée avant l'expiration des quarante jours qui auront suivi le crime, le coupable subira la peine de mort.

317. Quiconque, par aliments, breuvages, médicaments, violences, ou par tout autre moyen, aura procuré l'avortement d'une femme enceinte, soit qu'elle y ait consenti ou non, sera puni de la réclusion.

La même peine sera prononcée contre la femme qui se sera procuré l'avortement à elle-même, ou qui aura consenti à faire usage des moyens à elle indiqués ou administrés à cet effet, si l'avortement s'en est ensuivi.

Les médecins, chirurgiens et autres officiers de santé, ainsi que les pharmaciens qui auront indiqué ou administré ces moyens, seront condamnés à la peine des travaux forcés à temps, dans le cas où l'avortement aurait eu lieu.

318. Quiconque aura vendu ou débité des boissons falsifiées, contenant des mixtions nuisibles à la santé, sera puni d'un emprisonnement de six jours à deux ans, et d'une amende de seize francs à cinq cents francs.

Seront saisies et confisquées les boissons falsifiées trouvées appartenir au vendeur ou débitant.

SECTION III.

Homicide, Blessures et Coups involontaires ; Crimes et Délits excusables, et Cas ou ils ne peuvent être excusés; Homicide Blessures et Coups qui ne sont ni crimes ni délits.

§. Ier.

Homicide, Blessures et Coups involontaires.

319. Quiconque, par maladresse, imprudence, inattention, négligence ou inobservation des règlements, aura commis involontairement un homicide, ou en aura involontairement été la cause, sera puni d'un emprisonnement de trois mois à deux ans, et d'une amende de cinquante francs à six cents francs.

320. S'il n'est résulté du défaut d'adresse ou de précaution que des blessures ou coups, l'emprisonnement sera de six jours à deux mois, et l'amende sera de seize francs à cent francs.

§. II.

Crimes et Délits excusables, et Cas où ils ne peuvent être excusés.

321. Le meurtre, ainsi que les blessures et les coups, sont excusables s'ils ont été provoqués par des coups ou violences graves envers les personnes.

322. Les crimes et délits mentionnés au précédent article sont également excusables, s'ils ont été commis en repoussant pendant le jour l'escalade ou l'effraction des clôtures, murs ou entrée d'une maison ou d'un appartement habité ou de leurs dépendances.

Si le fait est arrivé pendant la nuit, ce cas est réglé par l'article 329.

3a3. Le parricide n'est jamais excusable.

3a4. Le meurtre commis par l'époux sur l'épouse, ou par celle-ci sur son époux, n'est pas excusable, si la vie de l'époux ou de l'épouse qui a commis le meurtre n'a pas été mise en péril dans le moment même où le meurtre a eu lieu.

Néanmoins, dans le cas d'adultère, prévu par l'article 336, le meurtre commis par l'époux sur son épouse, ainsi que sur le complice, à l'instant où il les surprend en flagrant délit dans la maison conjugale, est excusable.

3a5. Le crime de castration, s'il a été immédiatement provoqué par un outrage violent à la pudeur, sera considéré comme meurtre ou blessures excusables.

3a6. Lorsque le fait d'excuse sera prouvé,

S'il s'agit d'un crime emportant la peine de mort, ou celle des travaux forcés à perpétuité, ou celle de la déportation, la peine sera réduite à un emprisonnement d'un an à cinq ans;

S'il s'agit de tout autre crime, elle sera réduite à un emprisonnement de six mois à deux ans;

Dans ces deux premiers cas, les coupables pourront de plus être mis, par l'arrêt ou jugement, sous la surveillance de la haute police pendant cinq ans au moins, et dix ans au plus.

S'il s'agit d'un délit, la peine sera réduite à un emprisonnement de six jours à six mois.

§. III.

Homicide, Blessures et Coups non qualifiés crimes ni délits.

327. Il n'y a ni crime ni délit, lorsque l'homicide, les blessures et les coups étaient ordonnés par la loi et commandés par l'autorité légitime.

328. Il n'y a ni crime ni délit, lorsque l'homicide, les blessures et les coups étaient commandés par la nécessité actuelle de la légitime défense de soi-même ou d'autrui.

329. Sont compris dans les cas de nécessité actuelle de défense, les deux cas suivants :

1° Si l'homicide a été commis, si les blessures ont été faites, ou si les coups ont été portés en repoussant, pendant la nuit, l'escalade ou l'effraction des clôtures, murs ou entrée d'une maison ou d'un appartement habité ou de leurs dépendances ;

2° Si le fait a eu lieu en se défendant contre les auteurs de vols ou de pillages exécutés avec violence.

SECTION IV.

Attentats aux Mœurs.

330. Toute personne qui aura commis un outrage public à la pudeur, sera punie d'un emprisonnement de trois mois à un an, et d'une amende de seize francs à deux cents francs.

331. Quiconque aura commis le crime de viol, ou sera coupable de tout autre attentat à la pudeur, consommé ou tenté avec violence contre des individus de l'un ou de l'autre sexe, sera puni de la réclusion.

332. Si le crime a été commis sur la personne d'un

enfant au-dessous de l'âge de quinze ans accomplis, le coupable subira la peine des travaux forcés à temps.

333. La peine sera celle des travaux forcés à perpétuité, si les coupables sont de la classe de ceux qui ont autorité sur la personne envers laquelle ils ont commis l'attentat, s'ils sont ses instituteurs ou ses serviteurs à gages, ou s'ils sont fonctionnaires publics, ou ministres d'un culte, ou si le coupable, quel qu'il soit, a été aidé dans son crime par une ou plusieurs personnes.

334. Quiconque aura attenté aux mœurs, en excitant, favorisant ou facilitant habituellement la débauche ou la corruption de la jeunesse de l'un ou de l'autre sexe au-dessous de l'âge de vingt-un ans, sera puni d'un emprisonnement de six mois à deux ans, et d'une amende de cinquante francs à cinq cents francs.

Si la prostitution ou la corruption a été excitée, favorisée ou facilitée par leurs pères, mères, tuteurs ou autres personnes chargées de leur surveillance, la peine sera de deux ans à cinq ans d'emprisonnement, et de trois cents francs à mille francs d'amende.

335. Les coupables du délit mentionné au précédent article seront interdits de toute tutelle et curatelle, et de toute participation aux conseils de famille; savoir, les individus auxquels s'applique le premier paragraphe de cet article, pendant deux ans au moins et cinq ans au plus; et ceux dont il est parlé au second paragraphe, pendant dix ans au moins et vingt ans au plus.

Si le délit a été commis par le père ou la mère, le coupable sera de plus privé des droits et avantages à lui accordés sur la personne et les biens de l'enfant par le Code Napoléon, Livre Ier, titre IX, *De la Puissance paternelle*.

Dans tous les cas, les coupables pourront de plus être

mis, par l'arrêt ou le jugement, sous la surveillance de la haute police, en observant, pour la durée de la surveillance, ce qui vient d'être établi pour la durée de l'interdiction mentionnée au présent article.

336. L'adultère de la femme ne pourra être dénoncé que par le mari : cette faculté même cessera, s'il est dans le cas prévu par l'article 339.

337. La femme convaincue d'adultère subira la peine de l'emprisonnement pendant trois mois au moins et deux ans au plus.

Le mari restera le maître d'arrêter l'effet de cette condamnation, en consentant à reprendre sa femme.

338. Le complice de la femme adultère sera puni de l'emprisonnement pendant le même espace de temps, et, en outre, d'une amende de cent francs à deux mille francs.

Les seules preuves qui pourront être admises contre le prévenu de complicité, seront, outre le flagrant délit, celles résultant de lettres ou autres pièces écrites par le prévenu.

339. Le mari qui aura entretenu une concubine dans la maison conjugale, et qui aura été convaincu sur la plainte de la femme, sera puni d'une amende de cent francs à deux mille francs.

340. Quiconque étant engagé dans les liens du mariage en aura contracté un autre avant la dissolution du précédent, sera puni de la peine des travaux forcés à temps.

L'officier public qui aura prêté son ministère à ce mariage, connaissant l'existence du précédent, sera condamné à la même peine.

SECTION V.

Arrestations illégales et Séquestrations de personnes.

341. Seront punis de la peine des travaux forcés à temps ceux qui, sans ordre des autorités constituées et hors les cas où la loi ordonne de saisir des prévenus, auront arrêté, détenu ou séquestré des personnes quelconques;

Quiconque aura prêté un lieu pour exécuter la détention, ou séquestration, subira la même peine.

342. Si la détention ou séquestation a duré plus d'un mois, la peine sera celle des travaux forcés à perpétuité.

343. La peine sera réduite à l'emprisonnement de deux ans à cinq ans, si les coupables des délits mentionnés en l'article 341, non encore poursuivis de fait, ont rendu la liberté à la personne arrêtée, séquestrée ou détenue, avant le dixième jour accompli depuis celui de l'arrestation, détention ou séquestration. Ils pourront néanmoins être renvoyés sous la surveillance de la haute police depuis cinq ans jusqu'à dix ans.

344. Dans chacun des trois cas suivants,

1º Si l'arrestation a été exécutée avec le faux costume, sous un faux nom, ou sur un faux ordre de l'autorité publique;

2º Si l'individu arrêté, détenu ou séquestré, a été menacé de la mort;

3º S'il a été soumis à des tortures corporelles;

Les coupables seront punis de mort.

SECTION VI.

Crimes et Délits tendant à empêcher ou détruire la preuve de l'état civil d'un Enfant, ou à compromettre son existence. — Enlèvement de Mineurs. — Infraction aux lois sur les Inhumations.

§. Ier.

Crimes et Délits envers l'Enfant.

345. Les coupables d'enlèvement, de recélé ou de suppression d'un enfant, de substitution d'un enfant à un autre, ou de supposition d'un enfant à une femme qui ne sera pas accouchée, seront punis de la réclusion.

La même peine aura lieu contre ceux qui, étant chargés d'un enfant, ne le représenteront point aux personnes qui ont droit de le réclamer.

346. Toute personne qui, ayant assisté à un accouchement, n'aura pas fait la déclaration à elle prescrite par l'article 56 du Code Napoléon, et dans le délai fixé par l'article 55 du même Code, sera punie d'un emprisonnement de six jours à six mois, et d'une amende de seize francs à trois cents francs.

347. Toute personne qui, ayant trouvé un enfant nouveau-né, ne l'aura pas remis à l'officier de l'état civil, ainsi qu'il est prescrit par l'article 58 du Code Napoléon, sera punie des peines portées au précédent article.

La présente disposition n'est point applicable à celui qui aurait consenti à se charger de l'enfant, et qui aurait fait sa déclaration à cet égard devant la municipalité du lieu où l'enfant a été trouvé.

348. Ceux qui auront porté à un hospice un enfant au-dessous de l'âge de sept ans accomplis, qui leur aurait

été confié afin qu'ils en prissent soin ou pour toute autre cause, seront punis d'un emprisonnement de six semaines à six mois, et d'une amende de seize francs à cinquante francs.

Toutefois aucune peine ne sera prononcée, s'ils n'étaient pas tenus ou ne s'étaient pas obligés de pourvoir gratuitement à la nourriture et à l'entretien de l'enfant, et si personne n'y avait pourvu.

349. Ceux qui auront exposé et délaissé en un lieu solitaire un enfant au-dessous de l'âge de sept ans accomplis, ceux qui auront donné l'ordre de l'exposer ainsi, si cet ordre a été exécuté, seront, pour ce seul fait, condamnés à un emprisonnement de six mois à deux ans, et à une amende de seize francs à deux cents francs.

350. La peine portée au précédent article sera de deux ans à cinq ans, et l'amende de cinquante fr. à quatre cents francs, contre les tuteurs ou tutrices, instituteurs ou institutrices de l'enfant exposé et délaissé par eux ou par leur ordre.

351. Si, par suite de l'exposition et du délaissement prévus par les articles 349 et 350, l'enfant est demeuré mutilé ou estropié, l'action sera considérée comme blessures volontaires à lui faites par la personne qui l'a exposé et délaissé ; et si la mort s'en est ensuivie, l'action sera considérée comme meurtre : au premier cas, les coupables subiront la peine applicable aux blessures volontaires ; et, au second cas, celle du meurtre.

352. Ceux qui auront exposé et délaissé en un lieu non solitaire un enfant au-dessous de l'âge de sept ans accomplis, seront punis d'un emprisonnement de trois mois à un an, et d'une amende de seize francs à cent francs.

353. Le délit prévu par le précédent article sera puni

d'un emprisonnement de six mois à deux ans, et d'une amende de vingt-cinq francs à deux cents francs, s'il a été commis par les tuteurs ou tutrices, instituteurs ou institutrices de l'enfant.

§. II.

Enlevement de Mineurs.

354. Quiconque aura, par fraude ou violence, enlevé ou fait enlever des mineurs, ou les aura entraînés, détournés ou déplacés, ou les aura fait entraîner, détourner ou déplacer des lieux où ils étaient mis par ceux à l'autorité ou à la direction desquels ils étaient soumis ou confiés, subira la peine de la réclusion.

355. Si la personne ainsi enlevée ou détournée est une fille au-dessous de seize ans accomplis, la peine sera celle des travaux forcés à temps.

356. Quand la fille au-dessous de seize ans aurait consenti à son enlevement ou suivi volontairement le ravisseur, si celui-ci était majeur de vingt-un ans ou au-dessus, il sera condamné aux travaux forcés à temps.

Si le ravisseur n'avait pas encore vingt-un ans, il sera puni d'un emprisonnement de deux à cinq ans.

357. Dans le cas où le ravisseur aurait épousé la fille qu'il a enlevée, il ne pourra être poursuivi que sur la plainte des personnes qui, d'après le Code Napoléon, ont le droit de demander la nullité du mariage, ni condamné, qu'après que la nullité du mariage aura été prononcée.

§. III.

Infractions aux lois sur les Inhumations.

358. Ceux qui, sans l'autorisation préalable de l'officier public, dans le cas où elle est prescrite, auront fait

inhumer un individu décédé, seront punis de six jours à deux mois d'emprisonnement, et d'une amende de seize fr. à cinquante fr.; sans préjudice de la poursuite des crimes dont les auteurs de ce délit pourraient être prévenus dans cette circonstance.

La même peine aura lieu contre ceux qui auront contrevenu, de quelque manière que ce soit, à la loi et aux règlements relatifs aux inhumations précipitées.

359. Quiconque aura recélé ou caché le cadavre d'une personne homicidée ou morte des suites de coups ou blessures, sera puni d'un emprisonnement de six mois à deux ans, et d'une amende de cinquante francs à quatre cents francs; sans préjudice de peines plus graves, s'il a participé au crime.

360. Sera puni d'un emprisonnement de trois mois à un an, et de seize francs à deux cents francs d'amende, quiconque se sera rendu coupable de violation de tombeaux ou de sépultures, sans préjudice des peines contre les crimes ou les délits qui seraient joints à celui-ci.

SECTION VII.

Faux Témoignage, Calomnie, Injures, Révélation de secrets.

§. 1er.

Faux Témoignage.

361. Quiconque sera coupable de faux témoignage en matière criminelle, soit contre l'accusé, soit en sa faveur, sera puni de la peine des travaux forcés à temps.

Si néanmoins l'accusé a été condamné à une peine plus forte que celle des travaux forcés à temps, le faux témoin qui a déposé contre lui subira la même peine.

362. Quiconque sera coupable de faux témoignage en matière correctionnelle ou de police, soit contre le prévenu, soit en sa faveur, sera puni de la réclusion.

363. Le coupable de faux témoignage en matière civile sera puni de la peine portée au précédent article.

364. Le faux témoin en matière correctionnelle, de police ou civile, qui aura reçu de l'argent, une récompense quelconque ou des promesses, sera puni des travaux forcés à temps.

Dans tous les cas, ce que le faux témoin aura reçu sera confisqué.

365. Le coupable de subornation de témoins sera condamné à la peine des travaux forcés à temps, si le faux témoignage qui en a été l'objet emporte la peine de la réclusion; aux travaux forcés à perpétuité, lorsque le faux témoignage emportera la peine des travaux forcés à temps, ou celle de la déportation; et à la peine de mort, lorsqu'il emportera celle des travaux forcés à perpétuité ou la peine capitale.

366. Celui à qui le serment aura été déféré ou référé en matière civile, et qui aura fait un faux serment, sera puni de la dégradation civique.

§. II.

Calomnies, Injures, Révélation de secrets.

367. Sera coupable du délit de calomnie, celui qui, soit dans des lieux ou réunions publiques, soit dans un acte authentique et public, soit dans un écrit imprimé ou non qui aura été affiché, vendu ou distribué, aura imputé à un individu quelconque des faits qui, s'ils existaient, exposeraient celui contre lequel ils sont articulés à des poursuites criminelles ou correctionnelles, ou

même l'exposeraient seulement au mépris ou à la haine des citoyens..

La présente disposition n'est point applicable aux faits dont la loi autorise la publicité, ni à ceux que l'auteur de l'imputation était, par la nature de ses fonctions ou de ses devoirs, obligé de révéler ou de réprimer.

368. Est réputée fausse, toute imputation à l'appui de laquelle la preuve légale n'est point rapportée. En conséquence, l'auteur de l'imputation ne sera pas admis, pour sa défense, à demander que la preuve en soit faite : il ne pourra pas non plus alléguer comme moyen d'excuse que les pièces ou les faits sont notoires, ou que les imputations qui donnent lieu à la poursuite sont copiées ou extraites de papiers étrangers, ou d'autres écrits imprimés.

369. Les calomnies mises au jour par la voie de papiers étrangers pourront être poursuivies contre ceux qui auront envoyé les articles ou donné l'ordre de les insérer, ou contribué à l'introduction ou à la distribution de ces papiers en France.

370. Lorsque le fait imputé sera légalement prouvé vrai, l'auteur de l'imputation sera à l'abri de toute peine.

Ne sera considérée comme preuve légale que celle qui résultera d'un jugement, ou de tout autre acte authentique.

371. Lorsque la preuve légale ne sera pas rapportée, le calomniateur sera puni des peines suivantes :

Si le fait imputé est de nature à mériter la peine de mort, les travaux forcés à perpétuité ou la déportation, le coupable sera puni d'un emprisonnement de deux à cinq ans, et d'une amende de deux cents francs à cinq mille francs.

Dans tous les autres cas, l'emprisonnement sera d'un mois à six mois, et à l'amende de cinquante fr. à deux mille francs.

372. Lorsque les faits imputés seront punissables suivant la loi, et que l'auteur de l'imputation les aura dénoncés, il sera, durant l'instruction sur ces faits, sursis à la poursuite et au jugement du délit de calomnie.

373. Quiconque aura fait par écrit une dénonciation calomnieuse contre un ou plusieurs individus, aux officiers de justice ou de police administrative ou judiciaire, sera puni d'un emprisonnement d'un mois à un an, et d'une amende de cent francs à trois mille francs.

374. Dans tous les cas, le calomniateur sera, à compter du jour où il aura subi sa peine, interdit pendant cinq ans au moins, et dix ans au plus, des droits mentionnés en l'article 42 du présent Code.

375. Quant aux injures ou aux expressions outrageantes qui ne renfermeraient l'imputation d'aucun fait précis, mais celle d'un vice déterminé, si elles ont été proférées dans des lieux ou réunions publiques, ou insérées dans des écrits imprimés ou non, qui auraient été répandus et distribués, la peine sera une amende de seize francs à cinq cents francs.

376. Toutes autres injures ou expressions outrageantes qui n'auront pas eu ce double caractère de gravité et de publicité, ne donneront lieu qu'à des peines de simple police.

377. A l'égard des imputations et dès injures qui seraient contenues dans les écrits relatifs à la défense des parties, ou dans les plaidoyers, les juges saisis de la contestation pourront, en jugeant la cause, ou prononcer la suppression des injures ou des écrits injurieux, ou faire des injonctions aux auteurs du délit, ou les suspendre

de leurs fonctions , et statuer sur les dommages-intérêts.

La durée de cette suspension ne pourra excéder six
mois ; en cas de récidive , elle sera d'un an au moins, et
de cinq ans au plus.

Si les injures ou écrits injurieux portent le caractère
de calomnie grave, et que les juges saisis de la contestation
ne puissent connaitre du délit, ils ne pourront prononcer
contre les prévenus qu'une suspension provisoire de leurs
fonctions, et les renverront, pour le jugement du délit,
devant les juges compétents.

378. Les médecins, chirurgiens et autres officiers de
santé, ainsi que les pharmaciens, les sages-femmes, et
toutes autres personnes dépositaires, par état ou profession, des secrets qu'on leur confie, qui, hors le cas où la
loi les oblige à se porter dénonciateurs, auront révélé ces
secrets, seront punis d'un emprisonnement d'un mois
à six mois, et d'une amende de 100 francs à 500 francs.

CHAPITRE II.

(Décrété le 19 février 1810. Promulgué le 1er mars 1810.)

Crimes et Délits contre les Propriétés.

SECTION PREMIÈRE.

Vols.

379. Quiconque a soustrait frauduleusement une chose
qui ne lui appartient pas, est coupable de vol.

380. Les soustractions commises par des maris au préjudice de leurs femmes, par des femmes au préjudice de
leurs maris, par un veuf ou une veuve, quant aux
choses qui avaient appartenu à l'époux décédé, par des
enfants ou autres descendants au préjudice de leurs père
ou mères ou autres ascendants, par des pères et mères ou
autres ascendants au préjudice de leurs enfants ou autre

descendants, ou par des alliés aux mêmes degrés, ne pourront donner lieu qu'à des réparations civiles.

A l'égard de tous autres individus qui auraient recélé ou appliqué à leur profit tout ou partie des objets volés, ils seront punis comme coupables de vol.

381. Seront punis de la peine de mort, les individus coupables de vols commis avec la réunion des cinq circonstances suivantes :

1º Si le vol a été commis la nuit;

2º S'il a été commis par deux ou plusieurs personnes;

3º Si les coupables ou l'un d'eux étaient porteurs d'armes apparentes ou cachées;

4º S'ils ont commis le crime soit à l'aide d'effraction extérieure ou d'escalade ou de fausses clefs, dans une maison, appartement, chambre ou logement habités ou servant à l'habitation, ou leurs dépendances, soit en prenant le titre d'un fonctionnaire public ou d'un officier civil ou militaire, ou après s'être revêtus de l'uniforme ou du costume du fonctionnaire ou de l'officier, ou en alléguant un faux ordre de l'autorité civile ou militaire.

5º S'ils ont commis le crime avec violence ou menaces de faire usage de leurs armes.

382. Sera puni de la peine des travaux forcés à perpétuité, tout individu coupable de vol commis à l'aide de violence, et, de plus, avec deux des quatre premières circonstances prévues par le précédent article.

Si même la violence à l'aide de laquelle le vol a été commis, a laissé des traces de blessures ou de contusions, cette circonstance seule suffira pour que la peine des travaux forcés à perpétuité soit prononcée.

383. Les vols commis dans les chemins publics em-

porteront également la peine des travaux forcés à perpétuité.

384. Sera puni de la peine des travaux forcés à temps, tout individu coupable de vols commis à l'aide d'un des moyens énoncés dans le n° 4 de l'article 381, même quoique l'effraction, l'escalade et l'usage des fausses clefs aient eu lieu dans des édifices, parcs ou enclos non servant à l'habitation et non dépendants des maisons habitées, et lors même que l'effraction n'aurait été qu'intérieure.

385. Sera également puni de la peine des travaux forcés à temps tout individu coupable de vols commis, soit avec violence, lorsqu'elle n'aura laissé aucune trace de blessure ou de contusion, et qu'elle ne sera accompagnée d'aucune autre circonstance, soit sans violence, mais avec la réunion des trois circonstances suivantes :

1° Si le vol a été commis la nuit ;

2° S'il a été commis par deux ou plusieurs personnes;

3° Si le coupable, ou l'un des coupables, était porteur d'armes apparentes ou cachées.

386. Sera puni de la peine de la réclusion tout individu coupable de vols commis dans l'un des cas ci-après :

1° Si le vol a été commis la nuit, et par deux ou plusieurs personnes, ou s'il a été commis avec une de ces deux circonstances seulement, mais en même temps dans un lieu habité ou servant à l'habitation ;

2° Si le coupable, ou l'un des coupables, était porteur d'armes apparentes ou cachées, même quoique le lieu où le vol a été commis ne fût ni habité ni servant à l'habitation, et encore quoique le vol ait été commis le jour et par une seule personne :

3° Si le voleur est un domestique ou un homme de service à gages, même lorsqu'il aura commis le vol envers des personnes qu'il ne servait pas, mais qui se trouvaient soit dans la maison de son maître, soit dans celle où il l'accompagnait; ou si c'est un ouvrier, compagnon ou apprenti, dans la maison, l'atelier ou le magasin de son maître, ou un individu travaillant habituellement dans l'habitation où il aura volé;

4° Si le vol a été commis par un aubergiste, un hôtelier, un voiturier, un batelier ou un de leurs préposés, lorsqu'ils auront volé tout ou partie des choses qui leur étaient confiées à ce titre; ou enfin, si le coupable a commis le vol dans l'auberge ou l'hôtellerie dans laquelle il était reçu.

387. Les voituriers, bateliers ou leurs préposés, qui auront altéré des vins, ou toute autre espèce de liquide ou de marchandises dont le transport leur avait été confié, et qui auront commis cette altération par le mélange de substances malfaisantes, seront punis de la peine portée au précédent article.

S'il n'y a pas eu mélange de substances malfaisantes, la peine sera un emprisonnement d'un mois à un an, et une amende de seize francs à cent francs.

388. Quiconque aura volé, dans les champs, des chevaux, ou bêtes de charge, de voiture ou de monture, gros et menus bestiaux, des instruments d'agriculture, des récoltes ou meules de grains faisant partie de récoltes, sera puni de la réclusion.

Il en sera de même à l'égard des vols de bois dans les ventes, et de pierres dans les carrières, ainsi qu'à l'égard du vol de poisson en étang, vivier ou réservoir.

389. La même peine aura lieu, si, pour commettre un vol, il y a eu enlèvement ou déplacement de bornes, servant de séparation aux propriétés.

390. Est réputé maison habitée, tout bâtiment, loge-

ment, loge, cabane même mobile, qui, sans être actuellement habitée, est destinée à l'habitation, et tout ce qui en dépend, comme cours, basses-cours, granges, écuries, édifices qui y sont enfermés, quel qu'en soit l'usage, et quand même ils auraient une clôture particulière dans la clôture ou enceinte générale.

391. Est réputé *parc* ou *enclos*, tout terrain environné de fossés, de pieux, de claies, de planches, de haies vives ou sèches, ou de murs, de quelque espèce de matériaux que ce soit, quelles que soient la hauteur, la profondeur, la vétusté, la dégradation de ces diverses clôtures, quand il n'y aurait pas de porte fermant à clef ou autrement, ou quand la porte serait à claire-voie et ouverte habituellement.

392. Les parcs mobiles destinés à contenir du bétail dans la campagne, de quelque matière qu'ils soient faits, sont aussi réputés enclos; et lorsqu'ils tiennent aux cabanes mobiles ou autres abris destinés aux gardiens, ils sont réputés dépendants de maison habitée.

393. Est qualifié *effraction*, tout forcement, rupture, dégradation, démolition, enlevement de murs, toits, planchers, portes, fenêtres, serrures, cadenas, ou autres ustensiles ou instruments servant à fermer ou à empêcher le passage, et de toute espèce de clôture, quelle qu'elle soit.

394. Les effractions sont extérieures ou intérieures.

395. Les effractions extérieures sont celles à l'aide desquelles on peut s'introduire dans les maisons, cours, basses-cours, enclos ou dépendances, ou dans les appartements ou logements particuliers.

396. Les effractions intérieures sont celles qui, après l'introduction dans les lieux mentionnés en l'article pré-

cédent, sont faites aux portes ou clôtures du dedans, ainsi qu'aux armoires ou autres meubles fermés.

Est compris dans la classe des effractions intérieures, le simple enlevement des caisses, boites, ballots sous toile et corde, et autres meubles fermés, qui contiennent des effets quelconques, bien que l'effraction n'ait pas été faite sur le lieu.

397. Est qualifiée *escalade*, toute entrée dans les maisons, bâtiments, cours, basse-cours, édifices quelconques, jardins, parcs et enclos, exécutée par-dessus les murs, portes, toitures ou toute autre clôture.

L'entrée par une ouverture souterraine, autre que celle qui a été établie pour servir d'entrée, est une circonstance de même gravité que l'escalade.

398. Sont qualifiés *fausses clefs*, tous crochets, rossignols, passe-partout, clefs imitées, contrefaites, altérées, ou qui n'ont pas été destinées par le propriétaire, locataire, aubergiste ou logeur, aux serrures, cadenas ou aux fermetures quelconques auxquelles le coupable les aura employées.

399. Quiconque aura contrefait ou altéré des clefs, sera condamné à un emprisonnement de trois mois à deux ans, et à une amende de vingt-cinq francs à cent cinquante francs.

Si le coupable est un serrurier de profession, il sera puni de la réclusion ;

Le tout sans préjudice de plus fortes peines, s'il y échet, en cas de complicité de crime.

400. Quiconque aura extorqué par force, violence ou contrainte, la signature ou la remise d'un écrit, d'un acte, d'un titre, d'une pièce quelconque contenant ou opérant obligation, disposition ou décharge, sera puni de la peine des travaux forcés à temps.

401. Les autres vols non spécifiés dans la présente section, les larcins et filouteries, ainsi que les tentatives de ces mêmes délits, seront punis d'un emprisonnement d'un an au moins, et de cinq ans au plus, et pourront même l'être d'une amende qui sera de seize francs au moins, et de cinq cents francs au plus.

Les coupables pourront encore être interdits des droits mentionnés en l'article 42 du présent Code, pendant cinq ans au moins, et dix ans au plus, à compter du jour où ils auront subi leur peine.

Ils pourront aussi être mis, par l'arrêt ou le jugement, sous la surveillance de la haute police pendant le même nombre d'années.

SECTION II.

Banqueroutes, Escroqueries, et autres espèces de Fraude.

§. Ier.

Banqueroute et Escroquerie.

402. Ceux qui, dans les cas prévus par le Code de Commerce, seront déclarés coupables de banqueroute, seront punis ainsi qu'il suit :

Les banqueroutiers frauduleux seront punis de la peine des travaux forcés à temps ;

Les banqueroutiers simples seront punis d'un emprisonnement d'un mois au moins, et de deux ans au plus.

403. Ceux qui, conformément au Code de Commerce, seront déclarés complices de banqueroute frauduleuse, seront punis de la même peine que les banqueroutiers frauduleux.

404. Les agents de change et courtiers qui auront fait faillite, seront punis de la peine des travaux forcés à temps ; s'ils sont convaincus de banqueroute frauduleuse, la peine sera celle des travaux forcés à perpetuité.

405. Quiconque, soit en faisant usage de faux noms ou de fausses qualités, soit en employant des manœuvres frauduleuses pour persuader l'existence de fausses entreprises, d'un pouvoir ou d'un crédit imaginaire, ou pour faire naître l'espérance ou la crainte d'un succès, d'un accident ou de tout autre événement chimérique, se sera fait remettre ou délivrer des fonds, des meubles ou des obligations, dispositions, billets, promesses, quittances ou décharges, et aura, par un de ces moyens, escroqué ou tenté d'escroquer la totalité ou partie de la fortune d'autrui, sera puni d'un emprisonnement d'un an au moins et de cinq ans au plus, et d'une amende de 50 francs au moins et de 3000 francs au plus.

Le coupable pourra être, en outre, à compter du jour où il aura subi sa peine, interdit, pendant cinq ans au moins et dix ans au plus, des droits mentionnés en l'article 42 du présent Code : le tout sauf les peines plus graves, s'il y a crime de faux.

§. II.

Abus de confiance.

406. Quiconque aura abusé des besoins, des faiblesses ou des passions d'un mineur, pour lui faire souscrire, à son préjudice, des obligations, quittances ou décharges, pour prêt d'argent ou de choses mobilières, ou d'effets de commerce, ou de tous autres effets obligatoires, sous quelque forme que cette négociation ait été faite ou déguisée, sera puni d'un emprisonnement de deux mois au moins, de deux ans au plus, et d'une amende qui ne

pourra excéder le quart des restitutions et des dommages-intérêts qui seront dus aux parties lésées, ni être moindre de vingt-cinq francs.

La disposition portée au second paragraphe du précédent article pourra de plus être appliquée.

407. Quiconque abusant d'un blanc-seing qui lui aura été confié, aura frauduleusement écrit au-dessus une obligation ou décharge, ou tout autre acte pouvant compromettre la personne ou la fortune du signataire, sera puni des peines portées en l'article 405.

Dans le cas où le blanc-seing ne lui aurait pas été confié, il sera poursuivi comme faussaire, ou puni comme tel.

408. Quiconque aura détourné ou dissipé, au préjudice du propriétaire, possesseur ou détenteur, des effets, deniers, marchandises, billets, quittances ou tous autres écrits contenant ou opérant obligation ou décharge, qui ne lui auraient été remis qu'à titre de dépôt ou pour un travail salarié, à la charge de les rendre ou représenter, ou d'en faire un usage ou un emploi déterminé, sera puni des peines portées dans l'art. 406.

Le tout sans préjudice de ce qui est dit aux art. 254, 255 et 256, relativement aux soustractions et enlevemens de deniers, effets ou pièces, commis dans les dépôts publics.

409. Quiconque, après avoir produit dans une contestation judiciaire quelque titre, pièce ou mémoire, l'aura soustrait de quelque manière que ce soit, sera puni d'une amende de vingt-cinq fr. à trois cents fr.

Cette peine sera prononcée par le tribunal saisi de la contestation.

§. III.

Contravention aux Règlements sur les maisons de jeu,
les loteries, et les maisons de prêt sur gages.

410. Ceux qui auront tenu une maison de jeux de
hasard, et y auront admis le public, soit librement, soit
sur la présentation des intéressés ou affiliés, les banquiers
de cette maison, tous ceux qui auront établi ou tenu des
loteries non autorisées par la loi, tous administrateurs,
préposés ou agents de ces établissements, seront punis
d'un emprisonnement de deux mois au moins et de six
mois au plus, et d'une amende de cent francs à six mille
francs.

Les coupables pourront être de plus, à compter du
jour où ils auront subi leur peine, interdits, pendant
cinq ans au moins et dix ans au plus, des droits mentionnés en l'article 42 du présent Code.

Dans tous les cas, seront confisqués tous les fonds ou
effets qui seront trouvés exposés au jeu ou mis à la loterie, les meubles, instruments, ustensiles, appareils employés ou destinés au service des jeux ou des loteries, les
meubles et les effets mobiliers dont les lieux seront garnis
ou décorés.

411. Ceux qui auront établi ou tenu des maisons de
prêt sur gages ou nantissement, sans autorisation légale,
ou qui, ayant une autorisation, n'auront pas tenu un
registre conforme aux règlements, contenant de suite,
sans aucun blanc ni interligne, les sommes ou les objets
prêtés, les noms, domiciles et profession des emprun-
teurs, la nature, la qualité, la valeur des objets mis en
nantissement, seront punis d'un emprisonnement de
quinze jours au moins, de trois mois au plus, et d'une
amende de cent francs à deux mille francs.

§. IV.

Entraves apportées à la liberté des Enchères.

412. Ceux qui, dans les adjudications de la propriété, de l'usufruit ou de la location de choses mobilières ou immobilières, d'une entreprise, d'une fourniture, d'une exploitation ou d'un service quelconque, auront entravé ou troublé la liberté des enchères ou des soumissions, par voies de fait, violences ou menaces, soit avant, soit pendant les enchères ou les soumissions, seront punis d'un emprisonnement de quinze jours au moins, de trois mois au plus, et d'une amende de cent francs au moins et de cinq mille francs au plus.

La même peine aura lieu contre ceux qui, par dons ou promesses, auront écarté les enchérisseurs.

§. V.

Violation des Règlements relatifs aux manufactures, au commerce et aux arts.

413. Toute violation des règlements d'administration publique, relatifs aux produits des manufactures françaises qui s'exporteront à l'étranger, et qui ont pour objet de garantir la bonne qualité, les dimensions et la nature de la fabrication, sera punie d'une amende de deux cents francs au moins, de trois mille francs au plus, et de la confiscation des marchandises. Ces deux peines pourront être prononcées cumulativement ou séparément, selon les circonstances.

414. Toute coalition entre ceux qui font travailler des ouvriers, tendant à forcer injustement et abusivement l'abaissement des salaires, suivie d'une tentative ou d'un commencement d'exécution, sera punie d'un

emprisonnement de six jours à un mois, et d'une amende de deux cents francs à trois mille francs.

415. Toute coalition de la part des ouvriers pour faire cesser en même temps de travailler, interdire le travail dans un atelier, empêcher de s'y rendre et d'y rester avant ou après de certaines heures, et en général pour suspendre, empêcher, enchérir les travaux, s'il y a eu tentative ou commencement d'exécution, sera punie d'un emprisonnement d'un mois au moins et de trois mois au plus.

Les chefs ou moteurs seront punis d'un emprisonnement de deux ans à cinq ans.

416. Seront aussi punis de la peine portée par l'article précédent, et d'après les mêmes distinctions, les ouvriers qui auront prononcé des amendes, des défenses, des interdictions ou toutes proscriptions sous le nom de damnations et sous quelque qualification que ce puisse être, soit contre les directeurs d'ateliers et entrepreneurs d'ouvrages, soit les uns contre les autres.

Dans le cas du présent article et dans celui du précédent, les chefs ou moteurs du délit pourront, après l'expiration de leur peine, être mis sous la surveillance de la haute police pendant deux ans au moins et cinq ans au plus.

417. Quiconque, dans la vue de nuire à l'industrie française, aura fait passer en pays étranger des directeurs, commis ou des ouvriers d'un établissement, sera puni d'un emprisonnement de six mois à deux ans. et d'une amende de cinquante francs à trois cents francs.

418. Tout directeur, commis, ouvrier de fabrique qui aura communiqué à des étrangers ou à des Français résidant en pays étranger, des secrets de la fabrique où

il est employé, sera puni de la réclusion, et d'une amende de cinq cents francs à vingt mille francs.

Si ces secrets ont été communiqués à des Français résidant en France, la peine sera d'un emprisonnement de trois mois à deux ans, et d'une amende de seize francs à deux cents francs.

419. Tous ceux qui, par des faits faux ou calomnieux semés à dessein dans le public, par des sur-offres faites aux prix que demandaient les vendeurs eux-mêmes, par réunions ou coalitions entre les principaux détenteurs d'une même marchandise ou denrée, tendant à ne la pas vendre ou à ne la vendre qu'à un certain prix, ou qui, par des voies ou moyens frauduleux quelconques auront opéré la hausse ou la baisse du prix des denrées ou marchandises, ou des papiers et effets publics au-dessus ou au-dessous des prix qu'aurait déterminés la concurrence naturelle et libre du commerce, seront punis d'un emprisonnement d'un mois au moins, d'un an au plus, et d'une amende de cinq cents francs à dix mille francs. Les coupables pourront, de plus, être mis, par l'arrêt ou le jugement sous la surveillance de la haute police, pendant deux ans au moins et cinq ans au plus.

420. La peine sera d'un emprisonnement de deux mois au moins et de deux ans au plus, et d'une amende de mille francs à vingt mille francs, si ces manœuvres ont été pratiquées sur grains, grenailles, farines, substances farineuses, pain, vin, ou toute autre boisson.

La mise en surveillance qui pourra être prononcée, sera de cinq ans au moins et dix ans au plus.

421. Les paris qui auront été faits sur la hausse ou la baisse des effets publics, seront punis des peines portées par l'article 419.

422. Sera réputée pari de ce genre, toute convention

de vendre ou de livrer des effets publics qui ne seront
pas prouvés par le vendeur avoir existé à sa disposition
au temps de la convention, ou avoir dû s'y trouver au
temps de la livraison.

423. Quiconque aura trompé l'acheteur sur le titre
des matières d'or ou d'argent, sur la qualité d'une pierre
fausse vendue pour fine, sur la nature de toutes marchandises ; quiconque, par usage de faux poids ou de
fausses mesures, aura trompé sur la quantité des choses
vendues, sera puni de l'emprisonnement pendant trois
mois au moins, un an au plus, et d'une amende qui ne
pourra excéder le quart des restitutions et dommages-
intérêts, ni être au-dessous de cinquante francs.

Les objets du délit, ou leur valeur, s'ils appartiennent
encore au vendeur, seront confisqués : les faux poids et
les fausses mesures seront aussi confisqués, et de plus
seront brisés.

424. Si le vendeur et l'acheteur se sont servis, dans
leurs marchés, d'autres poids ou d'autres mesures que
ceux qui ont été établis par les lois de l'État, l'acheteur
sera privé de toute action contre le vendeur qui l'aura
trompé par l'usage de poids ou de mesures prohibés ;
sans préjudice de l'action publique pour la punition tant
de cette fraude que de l'emploi même des poids et des
mesures prohibés.

La peine, en cas de fraude, sera celle portée par l'article précédent.

La peine, pour l'emploi des mesures et poids prohibés, sera déterminée par le livre IV du présent Code,
contenant les peines de simple police.

425. Toute édition d'écrits, de composition musicale,
de dessin, de peinture ou de toute autre production,
imprimée ou gravée en entier ou en partie, au mépris

des lois et règlements relatifs à la propriété des auteurs, est une contrefaçon ; et toute contrefaçon est un délit.

426. Le débit d'ouvrages contrefaits, l'introduction sur le territoire français d'ouvrages qui, après avoir été imprimés en France, ont été contrefaits chez l'étranger, sont un délit de la même espèce.

427. La peine contre le contrefacteur, ou contre l'introducteur, sera une amende de 100 francs au moins, et de 2000 fr. au plus ; et contre le débitant une amende de 25 francs au moins et de 500 fr. au plus.

La confiscation de l'édition contrefaite sera prononcée tant contre le contrefacteur que contre l'introducteur et le débitant.

Les planches, moules ou matrices des objets contrefaits seront aussi confisqués.

428. Tout directeur, tout entrepreneur de spectacle, toute association d'artistes, qui aura fait représenter sur son théâtre des ouvrages dramatiques, au mépris des lois et règlements relatifs à la propriété des auteurs, sera puni d'une amende de 50 francs au moins, de 500 fr. au plus, et de la confiscation des recettes.

429. Dans les cas prévus par les quatre articles précédents, le produit des confiscations, ou les recettes confisquées, seront remis au propriétaire pour l'indemniser d'autant du préjudice qu'il aura souffert ; le surplus de son indemnité, ou l'entière indemnité, s'il n'y a eu ni vente d'objets confisqués ni saisie de recettes, sera réglé par les voies ordinaires.

§. VI.

Délits des Fournisseurs.

430. Tous individus chargés, comme membres de compagnie ou individuellement, de fournitures, d'entre-

prises ou régies pour le compte des armées de terre et de mer, qui, sans y avoir été contraints par une force majeure, auront fait manquer le service dont ils sont chargés, seront punis de la peine de la réclusion, et d'une amende qui ne pourra excéder le quart des dommages-intérêts, ni être au-dessous de 500 francs; le tout sans préjudice de peines plus fortes en cas d'intelligence avec l'ennemi.

431. Lorsque la cessation du service proviendra du fait des agents des fournisseurs, les agents seront condamnés aux peines portées par le précédent article.

Les fournisseurs et leurs agents seront également condamnés, lorsque les uns et les autres auront participé au crime.

432. Si des fonctionnaires publics ou des agents, préposés ou salariés du Gouvernement, ont aidé les coupables à faire manquer le service, ils seront punis de la peine des travaux forcés à temps; sans préjudice de peines plus fortes, en cas d'intelligence avec l'ennemi.

433. Quoique le service n'ait pas manqué, si, par négligence, les livraisons et les travaux ont été retardés, ou s'il y a eu fraude sur la nature, la qualité ou la quantité des travaux ou mains-d'œuvre ou des choses fournies, les coupables seront punis d'un emprisonnement de six mois au moins, et de cinq ans au plus, et d'une amende qui ne pourra excéder le quart des dommages-intérêts, ni être moindre de 100 francs.

Dans les divers cas prévus par les articles composant le présent paragraphe, la poursuite ne pourra être faite que sur la dénonciation du Gouvernement.

SECTION III.

Destructions, Dégradations, Dommages.

434. Quiconque aura volontairement mis le feu à des édifices, navires, bateaux, magasins, chantiers, forêts, bois taillis ou récoltes, soit sur pied, soit abattus, soit aussi que les bois soient en tas ou en cordes, et les récoltes en tas ou en meules, ou à des matières combustibles placées de manière à communiquer le feu à ces choses ou à l'une d'elles, sera puni de la peine de mort.

435. La peine sera la même contre ceux qui auront détruit, par l'effet d'une mine, des édifices, navires ou bateaux.

436. La menace d'incendier une habitation ou toute autre propriété sera punie de la peine portée contre la menace d'assassinat, et d'après les distinctions établies par les articles 305, 306 et 307.

437. Quiconque aura volontairement détruit ou renversé, par quelque moyen que ce soit, en tout ou en partie, des édifices, des ponts, digues ou chaussées ou autres constructions qu'il savait appartenir à autrui, sera puni de la réclusion et d'une amende qui ne pourra excéder le quart des restitutions et indemnités, ni être au-dessous de cent francs.

S'il y a eu homicide ou blessures, le coupable sera, dans le premier cas, puni de mort, et dans le second, puni de la peine des travaux forcés à temps.

438. Quiconque, par des voies de fait, se sera opposé à la confection de travaux autorisés par le Gouvernement, sera puni d'un emprisonnement de trois mois à deux ans, et d'une amende qui ne pourra excéder le quart des dommages-intérêts, ni être au-dessous de seize francs.

Les moteurs subiront le *maximum* de la peine.

439. Quiconque aura volontairement brûlé ou détruit d'une manière quelconque, des registres, minutes ou actes originaux de l'autorité publique, des titres, billets, lettres de change, effets de commerce ou de banque, contenant ou opérant obligation, disposition ou décharge, sera puni ainsi qu'il suit :

Si les pièces détruites sont des actes de l'autorité publique, ou des effets de commerce ou de banque, la peine sera la réclusion.

S'il s'agit de toute autre pièce, le coupable sera puni d'un emprisonnement de deux ans à cinq ans, et d'une amende de cent francs à trois cents francs.

440. Tout pillage, tout dégât de denrées ou marchandises, effets, propriétés mobilières, commis en réunion ou bande et à force ouverte, sera puni des travaux forcés à temps; chacun des coupables sera de plus condamné à une amende de deux cents francs à cinq mille francs.

441. Néanmoins, ceux qui prouveront avoir été entraînés par des provocations ou sollicitations à prendre part à ces violences, pourront n'être punis que de la peine de la réclusion.

442. Si les denrées pillées ou détruites sont des grains, grenailles ou farines, substances farineuses, pain, vin ou autre boisson, la peine que subiront les chefs, instigateurs ou provocateurs seulement, sera le *maximum* des travaux forcés à temps, et celui de l'amende prononcée par l'art. 440.

443. Quiconque, à l'aide d'une liqueur corrosive ou par tout autre moyen, aura volontairement gâté des marchandises ou matières servant à fabrication, sera puni d'un emprisonnement d'un mois à deux ans, et

d'une amende qui ne pourra excéder le quart des dommages-intérêts, ni être moindre de seize francs.

Si le délit a été commis par un ouvrier de la fabrique ou par un commis de la maison de commerce, l'emprisonnement sera de deux à cinq ans, sans préjudice de l'amende, ainsi qu'il vient d'être dit.

444. Quiconque aura dévasté des récoltes sur pied ou des plants venus naturellement ou faits de main d'homme, sera puni d'un emprisonnement de deux ans au moins, de cinq ans au plus.

Les coupables pourront, de plus, être mis, par l'arrêt ou le jugement, sous la surveillance de la haute police pendant cinq ans au moins et dix ans au plus.

445. Quiconque aura abattu un ou plusieurs arbres qu'il savait appartenir à autrui, sera puni d'un emprisonnement qui ne sera pas au-dessous de six jours, ni au-dessus de six mois, à raison de chaque arbre, sans que la totalité puisse excéder cinq ans.

446. Les peines seront les mêmes à raison de chaque arbre mutilé, coupé ou écorcé de manière à le faire périr.

447. S'il y a eu destruction d'une ou de plusieurs greffes, l'emprisonnement sera de six jours à deux mois, à raison de chaque greffe, sans que la totalité puisse excéder deux ans.

448. Le *minimum* de la peine sera de vingt jours dans les cas prévus par les articles 445 et 446, et de dix jours dans le cas prévu par l'article 447, si les arbres étaient plantés sur les places, routes, chemins, rues ou voies publiques ou vicinales, ou de traverse.

449. Quiconque aura coupé des grains ou des fourrages qu'il savait appartenir à autrui, sera puni d'un

emprisonnement qui ne sera pas au-dessous de six jours, ni au-dessus de deux mois.

450. L'emprisonnement sera de vingt jours au moins et de quatre mois au plus, s'il a été coupé du grain en vert.

Dans les cas prévus par le présent article et les six précédents, si le fait a été commis en haine d'un fonctionnaire public et à raison de ses fonctions, le coupable sera puni du *maximum* de la peine établie par l'article auquel le cas se référera.

Il en sera de même, quoique cette circonstance n'existe point, si le fait a été commis pendant la nuit.

451. Toute rupture, toute destruction d'instruments d'agriculture, de parcs de bestiaux, de cabanes de gardiens, sera punie d'un emprisonnement d'un mois au moins, d'un an au plus.

452. Quiconque aura empoisonné des chevaux ou autres bêtes de voiture, de monture ou de charge, des bestiaux à cornes, des moutons, chèvres ou porcs, ou des poissons dans des étangs, viviers ou réservoirs, sera puni d'un emprisonnement d'un an à cinq ans, et d'une amende de seize francs à trois cents francs. Les coupables pourront être mis, par l'arrêt ou le jugement, sous la surveillance de la haute police, pendant deux ans au moins, et cinq ans au plus.

453. Ceux qui, sans nécessité, auront tué l'un des animaux mentionnés au précédent article, seront punis ainsi qu'il suit :

Si le délit a été commis dans les bâtiments, enclos et dépendances, ou sur les terres dont le maître de l'animal tué était propriétaire, locataire, colon ou fermier, la peine sera un emprisonnement de deux mois à six mois.

S'il a été commis dans des lieux dont le coupable était propriétaire, locataire, colon ou fermier, l'emprisonnement sera de six jours à un mois.

S'il a été commis dans tout autre lieu, l'emprisonnement sera de quinze jours à six semaines.

Le *maximum* de la peine sera toujours prononcé en cas de violation de clôture.

454. Quiconque aura, sans nécessité, tué un animal domestique dans un lieu dont celui à qui cet animal appartient est propriétaire, locataire, colon ou fermier, sera puni d'un emprisonnement de six jours au moins et de six mois au plus.

S'il y a eu violation de clôture, le *maximum* de la peine sera prononcé.

455. Dans les cas prévus par les articles 444 et suiv., jusqu'au précédent article inclusivement, il sera prononcé une amende qui ne pourra excéder le quart des restitutions et dommages-intérêts, ni être au-dessous de seize francs.

456. Quiconque aura, en tout ou en partie, comblé des fossés, détruit des clôtures, de quelques matériaux qu'elles soient faites, coupé ou arraché des haies vives ou sèches; quiconque aura déplacé ou supprimé des bornes, ou pieds corniers, ou autres arbres plantés ou reconnus pour établir les limites entre différents héritages, sera puni d'un emprisonnement qui ne pourra pas être au-dessous d'un mois ni excéder une année, et d'une amende égale au quart des restitutions et des dommages-intérêts, qui, dans aucun cas, ne pourra être au-dessous de cinquante francs.

457. Seront punis d'une amende qui ne pourra excéder le quart des restitutions et des dommages-intérêts, ni être au-dessous de cinquante francs, les propriétaires ou

fermiers, ou toute personne jouissant de moulins, usines ou étangs, qui, par l'élévation du déversoir de leurs eaux au-dessus de la hauteur déterminée par l'autorité compétente, auront inondé les chemins ou les propriétés d'autrui.

S'il est résulté du fait quelques dégradations, la peine sera, outre l'amende, un emprisonnement de six jours à un mois.

458. L'incendie des propriétés mobilières ou immobilières d'autrui, qui aura été causé par la vétusté ou le défaut soit de réparation, soit de nettoyage des fours, cheminées, forges, maisons ou usines prochaines, ou par des feux allumés dans les champs à moins de cent mètres des maisons, édifices, forêts, bruyères, bois, vergers, plantations, haies, meules, tas de grains, pailles, foins, fourrages, ou de tout autre dépôt de matières combustibles, ou par des feux ou lumières portés ou laissés sans précaution suffisante, ou par des pièces d'artifice allumées ou tirées par négligence ou imprudence, sera puni d'une amende de cinquante francs au moins, et de cinq cents francs au plus.

459. Tout détenteur ou gardien d'animaux ou de bestiaux soupçonnés d'être infectés de maladie contagieuse, qui n'aura pas averti sur-le-champ le maire de la commune où ils se trouvent, et qui même, avant que le maire ait répondu à l'avertissement, ne les aura pas tenus renfermés, sera puni d'un emprisonnement de six jours à deux mois, et d'une amende de seize francs à deux cents francs.

460. Seront également punis d'un emprisonnement de deux mois à six mois, et d'une amende de cent francs à cinq cents francs, ceux qui, au mépris des défenses de l'administration, auront laissé leurs animaux ou bestiaux infectés communiquer avec d'autres.

461. Si, de la communication mentionnée au précédent article, il est résulté une contagion parmi les autres animaux, ceux qui auront contrevenu aux défenses de l'autorité administrative seront punis d'un emprisonnement de deux ans à cinq ans, et d'une amende de cent francs à mille francs ; le tout sans préjudice de l'exécution des lois et règlements relatifs aux maladies épizootiques, et de l'application des peines y portées.

462. Si les délits de police correctionnelle dont il est parlé au présent chapitre ont été commis par des gardes champêtres ou forestiers, ou des officiers de police, à quelque titre que ce soit, la peine d'emprisonnement sera d'un mois au moins, et d'un tiers au plus en sus de la peine la plus forte qui serait appliquée à un autre coupable du même délit.

DISPOSITION GÉNÉRALE.

463. Dans tous les cas où la peine d'emprisonnement est portée par le présent Code, si le préjudice causé n'excède pas vingt-cinq francs, et si les circonstances paraissent atténuantes, les tribunaux sont autorisés à réduire l'emprisonnement même au-dessous de six jours, et l'amende même au-dessous de 16 francs. Ils pourront aussi prononcer séparément l'une ou l'autre de ces peines, sans qu'en aucun cas elle puisse être au-dessous des peines de simple police.

FIN DU LIVRE TROISIÈME.

LIVRE IV.

CONTRAVENTIONS DE POLICE ET PEINES.

(Décrété le 20 février 1810. Promulgué le 2 mars.)

CHAPITRE PREMIER.

Des Peines.

464. Les peines de police sont,

L'emprisonnement,

L'amende,

Et la confiscation de certains objets saisis.

465. L'emprisonnement, pour contravention de police, ne pourra être moindre d'un jour, ni excéder cinq jours, selon les classes, distinctions et cas ci-après spécifiés.

Les jours d'emprisonnement sont des jours complets de vingt-quatre heures.

466. Les amendes pour contravention pourront être prononcées depuis 1 fr. jusqu'à 15 fr. inclusivement, selon les distinctions et classes ci-après spécifiées , et seront appliquées au profit de la commune où la contravention aura été commise.

467. La contrainte par corps a lieu pour le paiement de l'amende.

Néanmoins le condamné ne pourra être, pour cet objet, détenu plus de quinze jours, s'il justifie de son insolvabilité.

468. En cas d'insuffisance des biens, les restitutions et les indemnités dues à la partie lésée sont préférées à l'amende.

469. Les restitutions, indemnités et frais entraîneront la contrainte par corps; et le condamné gardera prison jusqu'à parfait paiement; néanmoins si ces condamnations sont prononcées au profit de l'État, les condamnés pourront jouir de la faculté accordée par l'article 467, dans le cas d'insolvabilité prévu par cet article.

470. Les tribunaux de police pourront aussi, dans les cas déterminés par loi, prononcer la confiscation, soit des choses saisies en contravention, soit des choses produites par la contravention, soit des matières ou des instruments qui ont servi ou étaient destinés à la commettre.

CHAPITRE II.

Contraventions et Peines.

SECTION PREMIÈRE.

Première classe.

471. Seront punis d'amende, depuis 1 franc jusqu'à 5 francs inclusivement,

1º Ceux qui auront négligé d'entretenir, réparer ou nettoyer les fours, cheminées ou usines où l'on fait usage du feu;

2º Ceux qui auront violé la défense de tirer, en certains lieux, des pièces d'artifice;

3º Les aubergistes et autres qui, obligés à l'éclairage, l'auront négligé; ceux qui auront négligé de nettoyer les rues ou passages, dans les communes où ce soin est laissé à la charge des habitants;

4° Ceux qui auront embarrassé la voie publique en y déposant ou y laissant, sans nécessité, des matériaux ou des choses quelconques qui empêchent ou diminuent la liberté ou la sûreté du passage; ceux qui, en contravention aux lois et règlements, auront négligé d'éclairer les matériaux par eux entreposés ou les excavations par eux faites dans les rues et places;

5° Ceux qui auront négligé ou refusé d'exécuter les règlements ou arrêtés concernant la petite voirie, ou d'obéir à la sommation émanée de l'autorité administrative, de réparer ou démolir les édifices menaçant ruine;

6° Ceux qui auront jeté ou exposé au-devant de leurs édifices, des choses de nature à nuire par leur chute ou par des exhalaisons insalubres;

7° Ceux qui auront laissé dans les rues, chemins, places, lieux publics, ou dans les champs, des coûtres de charrue, pinces, barres, barreaux ou autres machines, ou instruments ou armes dont puissent abuser les voleurs et autres malfaiteurs,

8° Ceux qui auront négligé d'écheniller dans les campagnes ou jardins où ce soin est prescrit par la loi ou les règlements;

9° Ceux qui, sans autre circonstance prévue par les lois, auront cueilli ou mangé, sur le lieu même, des fruits appartenant à autrui;

10° Ceux qui, sans autre circonstance, auront glané, râtelé ou grapillé dans les champs non encore entièrement dépouillés et vidés de leurs récoltes, ou avant le moment du lever ou après celui du coucher du soleil;

11° Ceux qui, sans avoir été provoqués, auront proféré contre quelqu'un des injures, autres que celles prévues depuis l'art. 367 jusques et compris l'art. 378;

12.º Ceux qui imprudemment auront jeté des immondices sur quelque personne ;

13º Ceux qui, n'étant ni propriétaires, ni usufruitiers, ni locataires, ni fermiers, ni jouissant d'un terrain ou d'un droit de passage, ou qui n'étant agents ni préposés d'aucune de ces personnes, seront entrés et auront passé sur ce terrain ou sur partie de ce terrain, s'il est préparé ou ensemencé ;

14º Ceux qui auront laissé passer leurs bestiaux ou leurs bêtes de trait, de charge ou de monture, sur le terrain d'autrui, avant l'enlevement de la récolte.

472. Seront, en outre, confisqués, les pièces d'artifice saisies dans le cas du nº 11 de l'article 471, les coûtres, les instruments et les armes mentionnés dans le nº VII du même article.

473. La peine d'emprisonnement, pendant trois jours au plus, pourra, de plus, être prononcée, selon les circonstances, contre ceux qui auront tiré des pièces d'artifice ; contre ceux qui auront glané, râtelé ou grapillé en contravention au nº 10 de l'art. 471.

474. La peine d'emprisonnement contre toutes les personnes mentionnées en l'art. 471, aura toujours lieu, en cas de récidive, pendant trois jours au plus.

SECTION II.

Deuxième classe.

475. Seront punis d'amende depuis six francs jusqu'à dix francs inclusivement,

1º Ceux qui auront contrevenu aux bans de vendanges ou autres bans autorisés par les règlements ;

2º Les aubergistes, hôteliers, logeurs ou loueurs de maisons garnies, qui auront négligé d'inscrire de suite, et sans aucun blanc, sur un registre tenu régulièrement, les noms, qualités, domicile habituel, dates d'entrée et de sortie de toute personne qui aurait couché ou passé une nuit dans leurs maisons; ceux d'entre eux qui auraient manqué à représenter ce registre aux époques déterminées par les règlements, ou lorsqu'ils en auraient été requis, aux maires, adjoints, officiers ou commissaires de police ou aux citoyens commis à cet effet; le tout sans préjudice des cas de responsabilité mentionnés en l'article 73 du présent Code, relativement aux crimes ou aux délits de ceux qui, ayant logé ou séjourné chez eux, n'auraient pas été régulièrement inscrits;

3.º Les rouliers, charretiers, conducteurs de voitures quelconques ou de bêtes de charge, qui auraient contrevenu aux règlements par lesquels ils sont obligés de se tenir constamment à portée de leurs chevaux, bêtes de trait ou de charge et de leurs voitures, et en état de les guider et conduire; d'occuper un seul côté des rues, chemins ou voies publiques; de se détourner ou ranger devant toutes autres voitures, et, à leur approche, de leur laisser libre au moins la moitié des rues, chaussées, routes et chemins;

4º Ceux qui auront fait ou laissé courir les chevaux, bêtes de trait, de charge ou de monture, dans l'intérieur d'un lieu habité, ou violé les règlements contre le chargement, la rapidité ou la mauvaise direction des voitures;

5º Ceux qui auront établi ou tenu dans les rues, chemins, places ou lieux publics, des jeux de loterie ou d'autres jeux de hasard;

6º Ceux qui auront vendu ou débité des boissons falsifiées, sans préjudice des peines plus sévères qui seront prononcées par les tribunaux de police correctionnelle, dans le cas où elles contiendraient des mixtions nuisibles à la santé ;

7° Ceux qui auraient laissé divaguer des fous ou des furieux étant sous leur garde, ou des animaux malfaisants ou féroces ; ceux qui auront excité ou n'auront pas retenu leurs chiens lorsqu'ils attaquent ou poursuivent les passants, quand même il n'en serait résulté aucun mal ni dommage ;

8° Ceux qui auraient jeté des pierres ou d'autres corps durs ou des immondices contre les maisons, édifices ou clôtures d'autrui, ou dans les jardins ou enclos, et ceux aussi qui auraient volontairement jeté des corps durs ou immondices sur quelqu'un.

9° Ceux qui, n'étant propriétaires, usufruitiers ni jouissant d'un terrain ou d'un droit de passage, y sont entrés et y ont passé dans le temps où ce terrain était chargé de grains en tuyau, de raisins ou autres fruits mûrs ou voisins de la maturité ;

10° Ceux qui auraient fait ou laissé passer des bestiaux, animaux de trait, de charge ou de monture sur le terrain d'autrui, ensemencé ou chargé d'une récolte, en quelque saison que ce soit, ou dans un bois taillis appartenant à autrui ;

11° Ceux qui auraient refusé de recevoir les espèces et monnaies nationales, non fausses ni altérées, selon la valeur pour laquelle elles ont cours ;

12° Ceux qui, le pouvant, auront refusé ou négligé de faire les travaux, le service, ou de prêter le secours dont ils auront été requis dans les circonstances d'accidents, tumultes, naufrages, inondation, incendie ou autres calamités, ainsi que dans les cas de brigandages, pillages, flagrant délit, clameur publique ou d'exécution judiciaire;

13º Les personnes désignées aux articles 284 et 288 du présent Code.

476. Pourra, suivant les circonstances, être prononcé, outre l'amende portée en l'article précédent, l'emprisonnement pendant trois jours au plus contre les rouliers, charretiers, voituriers et conducteurs en contravention, contre ceux qui auront contrevenu à la loi par la rapidité, la mauvaise direction ou le chargement des voitures ou des animaux; contre les vendeurs et débitants de boissons falsifiées; contre ceux qui auraient jeté des corps durs ou des immondices.

477. Seront saisis et confisqués, 1º les tables, instruments, appareils des jeux ou des loteries établies dans les rues, chemins et voies publiques, ainsi que les enjeux, les fonds, denrées, objets ou lots proposés aux joueurs, dans le cas de l'article 476; 2º les boissons falsifiées, trouvées appartenir au vendeur et débitant : ces boissons seront répandues; 3º les écrits ou gravures contraires aux mœurs : ces objets seront mis sous le pilon.

478. La peine de l'emprisonnement pendant cinq jours au plus, sera toujours prononcée, en cas de récidive, contre toutes les personnes mentionnées dans l'article 475.

SECTION III.

Troisième classe.

479. Seront punis d'une amende de onze à quinze francs inclusivement,

1° Ceux qui, hors les cas prévus depuis l'article 434 jusques et compris l'art. 462 , auront volontairement causé du dommage aux propriétés mobilières d'autrui ;

2° Ceux qui auront occasionné la mort ou la blessure des animaux ou bestiaux appartenant à autrui, par l'effet de la divagation des fous ou furieux, ou d'animaux malfaisants ou féroces, ou par la rapidité ou la mauvaise direction ou le chargement excessif des voitures, chevaux, bêtes de trait, de charge ou de monture ;

3° Ceux qui auront occasionné les mêmes dommages par l'emploi ou l'usage d'armes sans précaution ou avec maladresse, ou par jet de pierres ou d'autres corps durs.

4° Ceux qui auront causé les mêmes accidents par la vétusté, la dégradation, le défaut de réparation ou d'entretien des maisons ou édifices, ou par l'encombrement ou l'excavation, ou telles autres œuvres, dans ou près les rues, chemins, places ou voies publiques, sans les précautions ou signaux ordonnés ou d'usage ;

5° Ceux qui auront de faux poids ou de fausses mesures dans leurs magasins, boutiques, ateliers ou maisons de commerce, ou dans les halles, foires ou marchés, sans préjudice des peines qui seront prononcées par les tribunaux de police correctionnelle contre ceux qui auraient fait usage de ces faux poids ou de ces fausses mesures ;

6° Ceux qui emploieront des poids ou des mesures différents de ceux qui sont établis par les lois en vigueur ;

7° Les gens qui font le métier de deviner et pronostiquer ou d'expliquer les songes ;

8° Les auteurs ou complices de bruits ou tapages in-

jurieux ou nocturnes, troublant la tranquillité des habitants.

480. Pourra, selon les circonstances, être prononcée
la peine d'emprisonnement pendant cinq jours au plus,

1º Contre ceux qui auront occasionné la mort ou la
blessure des animaux ou bestiaux appartenant à autrui,
dans les cas prévus par le nº III du précédent article ; 2º
contre les possesseurs de faux poids et de fausses mesures ; 3º contre ceux qui emploient des poids ou des
mesures différents de ceux que la loi en vigueur a établis ; 4º contre les interprètes de songes ; 5º contre les
auteurs ou complices de bruits ou tapages injurieux ou
nocturnes.

481. Seront, de plus, saisis et confisqués, 1º les faux
poids, les fausses mesures, ainsi que les poids et les mesures différents de ceux que la loi a établis ; 2º les instruments, ustensiles et costumes servant ou destinés à l'exercice du métier de devin, pronostiqueur ou interprète de
songes.

482. La peine d'emprisonnement pendant cinq jours
aura toujours lieu, pour récidive, contre les personnes et
dans les cas mentionnés en l'art. 479.

Dispositions communes aux trois sections ci-dessus.

483. Il y a récidive dans tous les cas prévus par le
présent livre, lorsqu'il a été rendu contre le contreve-
nant, dans les douze mois précédents, un premier jugement pour contravention de police commise dans le ressort du même tribunal.

DISPOSITION GÉNÉRALE.

484. Dans toutes les matières qui n'ont pas été réglées par le présent Code, et qui sont régies par des lois et règlements particuliers, les cours et les tribunaux continueront de les observer.

FIN DU CODE PÉNAL.

DÉCRET *du* 13 *mars* 1810.

NAPOLÉON, EMPEREUR DES FRANÇAIS, ROI D'ITALIE, et PROTECTEUR DE LA CONFÉDÉRATION DU RHIN ;

Sur le rapport de notre grand-juge ministre de la justice ;

Considérant que le Code pénal présente des dispositions coordonnées avec celles du Code d'Instruction criminelle ; *

Notre Conseil d'État entendu,

Nous AVONS DÉCRÉTÉ et DÉCRÉTONS ce qui suit :

ART. Ier. Le Code pénal sera exécuté à l'époque fixée par notre décret du 17 décembre 1809, pour l'exécution du Code d'Instruction criminelle.

ART. II. Notre grand-juge ministre de la justice est chargé de l'exécution du présent décret.

Signé NAPOLÉON.

Par l'Empereur :

Le Ministre Secrétaire d'Etat, signé H. B.

DUC DE BASSANO.

---

* *Décret du* 17 *décembre* 1809.

ART. Ier. Nos cours et nos tribunaux continueront d'exécuter, comme par le passé, jusqu'au 1er janvier 1811, les lois relatives à la poursuite, à l'instruction et au jugement des affaires criminelles, de police correctionnelle et de simple police.

ART. II. Notre grand-juge ministre de la justice est chargé de l'exécution du présent décret.

TABLE
ANALYTIQUE ET RAISONNÉE
DES MATIÈRES

CONTENUES DANS CE CODE.

Les chiffres arabes indiquent les numéros des articles.

A.

ABUS D'AUTORITÉ. Peine pour abus d'autorité contre les particuliers, *article* 184 et suiv.; et contre la chose publique, 182 et suiv.

ABUS DE CONFIANCE. Peine contre celui qui aurait abusé, d'une manière quelconque, des besoins, des faiblesses ou des passions d'un mineur, 406. — Peine pour abus d'un blanc-seing, 407; pour soustraction d'effets et marchandises remis à titre de dépôt, 408; pour soustraction de titres, pièces ou mémoires produits dans une contestation judiciaire, 409. Voy. *Interdiction.*

ABUS DE POUVOIR. Voy. *Dons.*

ACCIDENTS. Voy. *Secours.*

ACCOUCHEMENT. Peine contre celui qui, ayant assisté à un accouchement, n'en a pas fait sa déclaration, 346.

ACCUSATION. Peine contre les procureurs-généraux, etc. qui auraient traduit un citoyen devant une cour d'assise ou spéciale sans qu'il ait été mis légalement en accusation, 122.

ACCUSÉ. Quand l'accusé a moins de seize ans, s'il est décidé qu'il a agi sans discernement, il est acquitté; mais il est remis à ses parents ou conduit dans une maison de correction, pour y être détenu pendant un temps qui ne peut excéder l'époque où il aura accompli sa vingtième année, 66. — Manière dont les peines doivent être prononcées lorsqu'il est décidé que l'accusé qui avait moins de seize ans a agi avec discernement, 67.

ACQUITTEMENT. Il a lieu à l'égard des accusés d'un âge au-dessous de seize ans qui sont reconnus avoir agi sans discernement, 66.

ACTES. Peines contre celui qui, sans titre, aurait fait les actes d'un office civil et militaire, 258.

ACTES ARBITRAIRES. Peine de la dégradation civique pour raison d'actes arbitraires et attentatoires à la liberté individuelle ou aux droits civiques, 114. — Exceptions, *ibid.* — Peine du bannissement pour actes arbitraires ordonnés ou faits par un ministre qui a négligé ou refusé de faire réparer ces actes, 115. — Les auteurs d'une fausse signature pour un acte contraire aux constitutions sont punis de la peine des travaux forcés à temps, 118. — Voy. *Constitution, Liberté, Ministres.*

ACTES DE BARBARIE. Voy. *Assassinats.*

ACTES DE L'ÉTAT CIVIL. Délits relatifs à la tenue de ces actes, 192 et suiv.

ACTION PUBLIQUE. Celle qui a lieu pour punir l'emploi de poids et mesures autres que ceux qui ont été établis par la loi de l'État, et la fraude qui en serait résultée, 424.

ADJOINTS DE MAIRE. Voy. *Aubergistes, Registres.*

ADJUDICATION. Voy. *Enchères.*

ADMINISTRATEURS. Peines contre ceux qui entreprendraient sur les fonctions judiciaires, et décideraient l'affaire malgré la réclamation des parties, 131. — Peines contre tout administrateur, fonctionnaire ou

officier public, pour destruction, suppression ou soustraction d'actes et de titres dont ils étaient dépositaires
en cette qualité, 173 ; pour violation d'un domicile,
174. — Voy. *Préfets, Déni de justice, Violences.*

ADMINISTRATION. Cas dans lesquels les tribunaux peuvent interdire l'exercice du droit d'être nommé aux
emplois de l'administration, 42 et 43.

ADULTÈRE. Cas dans lequel le meurtre commis par l'é-
poux sur son épouse adultère et sur son complice, est
excusable, 324. — L'adultère de la femme ne peut
être dénoncé que par le mari, qui n'a même pas
cette faculté s'il entretient une concubine dans la maison conjugale, 336 et 339. — Peine de la femme
convaincue d'adultère, et de son complice, 337 et 333.
— Seules preuves admissibles contre le prévenu de
complicité, 338.

AFFICHES. Lieux dans lesquels sont affichés les arrêts
portant peine de mort, des travaux forcés, de la déportation, du carcan, du bannissement et de la dégradation civique, 36. — Voy. *Ecrits.*

AFFICHEURS. Voy. *Crieurs.*

AGE. Commutation de peine à l'égard des individus qui,
âgés de moins de seize ans, ont encouru des peines afflictives et infamantes, ou simplement correctionnelles,
66 et 67. — Les individus âgés de soixante-dix ans, ne
peuvent être condamnés aux travaux forcés ni à la déportation, 70. — Par quelle peine celles-ci sont remplacées,
71. — Voy. *Accusé, Acquittement, Bannissement,
Carcan, Condamnation.*

AGENTS DE L'ADMINISTRATION DES POSTES. Voy. *Lettres.*

AGENTS DE LA POLICE. Voy. *Rebellion, Violences.*

AGENTS DU GOUVERNEMENT. Peine contre tout fonctionnaire public, tout agent du Gouvernement ou toute

autre personne qui aurait livré le secret d'une négociation ou d'une expédition aux agents de l'ennemi, ou lui aurait livré des plans de fortifications, etc., 80 et 81.—Voy. *Fonctionnaires publics*, *Force publique*, *Lettres*, *Préposés du Gouvernement*, *Soustraction*, *Violences*.

ALIMENTS. La confiscation générale des biens d'un condamné est grevée de la prestation des aliments à qui il en est dû de droit, 38.— Voy. *Avortement*.

ALTÉRATION D'ÉCRITURE. Peine encourue par un officier public pour altération des actes, écritures ou signatures, 143. — Voy. *Fausses-clefs*.

ALTÉRATION DE MONNAIE. Voy. *Contrefaçon*, *Fausse monnaie*.

AMENDES en matière correctionnelle, 9. — Communes aux matières correctionnelles et criminelles, 11. — Mode de poursuite de l'exécution des condamnations à l'amende, 52. — Durée de l'emprisonnement pour acquit des amendes au profit de l'État, après laquelle le condamné insolvable obtient sa liberté provisoire, 53.—Amende prononcée contre ceux qui ne font point la déclaration des crimes et des complots dont ils ont connaissance, 105. — Amende contre le vendeur ou l'acheteur d'un suffrage, 113. — Contre le concierge d'une maison d'arrêt, etc., pour détention arbitraire, 120; contre les juges qui auraient jugé une affaire portée devant eux, ou décerné des mandats contre des préposés du Gouvernement avant la décision de l'autorité supérieure, 128 et 129; et contre les administrateurs qui entreprendraient sur les fonctions judiciaires, 131. — Amendes contre ceux qui, avec connaissance de cause, font usage de monnaies contrefaites ou altérées, 135 ; contre les dépositaires publics condamnés pour soustractions par eux commises, 172 ; contre des fonctionnaires publics qui se seraient ingérés dans un commerce incompatible avec leur qualité, 175. — Amendes prononcées pour fait de corruption, 171, 179 et 181 ; pour abus d'autorité, déni de justice et suppression ou ouverture de lettres confiées à la poste, 184, 185 et 187 ; contre le fonctionnaire public entré en exercice de ses fonctions sans avoir prêté serment, 196 ; contre celui qui aurait illégalement continué cet exercice après révocation, destitution, etc., 197 ; contre le ministre d'un culte qui aurait procédé aux cérémonies d'un mariage sans s'être fait représenter l'acte reçu par les officiers de l'état civil, 199. Cas dans lesquels il y a lieu à l'amende pour outrage envers les dépositaires de l'autorité et de la force publique, 224 et 227. — Amendes pour la non-comparution des témoins et des jurés, 236 ; et pour soustraction et enlèvement de papiers ou registres commis à la garde de dépositaires négligents, 254 ; pour dégradation de monuments, 257 ; pour entraves au libre exercice des cultes, 260 et suiv. ; pour exposition ou distribution de pamphlets ou images contraires aux bonnes mœurs, 287 ; pour réunion illicite de sociétés littéraires, etc., 292 et 294 ; pour menaces écrites ou verbales, 306 et 307 ; pour blessures et coups volontaires, 311 et suiv. ; pour port d'armes prohibées, 314 ; pour vente ou débit de boissons falsifiées, 318 ; pour homicide ou blessures involontaires, 319 et 320 ; pour attentats aux mœurs, 330 et suiv. ; pour crimes et délits tendant à empêcher ou détruire la preuve de l'état civil d'un enfant, ou à compromettre son existence, 346 et suiv.; pour infraction aux lois sur les inhumations, 358 et suiv.; pour calomnies, injures ou révélation de secrets, 373 et suiv.; pour contrefaçon ou altération de clefs, 399; pour larcins et filouteries, 401; pour escroquerie, 405; pour abus de confiance, 406 et suiv.; pour contravention aux règlements sur les maisons de jeu et de prêt, 410 et 411; pour entraves apportées à la liberté des enchères, 412; pour violation des règlements relatifs aux manufactures, au commerce et aux arts, 413 et suiv.; pour délits des fournisseurs, 430 et suiv.; pour destruction, dégradation et dommage de différentes sortes, 437 et suiv.—Circonstances dans lesquelles les tribunaux sont autorisés à réduire la quotité de l'amende, 463. — Les amendes pour contraventions peuvent être prononcées depuis 1 fr. jusqu'à 15, 466. — Commune au profit de laquelle s'en fait l'application, *ibid.* — Comment se poursuit le paiement de l'amende, 467. — Terme après lequel le condamné obtient sa liberté en justifiant de son insolvabilité, *ibid.* — En cas d'insuffisance des biens, les restitutions et indemnités dues à la partie lésée sont préférées à l'amende, 468. — Contraventions qui sont punies d'amende : première classe, 471 et suiv.; deuxième classe, 475 et suiv.; troisième classe, 479 et suiv. — Voy. *Condamnation, Indemnités, Restitution, Solidarité.*

Animaux domestiques. Peines pour en avoir tué dans des lieux appartenant aux propriétaires de ces animaux, 455.

Animaux malfaisants. Amende contre ceux qui en auraient laissé divaguer, 477. — Peine contre ceux qui auraient occasionné la mort ou la blessure de bestiaux par l'effet de la divagation d'animaux malfaisants ou féroces, 479. — Emprisonnement suivant les circonstances, 480.

APPRENTIS. Voy. *Ouvriers.*

ARBRES. Peines pour avoir abattu, mutilé, coupé ou écorcé des arbres appartenant à autrui, 445 et 446. — Pour avoir détruit des greffes, 447. — Peine plus forte si les arbres étaient plantés sur les places, routes, chemins, rues ou voies publiques, ou vicinales, ou de traverse, 448. — Peine qui s'inflige dans le cas où les délits ci-dessus ont été commis en haine d'un fonctionnaire public et pendant la nuit, 450. — Voy. *Limites.*

ARCHIVISTES. Voy. *Scellés.*

ARGENT. Voy. *Matières d'or et d'argent, Récompenses.*

ARMES. Tout Français qui a porté les armes contre la France est puni de mort, 75. — Ses biens sont confisqués, *ibid.* — Peines contre ceux dont les manœuvres auraient eu pour objet de seconder les progrès des armes des ennemis sur les possessions ou contre les forces françaises de terre ou de mer, 77; contre l'attentat ayant pour but d'armer les citoyens les uns contre les autres, 91. — Peine pour avoir fourni ou procuré des armes à des bandes illégalement formées, 96. — On désigne par le mot *arme,* toute machine, tous instruments ou ustensiles tranchants, perçants ou contondants, 101. Les couteaux, les ciseaux de poche, les cannes simples ne sont réputés *armes* qu'autant qu'il en serait fait usage pour tuer, blesser ou frapper, *ibid.* — Peines pour fournitures d'armes destinées à favoriser l'évasion des détenus ou les crimes des malfaiteurs, 243 et 268. —

Pour fabrication ou débit de stylets, tromblons ou autres armes prohibées par la loi ou par des règlements d'administration publique, 314; et contre les porteurs de ces armes, *ibid.* — Peines contre les voleurs munis d'armes apparentes ou cachées, 381 382 et 385. — Amende contre ceux qui auraient laissé dans les rues, chemins, places et lieux publics, des armes ou instruments dont les voleurs et les malfaiteurs pourraient abuser, 471.—Confiscation de ces armes, 472. — Peine contre ceux qui auraient occasionné la mort ou la blessure de bestiaux par l'usage d'armes sans précaution ou avec maladresse, 479.—Voy. *Instruments, Port d'armes, Réunion armée, Vagabondage.*

ARRESTATION. Peines contre les individus qui, sans ordre légal, et hors les cas prévus où la loi ordonne de saisir des prévenus, auraient arrêté des personnes quelconques, 341. — Diminution de la peine en cas de relaxation de l'individu arrêté, 343. — Circonstances qui donnent lieu contre les coupables à la peine de mort, 344.

ARRÊT. Un huissier lit au peuple l'arrêt de condamnation du parricide pendant que celui-ci est exposé sur l'échafaud, 13. — Les arrêts portant peine de mort, des travaux forcés, de la déportation, du carcan, du bannissement ou de la dégradation civique, sont imprimés par extrait, 36. — Communes dans lesquelles ils sont affichés, *ibid.* — Voy. *Bannissement, Exécution, Place publique.*

ARSENAUX. Voy. *Bandes armées, Incendie, Mine, Places, Plans.*

ARTIFICE. Amende contre ceux qui auraient violé la défense de tirer en certains lieux des pièces d'artifice,

471. — Peine de l'emprisonnement et confiscation des pièces d'artifice, 472 et 475. — Voy. *Incendie.*

ARTIFICES. Voy. *Dons.*

ARTS. Voy. *Manufactures.*

ASSASSINAT. Dans quel cas le meurtre est ainsi qualifié, 296.—De quelle peine l'assassinat est puni, 302. — Les malfaiteurs qui emploient des tortures, ou commettent des actes de barbarie, sont punis comme coupables d'assassinat, 305.

ASSOCIATION DE MALFAITEURS. Cette sorte d'association envers les personnes et les propriétés est un crime contre la paix publique, 265. — Peines contre les directeurs et commandants de ces bandes, 267 ; et contre ceux qui leur auraient fourni des armes, munitions, instruments, logements, retraites ou lieux de réunion, 268. — Voy. *Sociétés.*

ATELIERS PARTICULIERS. Voy. *Boutiques.*

ATELIERS PUBLICS. Voy. *Ouvriers.*

ATTAQUE. Voy. *Rebellion.*

ATTENTAT. L'attentat ou complot contre la vie ou contre la personne de l'Empereur est crime de lèse-majesté, 86. — De quelle peine ce crime est puni, *ibid.* — Peines encourues pour attentat ou complot contre les membres de la famille impériale, ou dont le but serait de détruire le Gouvernement, de changer l'ordre de successibilité du trône, ou d'exciter les citoyens à s'armer contre l'exercice de l'autorité impériale, 87. — Circonstances dans lesquelles il y a attentat, 88. — Attentat à la liberté, 114 et suiv. — Voy. *Menaces, Mœurs.*

ATTROUPEMENT. Peines contre ceux qui, par attroupement, voies de fait ou menaces, auraient empêché des citoyens d'exercer leurs droits civiques, 109.

AUBERGISTES et HÔTELIERS. Leur responsabilité, dans le cas où ils ont logé, sans l'avoir inscrit sur leurs registres, le coupable d'un crime ou délit commis pendant son séjour, 73. — Peines pour inscription sur leurs registres, avec connaissance de cause, de noms faux et supposés, 154; pour vols de choses à eux confiées à ce titre, ou pour vols commis dans l'auberge ou l'hôtellerie où le coupable était reçu, 386. — Amende contre les aubergistes, hôteliers, logeurs ou loueurs de maisons garnies, pour défaut d'inscription sur leurs registres des personnes qui auraient couché ou passé une nuit dans leurs maisons, ou de représentation de ces registres aux maires, adjoints, officiers ou commissaires de police, sans préjudice de la responsabilité pour les délits commis par les personnes qui auraient séjourné chez eux, 475; pour préjudice causé aux propriétés mobilières de ceux qu'ils logent, 479. — Emprisonnement qui peut être prononcé pour cette dernière contravention, 481.

AUTEURS. Ils encourent le *maximum* des peines attachées aux délits commis par la voie d'écrits ou images anonymes, 288. — Voy. *Contrefaçon, Écrits, Indemnités.*

AUTORISATION. Celle du Gouvernement est nécessaire pour la réunion des sociétés littéraires, etc., 290 et 294.

AUTORITÉ. Voy. *Abus d'autorité.*

AUTORITÉ ADMINISTRATIVE. Peines contre les juges, les procureurs généraux ou impériaux, leurs substituts et les officiers de police judiciaire qui se seraient immiscés dans les affaires attribuées aux autorités administratives, 127. Voy. *Déni de justice.*

AUTORITÉ PUBLIQUE. Peines pour exercice de l'autorité publique illégalement anticipé ou prolongé, 196 et 197; pour résistance, désobéissance et autres manquements envers l'autorité publique, 209 et suiv.; pour outrages et violences envers les dépositaires de l'autorité et de la force publique, 222 et suiv. Voy. *Bannissement, Censure de l'autorité publique, Déportation.*

AVIS DE FAMILLE. Voy. *Tuteur.*

AVORTEMENT. Peines contre ceux qui, par aliments, breuvages, médicaments, violences, etc., auraient procuré l'avortement d'une femme enceinte, 317. — Même peine contre la femme, *ibid.* — Peines contre les officiers de santé, médecins, chirurgiens et pharmaciens qui auraient indiqué ou administré ces moyens, *ibid.*

B.

BAISSE. Voy. *Effets publics.*

BAN. Amende contre ceux qui auraient contrevenu aux bans de vendange ou autres bans autorisés par les règlements, 475.

BANDES ARMÉES. Peine de mort, avec confiscation de biens, contre ceux qui, pour envahir des domaines, propriétés ou deniers publics, places, villes, forteresses, postes, magasins, arsenaux, ports, vaisseaux ou bâtiments appartenant à l'État, ou pour piller des propriétés publiques, se seraient mis à la tête de bandes armées ou auraient contribué à leur levée et à leur entretien, 96. — Autres crimes pour lesquels chacun des individus faisant partie d'une bande armée encourt la même peine, 97. — Peines contre ceux qui, connaissant le but et le caractère des bandes, leur auraient volontairement fourni des logements ou lieux de retraite, 99. — Peines pour tout pillage ou dégât de marchandises, effets et propriétés mobilières, commis en réunion ou bande et à force ouverte, 440 à 442.

BANNISSEMENT. C'est une peine infamante, 8. — Lieu où les condamnés au bannissement sont transportés, 32. — Durée du bannissement, *ibid.* — Peine de la déportation encourue par le banni rentré sur le territoire de l'Empire, 33. — La durée du bannissement se compte du jour où l'arrêt est devenu irrévocable, 35. — Temps pendant lequel les coupables condamnés au bannissement sont de plein droit sous la surveillance de la haute police de l'État, 48. — Celui qui, ayant été condamné pour crime, en a commis un second emportant la peine du bannissement, doit être condamné à la peine de la réclusion, 56. — L'individu, ayant moins de seize ans, qui, agissant avec discernement, a encouru la peine du bannissement, doit être condamné à être enfermé d'un an à cinq dans une maison de correction, 67. — Tout agent public subit la peine du bannissement encourue pour avoir livré à une puissance neutre ou alliée des plans de fortifications, arsenaux, etc., 81 ; pour avoir, par des actions hostiles non approuvées du Gouvernement, exposé l'État à une déclaration de guerre, ou des Français à éprouver des représailles, 84 et 85 ; pour provocation au pillage, 102. — Elle est prononcée contre ceux qui auraient empêché l'exercice des droits civiques par suite d'un plan concerté pour tout l'Empire, 110. — Contre les ministres qui auraient ordonné des actes attentatoires à la liberté individuelle, aux droits civiques ou aux constitutions de l'Empire, et auraient refusé, malgré les invitations légales, de réparer ces actes, 115. — Bannissement pour mesures concertées entre des fonctionnaires contre l'exécution des lois ou contre les ordres du Gouvernement, 124. — La peine du bannissement se prononce contre l'officier public qui, instruit de la supposition du nom, a cependant délivré le passe-port sous le nom supposé, 153. — Cas où des paiements de frais de route, faits par le trésor public sur une fausse feuille, donnent lieu au bannissement ou à la réclusion, 154. — Circonstances qui font encourir la peine du bannissement, de la réclusion ou des travaux forcés à temps, à l'officier public qui a délivré une feuille de route contenant supposition de nom, 156; au ministre des cultes qui, dans un discours pastoral, aurait donné lieu à la désobéissance aux lois, 202 et 203; à l'individu qui a enfreint l'ordre de s'éloigner de l'habitation du magistrat par lui insulté, 229.

BANQUEROUTES. Peines des banqueroutiers frauduleux, 402; des banqueroutiers simples, *ibid.*; des complices de banqueroutes frauduleuses, 403; des agents de change et courtiers qui ont fait faillite ou banqueroute, 404.

BATEAUX. Vols commis par les bateliers, 386. — *Voy. Incendie, Mine.*

BATIMENTS. Voy. *Destruction, Incendie, Mine.*

BATIMENTS DE GUERRE. Voy. *Bandes armées, Commandement militaire.*

BESTIAUX. Peines contre ceux qui auraient tué sans nécessité des bestiaux à cornes, des moutons, chèvres ou porcs, 452. — Voy. *Animaux malfaisants, Champs, Empoisonnement, Epizootie, Fous, Voituriers.*

BÊTES DE CHARGE ET DE MONTURE. Voy. *Champs.*

BIENS. Voy. *Confiscation, Curateur, Interdiction, Provisions.*

BILLETS. Voy. *Destruction,*

BILLETS DE BANQUE. Voy. *Effets publics.*

BLANC-SEING. Voy. *Abus de confiance.*

BLESSURES. Peines pour blessures de fonctionnaires publics dans le cas où la mort s'en serait suivie, et dans celui où les blessures porteraient le caractère de meurtre, 230 et 232.—Peines pour blessures ou coups volontaires qui ont occasionné une maladie ou incapacité de travail personnel pendant plus de vingt jours, 309. — Aggravation de la peine lorsque le crime a été commis envers des ascendants, *ibid.;* et lorsqu'il a eu lieu avec préméditation ou guet-apens, 310 et 312. — Punition des coupables quand les blessures ou les coups n'ont occasionné ni maladie ni incapacité de travail, 311.—Ces délits imputables aux chefs, auteurs, instigateurs et provocateurs, lorsqu'il y a eu réunion séditieuse avec rebellion ou pillage, 313.—Peines de ceux qui auraient causé des blessures ou exposé à recevoir des coups par défaut d'adresse ou de précaution, 320. — Provocation ou violences graves qui rendent les blessures et les coups excusables, 321. — Autres circonstances dont il résulte une pareille excuse, 322. — Cas dans lesquels les blessures et les coups ne sont qualifiés ni crimes ni délits, 327 et suiv. — L'exposition d'un enfant qui, par suite de délaissement, est demeuré mutilé ou estropié, est punie des peines applicables aux blessures volontaires, 351. — Voy. *Destruction, Violences.*

BOIS. Voy. *Champs, Incendie, Terrains.*

BOISSONS FALSIFIÉES. Peines pour vente ou débit de ces boissons, contenant des mixtions nuisibles à la santé.

318 et 476 ; contre les voituriers, bateliers et autres qui ont alléré celle dont le transport leur est confié, 387. — Emprisonnement encouru par les vendeurs et débitants de boissons falsifiées, 475. — Ces boissons sont confisquées et répandues, 477.

BORNES. Peines pour vols accompagnés d'enlèvement ou de déplacement de bornes séparant les propriétés, 389. — Voy. *Limites.*

BOULET. Les individus condamnés aux travaux forcés en traînent un à leurs pieds, 15.

BOUTIQUES. Peines contre ceux qui auraient empêché d'ouvrir ou de fermer pendant certains jours les ateliers, boutiques ou magasins, et de faire quitter des travaux, 260.

BREUVAGES. Voy. *Avortement, Boissons falsifiées.*

BRIGANDAGE. Voy. *Etat, Secours.*

BRIS DE PRISON. Peines contre ceux qui auraient favorisé une évasion avec violence ou bris de prison, 241 et 243 ; et contre les détenus qui se seraient évadés par ces moyens, 244.

BRIS DE SCELLÉS. Voy. *Scellés.*

BRUITS NOCTURNES. Amende contre les auteurs ou complices de bruits ou tapages injurieux ou nocturnes troublant la tranquillité des habitants, 479.—Circonstances qui peuvent donner lieu à leur emprisonnement, 480.

BULLETINS. Voy. *Ecrits.*

C.

CABANES DES GARDIENS. Voy. *Champs.*

CADENAS. Voy. *Effraction, Fausses-clefs.*

CALOMNIE. Cas dans lesquels ce délit a lieu, 367. — Exception, *ibid.* — Toute imputation à l'appui de laquelle on ne rapporte point de preuve légale est ré-

putée fausse, 368. — Contre qui peuvent être poursuivies les calomnies mises au jour par la voie des
papiers étrangers, 369.—Comment se fait la preuve
légale de l'imputation, 370. — Peines du calomniateur à défaut de preuves, 371. — Faits imputés qui
donnent lieu à surseoir à la poursuite et au jugement
du délit de calomnie, 372.—Peines des dénonciations
calomnieuses faites par écrit aux officiers de justice ou
de police, 373.—Droits dont le calomniateur est tem·
porairement interdit après avoir subi sa peine, 374.

Cannes. Voy. *Armes.*

Carcan. C'est une peine infamante, 8.—Durée du temps
pendant lequel on attache au carcan les individus
condamnés aux travaux forcés ou à la réclusion, 22.
— La durée de ces deux peines se compte du jour de
l'exposition, 23. — Celui qui, ayant été condamné
pour crime, en a commis un second emportant la
peine du carcan, doit être condamné à la peine de la
réclusion, 56. — L'individu ayant moins de seize
ans, qui, agissant avec discernement, a encouru la
peine du carcan, est condamné à être enfermé d'un
an à cinq dans une maison de correction, 67.—Peine
du carcan contre tout citoyen qui aurait falsifié des
billets contenant les suffrages des citoyens, 111; contre
ceux qui auront falsifié, contrefait et employé les
sceaux, timbres, poinçons de l'État, etc, 143. —
La peine du carcan a lieu contre les fonctionnaires
publics ou agents d'une administration qui se seraient
laissé corrompre, 177. — Cas dans lequel des voies
de fait contre un magistrat emportent la peine du
carcan, 228.—Même peine pour avoir frappé le ministre d'un culte dans ses fonctions, 263. — Voyez
Arrêt, Condamnés, Dégradation civique.

Castration. Peines encourues par les coupables de ce

crime, 316. —— Cas dans lequel le crime de castration peut être considéré comme meurtre ou blessures excusables, 325.

CAUTIONNEMENT pour un condamné mis sous la surveillance de la haute police de l'État, 44 et 45. Les personnes qui ont souscrit un acte de cautionnement d'un individu mis sous la surveillance spéciale de l'État sont contraintes, même par corps, au paiement des sommes portées dans l'acte, lorsque l'individu cautionné a été condamné pour crimes ou délits commis dans l'intervalle déterminé, 46. ——Paiements auxquels les sommes recouvrées sont affectées de préférence, *ibid.* —— Voy. *Surveillance de la haute police*, *Vagabondage.*

CENSURE DE L'AUTORITÉ PUBLIQUE. Peines pour critiques, censures ou provocations dirigées contre l'autorité publique dans un discours pastoral prononcé publiquement, 201 et suiv.

CERTIFICAT DE BONNE CONDUITE. Voy. *Certificat d'indigence.*

CERTIFICAT D'INDIGENCE. Peine pour fabrication, sous le nom d'un fonctionnaire ou officier public, d'un certificat de bonne conduite, d'indigence, etc., 159. —— Même peine pour falsification d'un certificat de cette espèce auparavant véritable, et pour emploi de ce certificat falsifié, *ibid.*

CERTIFICAT DE MALADIE. Peine encourue par ceux qui délivrent un certificat de maladie ou d'infirmité sous le nom d'un officier de santé, 159; et par l'officier de santé qui a délivré lui-même le faux certificat, 160

CERTIFICATS. Punition encourue pour faux certificats dont il pourrait résulter lésion envers des tiers ou préjudice envers le trésor public, 162. —— Les peines pour faux certificats sont portées au *maximum* quand on les applique aux mendiants ou vagabonds, 281.

CHAINE. Celle à laquelle on attache deux à deux les individus condamnés aux travaux-forcés, 15.

CHAMPS. Peines pour vols commis dans les champs, de chevaux ou bêtes de charge, de bestiaux, d'instruments d'agriculture, de récoltes ou meules de grains, de bois dans les ventes, de pierres dans les carrières, et de poissons dans les étangs, viviers, et réservoirs, 388.

—— Peines pour rupture ou destruction d'instruments d'agriculture, de parcs de bestiaux, de cabanes de gardiens, 451.

CHANSONS. Voy. *Gravures.*

CHANTIERS. Voy. *Incendie.*

CHARRETIERS. Voy. *Voituriers.*

CHARRUES. (coûtres des) Voy. *Instruments.*

CHAUSSÉES. Voy. *Destruction.*

CHEMINÉES. Voy. *Fours, Incendie.*

CHEMINS, (peines des vols commis sur les grands) 383.

—— Voy. *Arbres, Jeux de hasard, Voituriers.*

CHEVAUX. Peines contre ceux qui en auraient tué sans nécessité, 454. —— Voy. *Champs, Empoisonnement.*

CHÈVRES. Voy. *Empoisonnement.*

CHIENS (peines contre ceux qui excitent leurs) contre les passants, 475.

CHIRURGIENS. Voy. *Avortement, Certificats et Secrets.*

CHUTE. Amende contre ceux qui exposeraient sur leurs fenêtres des choses propres à nuire par leur chute, 471.

CISEAUX. Voy. *Armes.*

CLAMEUR PUBLIQUE. Voy. *Forfaiture, Mandat, Secours.*

CLEFS. Voy. *Fausses clefs.*

CLÔTURES. Peines pour avoir tué sans nécessité un animal domestique, en violant la clôture du propriétaire, 454 et 455; pour avoir détruit des clôtures, 456. —— Voy. *Effraction, Enclos, Escalade.*

COALITION. Peines auxquelles peuvent donner lieu de

coalitions formées par des fonctionnaires publics, 123
et suiv. — Peines pour coalition contre ceux qui font
travailler des ouvriers pour l'abaissement injuste de
leurs salaires, 414 ; et pour coalition de la part
des ouvriers tendant à une cessation de travail par
amendes, défenses, proscriptions, etc., 415 et 416.

COLLUSION. Voy. *Officiers de l'état civil.*

COMMANDANT MILITAIRE. Voy. *Commerce, Détenus,
Force publique.*

COMMANDEMENT MILITAIRE. Peine de mort, avec confiscation de biens, contre ceux qui, sans droit ou motif légitime, auraient pris un commandement militaire, ou tenu une troupe rassemblée après l'ordre
de licenciement ou de séparation, 93.

COMMERCE. Amende et confiscation encourues par tout
commandant des divisions militaires, des départements ou des places et villes, tout préfet ou sous-
préfet qui, dans l'étendue des lieux soumis à son autorité, aurait fait ouvertement, ou par interposition
de personnes, le commerce des grains, farines ou
boissons, 176. — Voy. *Manufactures.*

COMMIS. Voy. *Concussion, Soustraction.*

COMMISSAIRES DE POLICE. Voy. *Aubergistes, Registres.*

COMMUNES. Voy. *Dévastation.*

COMPAGNONS. Voy. *Ouvriers.*

COMPLICES. Ceux d'un crime ou d'un délit sont punis de
la même peine que les auteurs, 59. — Circonstances
qui établissent la complicité, 60 à 62.—Voy. *Adultère.*

COMPLOT. Circonstances dans lesquelles il y a complot,
89. — Peines lorsqu'il y a proposition de complot faite
et non agréée, pour arriver au crime de lèse-majesté,
90.—Peine contre ceux qui, par des discours tenus dans
les lieux publics, par des placards ou des écrits imprimés, auraient excité les citoyens à commettre des

crimes et complots contre la sûreté de l'État, 102.
—— Peine pour non-révélation de complots formés contre la sûreté de l'État, 103 et 125.—— Voy. *Attentat.*

CONCIERGES. Cas dans lesquels les gardiens et concierges des maisons de dépôt, d'arrêt, de justice ou de peine, sont coupables de détention arbitraire, 120,
—— Peines à eux infligées, *ibid.* ——Voy. *Détenus.*

CONCUBINAGE. Peines contre le mari qui, sur la plainte de sa femme, a été convaincu d'entretenir une concubine dans sa maison, 339.

CONCUSSION. Peines encourues par les fonctionnaires, officiers publics, leurs commis ou préposés, les percepteurs des droits, taxes, contributions, etc., qui se seraient rendus coupables du crime de concussion, 174. —— Amende, restitution et dommages-intérêts auxquels sont en outre condamnés les coupables, *ibid.*

CONDAMNATION. La condamnation aux peines établies par la loi est toujours prononcée sans préjudice des restitutions et dommages-intérêts qui peuvent être dus aux parties, 10. —— Les condamnations aux travaux forcés à perpétuité, et à la déportation, emportent mort civile, 18. —— Aucune condamnation ne peut être exécutée les jours de fêtes nationales ou religieuses, ni les dimanches, 25. —— Condamnations qui peuvent être prononcées pour crimes ou délits, 44 et suiv. —— L'exécution des condamnations à l'amende, aux restitutions, aux dommages-intérêts et aux frais, peut être poursuivie par la voie de la contrainte par corps, 52.—— Condamnations qui se prononcent pour crimes et délits commis par récidive, 56 à 58. —— Diminution des peines pour crimes et délits commis par des individus âgés de moins de seize ans, 66. Voy. *Arrêt, Restitution, Solidarité.*

CONDAMNÉS. Droits dont sont déchus les individus condamnés à la peine des travaux forcés à temps, du bannissement, de la réclusion et du carcan, 28 et 29. — Époque à laquelle les biens sont remis aux condamnés qui, pendant la durée de leur peine, étaient en état d'interdiction légale, 30. — Comptes qui doivent leur être rendus par leurs curateurs, *ib.* — Pendant la durée de la peine, il ne peut leur être remis aucune somme, ni provision, ni portion de revenus, 31.

CONDUCTEURS DE PRISONNIERS. Voy. *Détenus.*

CONFIANCE. Voy. *Abus de confiance.*

CONFISCATION. On peut, dans certains cas, prononcer la confiscation générale concurremment avec une peine afflictive, 7. — La confiscation spéciale du corps du délit, des choses produites par le délit, ou de celles qui ont servi à le commettre, est une peine commune aux matières criminelles et correctionnelles, 11. — En quoi consiste la confiscation générale, 37. — Elle n'a lieu que dans le cas où la loi la prononce expressément, *ibid.* — Charges dont la confiscation générale est grevée, 38 — Personnes en faveur desquelles l'Empereur peut disposer des biens confisqués, 39. — Les biens de tout Français qui a porté les armes contre la France sont confisqués, 75. — Pareille confiscation pour intelligences avec les puissances étrangères et ennemies de l'État, 76 et 77. — Confiscation de biens pour crime de lèse-majesté, 86. — Pour complot contre les membres de la famille impériale ou contre le Gouvernement, 87. — Pour crime tendant à troubler l'État par la guerre civile, 91 et suiv. — Pour crime de fausse monnaie, 132. — Pour contrefaçon des sceaux de l'État, des billets de banque et des effets publics, 139. — Confiscation des denrées appartenant à un commerce interdit aux fonctionnaires publics, 176. — Les exemplaires d'écrits imprimés et distribués sans nom d'auteur ni d'imprimeur sont confisqués en cas de saisie, 286. — Il en est de même des images et gravures obscènes, 287; des armes prohibées, 314; des boissons falsifiées, 318; de l'argent reçu par un faux témoin, 364; des fonds ou effets qui seraient trouvés exposés aux jeux de hasard ou mis à des loteries non autorisées, et des meubles garnissant les maisons, 410; des marchandises à l'égard desquelles on a violé les règlements d'administration destinés à garantir leur bonne qualité, 413; De celles qui ont été vendues à faux poids ou fausses mesures, 423; des éditions contrefaites et des matières qui ont servi à la contrefaçon, 427; des recettes de représentations théâtrales faites au mépris des lois sur la propriété des auteurs, 428. — Remise aux propriétaires du produit des confiscations et des recettes confisquées, pour les indemniser d'autant du préjudice par eux souffert, 429. — Cas dans lesquels les tribunaux de police peuvent prononcer la confiscation des choses saisies en contravention, des choses produites par la contravention, ou des matières et instruments qui y ont servi, 470. — Contraventions de police pour lesquelles la confiscation a lieu, 477. — Divers cas de confiscation 481. — Voy. *Restitution.*

CONFLIT. Peines contre les juges, etc., qui, malgré la notification d'un conflit, auraient persisté dans l'exécution de jugements ou ordonnances rendus contre des administrateurs pour raison de l'exercice de leurs fonctions, 127. — Voy. *Revendication.*

CONNIVENCE. Peines contre les préposés à la garde ou à la conduite des détenus dont l'évasion aurait eu lieu par suite de connivence, 238 et suiv. — Dommages-intérêts pour la partie civile contre ceux qui auraient connivé à l'évasion d'un détenu, 244.

CONSCRIPTION MILITAIRE. Maintien des lois pénales et des règlements qui la concernent, 235.

CONSEIL D'ÉTAT. Voy. *Forfaiture, Mandat.*

CONSEIL DE FAMILLE. Attentats aux mœurs qui font interdire aux coupables toute participation aux conseils de famille, 335.

CONSTITUTIONS (crimes et délits contre les) de l'Empire, 109 et suiv.

CONSTRUCTIONS publiques et particulières, (destruction de) 437.

CONTRAINTE. Celle qui est exercée contre les cautions d'un individu mis sous la surveillance spéciale du Gouvernement et condamné pour crimes et délits, 46. — Contrainte par corps pour l'exécution des condamnations à l'amende, aux restitutions, aux dommages-intérêts et aux frais, 52, 467 et 469. — Voy. *Condamnation, Corruption.*

CONTRAVENTION. On appelle ainsi l'infraction des lois de police, 1. — Les dispositions du Code ne sont pas applicables aux contraventions militaires, 5. — Contraventions de police, 464 et suiv. — Voy. *Peines.*

CONTREFAÇON. Peines pour contrefaçon ou altération des monnaies, 132 et suiv. — Pour contrefaçon des sceaux de l'État, des billets de banque, des effets publics et des poinçons, timbres et marques, 139 et suiv. — Toute édition d'écrits, de composition musicale, de dessin, de peinture ou de toute autre production imprimée ou gravée au mépris des lois relatives à la propriété des auteurs, est une contrefaçon, 425. — Le débit d'ouvrages contrefaits et l'introduction de ceux qui, après avoir été imprimés en France, ont été contrefaits chez l'étranger, sont un délit de la même espèce, 426. — Peines contre le contrefacteur, l'introducteur et le débitant, 427. — Confiscation de l'édition contrefaite, des planches, des moules et des matrices, *ibid.* — Voy. *Fausse monnaie, Fausses clefs.*

CONTRIBUTIONS. Voy. *Force publique.*

CORPS LÉGISLATIF. Voy. *Forfaiture, Mandat.*

CORRESPONDANCE criminelle avec les ennemis de l'État, 77 et suiv.; dans l'intérieur, 123. — Voy. *Ministres des cultes.*

CORRUPTION. Peines qui se prononcent contre des individus convaincus d'avoir soustrait, par corruption, fraude ou violence, des plans de fortification etc., et de les avoir livrés à l'ennemi, 82. — Peines contre tout fonctionnaire public, tout agent ou préposé d'une administration publique qui aurait agréé des offres ou promesses, ou reçu des dons ou présents pour faire un acte de sa fonction non sujet à salaire, ou s'être abstenu d'en faire un qui entrait dans l'ordre de ses devoirs, 177 et 178. — Celui qui aurait contraint, séduit, corrompu, ou tenté de contraindre ou corrompre un fonctionnaire ou préposé d'une administration publique, serait puni des mêmes peines, 179. — Cas où la peine de ces tentatives serait moindre, *ibid.* — Confiscation des choses livrées au profit des hospices des lieux, 180. — Peine contre le juge prononçant en matière criminelle ou le juré qui se serait laissé séduire, 181 et 182. — Voy. *Mœurs.*

COSTUME. Peines contre toute personne qui aurait publiquement porté un costume, un uniforme ou une décoration qui ne lui appartenait pas, 259. — Peine pour arrestation illégale faite sous un faux costume, 344. — Pour vols commis sous l'uniforme ou

le costume d'un fonctionnaire public ou d'un officier
civil ou militaire, 381 et 384.

COUPABLES. Voy. *Excuses.*

COUPS. Voy. *Blessures.*

COUTEAUX. Voy. *Armes.*

COUTRES DE CHARRUE. Voy. *Instruments.*

CRIEURS. Réduction des peines encourues par les crieurs,
afficheurs, vendeurs ou distributeurs d'écrits publiés
sans nom d'auteur et d'imprimeur, lorsqu'ils les auront fait connaître, 284. — A défaut de révélation,
et lorsque l'écrit imprimé contient quelque provocation à des crimes, les crieurs, etc., sont punis comme
complices, 285. — Même disposition relativement à
la distribution de figures ou images obscènes, 287.—
Peines contre tout individu qui ferait le métier de
crieur ou d'afficheur sans autorisation de la police, 290.

CRIME. Quelle sorte d'infraction aux lois est ainsi qualifiée, 1. — En quel cas la tentative de crime est
considérée comme le crime même, 2. — Les dispositions du Code ne sont pas applicables aux crimes
militaires, 5. — Il n'y a ni crime ni délit lorsque
le prévenu a été contraint par une force à laquelle
il n'a pu résister, 64. — Crimes et délits contre la
chose publique, 75 et suiv.; contre les personnes,
295 et suiv.; contre les propriétés, 380 et suiv. —
Voy. *Complices, Complot, Délit, Démence, Excuses, Peines, Recéleurs, Récidive, Révélation.*

CRIMINELS. (recèlement de) Voy. *Recèlement.*

CRITIQUES. Voy. *Censure de l'autorité publique.*

CROCHETS. Voy. *Fausses clefs.*

CULTES. Peines pour troubles apportés à l'ordre public
par les ministres des cultes dans l'exercice de leur mi-

nistère, 199 et suiv. — Peines pour entraves apportées au libre exercice des cultes, 260. et 261. — Pour outrages faits, par paroles ou gestes, aux objets d'un culte dans les lieux servant à son exercice, ou aux ministres de ce culte dans leurs fonctions, 262 et 263. — Peine encourue par tout individu qui, sans l'autorisation du Gouvernement, aurait accordé sa maison pour l'exercice d'un culte, 294. — Voy. *Boutiques, Fêtes religieuses.*

CURATELLE. Voy. *Tutelle.*

CURATEUR. Il en est nommé un pour gérer et administrer les biens de celui qui est en état d'interdiction pendant la durée d'une condamnation aux travaux forcés à temps ou à la réclusion, 29. — Compte à rendre par ce curateur après l'expiration de la peine, 30. — Voy. *Tutelle.*

CURÉS et VICAIRES. Voy. *Ministres des cultes.*

D.

DÉBAUCHE. Voy. *Mœurs.*

DÉBITANTS. Voy. *Boissons falsifiées, Crieurs.*

DÉCHARGE. Voy. *Extorsion.*

DÉCLARATIONS. Celles que doivent faire les personnes qui ont connaissance de complots formés ou de crimes projetés contre la sûreté de l'État, 103 et suiv.; les personnes qui ont assisté à un accouchement, 346; ou qui consentent à se charger d'un enfant trouvé, 347.

DÉCORATION. Voy. *Costume.*

DÉGATS. Voy. *Pillage.*

DÉGRADATION. Voy. *Destruction, Monuments.*

DÉGRADATION CIVIQUE. Cette peine est infamante, art. 8.

— En quoi elle consiste, 34. — Celui qui, ayant été condamné pour un crime, en a commis un second emportant la peine de la dégradation civique, doit être condamné à la peine du carcan, 56. — Peine de la dégradation civique contre les fonctionnaires publics qui n'auraient pas déféré à une réquisition tendant à constater des détentions illégales et arbitraires, 119. — Autres cas où ils encourent la même peine, 121, 122, 127 et 130. — Cas dans lesquels la dégradation civique est la peine infligée aux fonctionnaires publics pour forfaiture, 167. — Cette peine est prononcée contre le juge ou l'administrateur qui se serait décidé, par faveur ou par inimitié pour une partie, 183 ; contre celui à qui le serment aurait été déféré en matière civile, et qui se serait rendu faussaire, 366. — *Voy. Arrêt, Forfaiture.*

DÉLAISSEMENT. *Voy. Exposition d'enfant.*

DÉLIBÉRATIONS DE FAMILLE. Cas dans lesquels les tribunaux peuvent interdire temporairement l'exercice du droit de vote et de suffrage dans ces délibérations, 42 et 43.

DÉLITS. Quelle sorte d'infraction aux lois est ainsi qualifiée, 1. — Cas dans lesquels la tentative de délits est considérée comme le délit lui-même, 3. — Les dispositions du Code ne sont pas applicables aux délits militaires, 5. — Peine encourue par celui qui, ayant été condamné pour un crime, a commis un délit de nature à être puni correctionnellement, 57 ; et par les coupables condamnés à un emprisonnement de plus d'une année qui commettent un nouveau délit, 58. — *Voy. Complices, Crimes, Démence, Excuses, Peines, Recéleurs.*

DÉMENCE. Il n'y a ni crime ni délit, lorsque le prévenu était en état de démence au temps de l'action, 64.

DÉNI DE JUSTICE. Peines contre tout juge ou tribunal, tout administrateur ou autorité administrative qui, sous quelque dénomination que ce soit, même du silence ou de l'obscurité de la loi, aurait dénié de rendre justice aux parties, 184.

DENIERS PUBLICS. Voy. *Bandes armées.*

DÉNONCIATION. Voy. *Calomnie, Détention arbitraire, Fournisseurs.*

DENRÉES. Voy. *Marchandises.*

DÉPORTATION. Elle est au nombre des peines afflictives et infamantes, 7. — En quoi elle consiste, 17. — A quoi est condamné, et sur la seule preuve d'identité, le déporté qui rentre sur le territoire de l'Empire, *ibid*; et celui que l'on saisit dans les pays occupés par les armées françaises, *ibid.* — Le Gouvernement peut accorder aux déportés, dans le lieu de la déportation, l'exercice des droits civils, 18. — Celui qui, ayant été condamné pour crime, en a commis un second emportant la peine de la déportation, doit être condamné aux travaux forcés à perpétuité, 56. — L'accusé ayant moins de vingt ans qui, agissant avec discernement, a encouru la peine de la déportation, ne doit être condamné qu'à celle de dix à vingt ans d'emprisonnement dans une maison de correction, 67. — Peine de la déportation contre ceux qui, par des actions hostiles, auraient occasionné une guerre, 85; contre les personnes faisant partie de bandes armées qui auraient été saisies sur les lieux de réunion, 96; contre les membres d'autorités civiles ou militaires qui auraient provoqué des mesures propres à empêcher l'exécution des lois ou des ordres du Gouvernement, 124; contre les fonctionnaires qui auraient requis ou ordonné l'emploi de la force publique pour empêcher l'exécution d'une loi, etc., lorsque la réquisition ou l'ordre ont été suivis de leur effet, 189 ; contre les ministres des cultes qui, pour la troisième fois, auraient procédé aux cérémonies d'un mariage sans justification de l'acte préalablement reçu par les officiers de l'état civil, 198 ; et contre celui qui, dans un écrit pastoral, aurait provoqué la désobéissance aux lois ou autres actes de l'autorité publique, 204 et 206. — Les étrangers déclarés vagabonds par jugement peuvent être conduits hors du territoire de l'Empire, 272. — Voy. *Age*, *Arrêt*, *Bannissement*, *Condamnation*.

DÉPOSITAIRES DE L'AUTORITÉ ET DE LA FORCE PUBLIQUES. Voy. *Autorité publique*, *Force publique*, *Outrages*, *Violences*.

DÉPOSITAIRES PUBLICS. Voy. *Scellés*, *Soustraction*.

DÉPÔTS DE MENDICITÉ. Voy. *Mendicité*.

DÉSORDRES. Voy. *Cultes*.

DESSIN. Voy. *Contrefaçon*.

DESSINATEURS. Voy. *Crieurs*, *Gravures*.

DESTITUTION. Voy. *Fonctionnaires publics*.

DESTRUCTION. Peine de quiconque aurait volontairement détruit ou renversé des bâtiments, maisons, édifices, ponts, digues, chaussées ou autres choses immobilières appartenant à autrui, 437. — Augmentation de peine, s'il y eu homicide ou blessure, *ibid*. — Peines encourues par ceux qui auraient volontairement brûlé ou détruit des registres, minutes ou actes originaux de l'autorité publique; des titres, billets, lettres de change, effets de commerce ou de banque, contenant obligation ou opérant décharge, 439. — Voy. *Monuments*, *Scellés*.

DÉTENTIONS ARBITRAIRES. Peine encourue par les fonctionnaires publics chargés de la police administrative ou judiciaire qui auraient refusé ou négligé de déférer à une réclamation tendant à constater des détentions illégales et arbitraires, 119.

DÉTENUS. Les détenus dans des maisons de correction y sont employés à des travaux dont le produit est en partie appliqué à leur profit, 41.——Peines encourues, suivant les circonstances qui ont accompagné une évasion de détenus, par les huissiers, les commandans en chef ou en sous-ordre, soit de la gendarmerie, soit de la force armée servant d'escorte ou garnissant les postes, les concierges, gardiens, geôliers, et tous autres préposés à la conduite, au transport ou à la garde des détenus, 237.

DETTES. La confiscation générale des biens d'un condamné est grevée de toutes les dettes légitimes jusqu'à concurrence de la valeur de ses biens, 33.

DÉVASTATION. Peine contre les complots ayant pour but de porter la dévastation, le massacre et le pillage dans une ou plusieurs communes, 91.

DEVIN. Voy. *Songes.*

DIGUES. Voy. *Destruction.*

DIMANCHES. Voy. *Condamnation.*

DISCERNEMENT. Ce qui est prononcé à l'égard d'individus âgés de moins de seize ans qui ont commis des crimes ou délits avec ou sans discernement, 66 et 67.

DISCOURS. Sont punis comme coupables de rebellion ceux qui l'ont provoquée par des discours tenus dans des réunions ou lieux publics, par des placards affichés ou par des écrits imprimés, 217.

DISCOURS PASTORAL. Voy. *Censure de l'autorité publique.*

DISTRIBUTEURS. Voy. *Crieurs,*

DOMESTIQUE. Peine pour vols commis par un domestique ou un homme de service à gages, 386.

DOMICILE. Peines contre tout juge, tout procureur général ou impérial, tout substitut ou tout autre officier de justice ou de police qui se serait introduit dans le domicile d'un citoyen hors les cas prévus et sans les formalités prescrites, 184.

DOMMAGE. Amende contre ceux qui auraient volontairement causé du dommage aux propriétés mobilières d'autrui, 479.—Voy. *Destruction.*

DOMMAGES-INTÉRÊTS. Les sommes provenant des paiements faits par les cautions d'un individu condamné pour crimes ou délits sont affectées de préférence aux dommages-intérêts envers les parties lésées, 46. — Mode de demande et de règlement des dommages-intérêts prononcés à raison d'attentat contre la liberté, 117. — Dommages-intérêts dont sont tenus les fonctionnaires publics qui n'auraient pas constaté des détentions illégales et arbitraires, et ne les auraient pas dénoncées aux autorités supérieures, 119. — Dommages-intérêts dans le cas d'abus de confiance, 406; et lorsque le service des armées a manqué par la négligence des fournisseurs, ou a été compromis par fraude sur la qualité des choses fournies, 430 et 433. — Voy. *Condamnation, Injonction, Restitution, Solidarité.*

DONS. Ceux qui, par dons, promesses, menaces, abus d'autorité ou de pouvoir, machinations ou artifices coupables, ont provoqué à un crime ou délit, en sont réputés complices, 60. — Voy. *Corruption.*

DROITS. Voy. *Interdiction.*

DROITS CIVILS. Ceux dont sont privés les individus con-

damnés aux travaux forcés à temps, au bannissemen,,
à la réclusion, au carcan ou à la dégradation civique.
28 et 34. — Les tribunaux peuvent, dans certains
cas, interdire temporairement l'exercice des droits civils, 42 et 43. — Voy. *Déportation*.

DROITS CIVIQUES. On peut prononcer l'interdiction temporaire de l'exercice des droits de vote, d'élection et
d'éligibilité, 42. — Crimes et délits relatifs à l'exercice des droits civiques, 109 à 113. — L'interdiction
des droits civiques est une des peines encourues par
les fonctionnaires qui concertent des mesures non autorisées par les lois, 123. — Voy. *Interdiction*.

DROITS DE FAMILLE. Voy. *Interdiction*.

E.

ÉCHAFAUD. Le coupable condamné à mort pour parricide,
y est exposé pendant la lecture de l'arrêt, 13.

ÉCHENILLAGE. Peines contre ceux qui auraient négligé
d'écheniller dans les campagnes ou jardins où ce soin
est prescrit, 471.

ÉCLAIRAGE. Amende contre les aubergistes et autres qui
auront négligé d'éclairer les matériaux par eux entreposés, ou les excavations par eux faites dans les rues
et places, 471.

ÉCRITEAU. Ce que doit porter celui qu'on place sur la tête
des individus attachés au carcan sur la place publique, 22.

ÉCRITS. Délits commis par la voie d'écrits, bulletins, affiches, journaux, etc., distribués sans nom d'auteurs
ou imprimeurs, 283 et suiv. — Voy. *Contrefaçon,
Discours, Extorsion, Menaces*.

ÉCRITURES. Voy. *Faux*.

ÉDIFICES. Voy. *Destruction, Incendie, Mine*.

ÉDITION. Voy. *Contrefaçon*.

Effet. Différence des peines prononcées quand les crimes ou délits ont ou n'ont pas reçu d'effet, 84 et 94.

Effets publics. Ceux qui ont contrefait ou falsifié des effets émis par le trésor public, ou des billets de banque autorisés par la loi, ou qui ont fait usage de ces eff ts falsifiés, sont punis de mort, et leurs biens confisqués, 139. — Peine contre ceux qui, par des voies ou moyens frauduleux, auraient opéré la hausse ou la baisse des effets publics, 419. — Les paris qui auraient été faits sur cette hausse ou baisse, sont punis des mêmes peines, 421. — Quelle convention est réputée pari de ce genre, 422. Voy. *Destruction*.

Effraction. Ce qui est qualifié effraction, 392. — Division des effractions en extérieures ou intérieures, 393 à 396. — Voy. *Escalade*.

Élection. Cas dans lesquels les tribunaux peuvent interdire l'exercice des droits d'élection, 42.

Éligibilité. Cas dans lesquels les tribunaux peuvent interdire pour un temps l'exercice du droit d'éligibilité, 42 et 43. — Interdiction temporaire du droit d'éligibilité contre ceux qui auraient empêché l'exercice des droits civiques, 109.

Éloignement. Cas dans lesquels un coupable de voies de fait envers un magistrat est tenu de s'éloigner du lieu où il habite, 229.

Empereur. Voy. *Attentat*, *Lèse-majesté*.

Empiètements. Peines auxquelles donnent lieu les empiétements des autorités administratives et judiciaires, 127 et suiv.

Emplois publics. Les fonctionnaires qui concertent entre eux des mesures non autorisées par les lois encourent la peine d'une interdiction temporaire de tout emploi public, 123. — Voy. *Administration , Fonctions publiques.*

EMPOISONNEMENT. Quel attentat est ainsi qualifié, 301. — De quelle peine ce crime est puni, 302. — Peines contre quiconque aurait empoisonné des chevaux ou autres bêtes de voiture, de monture ou de charge, des bestiaux à cornes, des moutons, chèvres ou porcs, ou des poissons dans les étangs, viviers et réservoirs, 452.

EMPRISONNEMENT. L'emprisonnement à temps dans un lieu de correction est une des peines qui se prononcent en matière correctionnelle, 9. — Maison dans laquelle sont renfermés ceux qui ont été condamnés à la peine d'emprisonnement, et travaux auxquels ils y sont employés, 40. — Durée de la peine d'emprisonnement, *ibid.* — Emprisonnement pour acquit des condamnations pécuniaires, 53. — Conversion de la peine des travaux forcés à perpétuité en un emprisonnement de dix à vingt ans à l'égard des individus âgés de moins de seize ans, 67. — Peine d'un emprisonnement de plus ou de moins de durée pour non-révélation de crimes ou complots contre la sûreté de l'État, 105. — Même peine contre ceux qui auraient empêché l'exercice des droits civiques, 109; contre le concierge d'une maison d'arrêt, etc.; pour détention arbitraire, 120. — Les fonctionnaires qui se seraient coalisés pour concerter des mesures non autorisées par les lois encourent la peine d'un emprisonnement, 123. — Emprisonnement pour non-révélation d'une fabrique ou d'un dépôt connu de monnaies contrefaites, 136. — Divers cas de faux commis dans un passe-port qui donnent lieu à un emprisonnement plus ou moins long, 153 et suiv.;

autres pour les feuilles de route, 156. — Peine de
l'emprisonnement pour faux certificat de maladie ou
d'infirmité, 159 et 160; et pour certificat de bonne
conduite, indigence, etc., 161. — Cas dans lesquels
un emprisonnement de deux à cinq ans a lieu pour
soustractions commises par des dépositaires publics,
171. — Commis et préposés de fonctionnaires publics
qui encourent l'emprisonnement pour concussion par
eux commise, 174. — Emprisonnement pour tentative de contrainte, séduction ou corruption 179;
pour délits relatifs à la tenue des actes de l'état civil,
193. — Durée de l'emprisonnement encouru par le fonctionnaire public révoqué, suspendu, etc. qui aurait continué l'exercice de ses fonctions, 197. — Peine de
l'emprisonnement contre tout ministre de culte qui,
pour la seconde fois, aurait procédé aux cérémonies
religieuses d'un mariage sans justification de l'acte
préalablement reçu par les officiers de l'état civil,
200; et contre celui qui se serait permis de censurer,
dans un discours pastoral, l'autorité publique, 201
et 202. — Cas de rébellion pour lesquels cette peine
est prononcée, 211, 212, 217 et 218. — Cas dans
lesquels les outrages et violences envers les dépositaires de l'autorité et de la force publique donnent
lieu à un emprisonnement, 222 et suiv. — Celui
auquel peut donner lieu le refus d'un service dû légalement, 234 et 236. — Divers cas où cette peine
se prononce contre les gardiens ou préposés à la
conduite de détenus qui s'évadent, 238 et suiv. —
Contre les personnes qui ont laissé commettre un
bris de scellés et un enlèvement de pièces dans les
dépôts publics, 249 et suiv.; pour recèlements de
criminels et dégradation de monuments, 248 et

257 ; pour entraves au libre exercice des cultes, 260 et suiv. ; pour vagabondage et mendicité, 271 et suiv. ; pour publication, distribution ou affiche d'écrits, pamphlets et gravures obscènes publiés, sans nom d'auteur, d'imprimeur ou de graveur, 283 et suiv. ; pour menaces écrites ou verbales, 3o6 et 3o7 ; pour blessures et coups volontaires, 311 et suiv. ; pour fabrication ou débit d'armes prohibées, 314 ; pour vente ou débit de boissons falsifiées, 318 ; pour homicide ou blessures involontaires, 319 et 320. — Durée de l'emprisonnement pour les cas dans lesquels les auteurs de crimes et délits emportant plus forte peine ont été reconnus excusables, 326. — Emprisonnement pour attentats aux mœurs, 330 et suiv. ; pour crimes et délits tendant à empêcher ou détruire la preuve de l'état civil d'un enfant, ou à compromettre son existence, 346 et suiv. ; pour enlèvement d'une fille mineure quand le ravisseur n'a pas atteint vingt-un ans. 356 ; pour infraction aux lois sur les inhumations, 358 et suiv. ; pour calomnies, injures et révélation de secrets, 373 et suiv. ; pour contrefaçon ou altération de cléfs, 399 ; pour larcin et filouterie, 401 ; pour banqueroute simple, 402 ; pour escroquerie, 405 ; pour abus de confiance, 406 et suiv. ; pour contravention aux règlements sur les maisons de jeu et de prêt, 410 et 411 ; pour entraves apportées à la liberté des enchères, 412 ; pour violation des règlements relatifs aux manufactures, au commerce et aux arts, 414 ; pour délits des fournisseurs, 433 ; pour destruction, dégradation et dommages de différentes sortes, 438 et suiv. — Circonstances qui permettent aux tribunaux de réduire la durée de l'emprisonnement pour les délits de police

correctionnelle, 463. — Durée que ne peut excéder l'emprisonnement pour contraventions de police, 465. — Comment se calculent les jours d'emprisonnement, *ibid.* — Contraventions de police pour lesquelles la peine d'emprisonnement peut être prononcée, 472.— Délits pour lesquels elle a toujours lieu en cas de récidive, 474.— Circonstances suivant lesquelles l'emprisonnement peut avoir lieu pour délits de police correctionnelle, 482.— Cas dans lesquels cette peine est toujours prononcée pour récidive, 483. — Voy. *Age, Amendes, Excuses.*

ENCHÈRES. Peines pour entraves apportées à la liberté des enchères ou des soumissions, 412.

ENCLOS. Quelle espèce de clôture constitue un terrain réputé parc ou enclos, 391. — Les parcs mobiles destinés à contenir du bétail dans la campagne sont aussi réputés enclos, 392. — Lorsqu'ils tiennent aux cabanes mobiles des gardiens, ils sont réputés dépendants de la maison habitée, *ibid.*

ENFANT. Peines contre les coupables d'enlèvement, de recélé ou de suppression d'un enfant, de substitution d'un enfant à un autre, ou de supposition d'un enfant à une femme qui ne serait pas accouchée, 345. — Mêmes peines contre ceux qui, étant chargés d'un enfant, ne le représenteraient pas aux personnes ayant droit de le réclamer, *ibid.* — Peine contre toute personne qui, ayant trouvé un enfant nouveau-né, ne l'aurait pas remis à l'officier de l'état civil, à moins qu'elle n'ait déclaré consentir à se charger de cet enfant, 347. — Peines contre ceux qui auraient porté à un hospice un enfant au dessous de l'âge de sept ans à eux confié, 348.— Exception, *ibid.*— Peines contre ceux qui l'auraient exposé et délaissé, 349 et suiv.

ENGAGEMENT. Voy. *Enrôlement.*

ENLÈVEMENT. Voy. *Enfants, Mineurs.*

ENLÈVEMENT DE PROCÉDURES CRIMINELLES. Voy. *Scellés.*

ENNEMI. Voy. *Intelligence.*

ENRÔLEMENT. Peine de mort, avec confiscation de biens, pour engagement ou enrôlement de soldats sans ordre du pouvoir légitime, 92.

ENTREPRISES. Voy. *Fournisseurs.*

ÉPIZOOTIE. Peines contre tout détenteur ou gardien d'animaux ou de bestiaux soupçonnés d'être infectés de maladies contagieuses, qui n'aurait pas averti sur-le-champ le maire de la commune où il se trouve, et ne les aurait pas tenus provisoirement renfermés, 459. —— Même peine contre ceux qui, au mépris des défenses de l'administration, auraient laissé leurs animaux ou bestiaux infectés communiquer avec d'autres. 460. —— Peine plus grave dans le cas où il serait résulté de la communication une contagion parmi les autres animaux, 461.

ÉPOUX. Voy. *Adultère, Meurtre.*

ESCALADE. Les crimes ou délits sont excusables lorsqu'ils ont été commis en repoussant, pendant le jour, l'escalade ou l'effraction des clôtures, 322. —— Il n'y a ni crime ni délit quand l'homicide a été commis ou les coups portés en repoussant pendant la nuit l'escalade ou l'effraction des clôtures, 329. —— Peines pour vols commis à l'aide d'effraction extérieure ou d'escalade dans un logement habité, 384 et 386. —— Ce qui est qualifié escalade, 397.

ESCROQUERIE. Diverses espèces d'escroqueries, et peines dont elles sont punies, 405. —— Voy. *Interdiction.*

ESPIONNAGE. Peine encourue pour des instructions données aux ennemis de l'État, et qui auraient été la suite d'un concert constituant un fait d'espionnage, 78. — Peine de mort contre ceux qui auraient recélé les espions ou les soldats ennemis envoyés à la découverte, 83.

ÉTANGS. Voy. *Champs, Emprisonnement, Inondations*.

ÉTAT. Ceux qui ont été condamnés pour crimes ou délits qui intéressent la sûreté intérieure ou extérieure de l'État, doivent être renvoyés sous la surveillance de la haute police, 49. — Ceux qui connaissent la conduite criminelle des malfaiteurs exerçant des brigandages ou des violences contre la sûreté de l'État et la paix publique, leur fournissent habituellement logement, lieu de retraite ou de réunion, sont punis comme leurs complices, 61. — Crimes contre la sûreté extérieure de l'État, 86 et suiv. — Crimes tendant à troubler l'État par la guerre civile, l'illégal emploi de la force armée, la dévastation et le pillage publics, 91 et suiv. — Dispositions relatives à la révélation ou non-révélation des crimes qui compromettent la sûreté intérieure ou extérieure de l'État, 103 et suiv. — Voy. *Intelligences*.

ÉTAT CIVIL. Peines pour contraventions propres à compromettre l'état civil des personnes, 199 et suiv. — Crimes et délits tendant à empêcher ou détruire la preuve de l'état civil d'un enfant, 345. — Voy *Actes de l'état civil*.

ÉTRANGERS. Voy. *Transportation*.

ÉVASION DES DÉTENUS. Peines encourues par les individus non chargés de la garde ou conduite des détenus, qui auraient procuré ou facilité leur évasion, 237, 238 et suiv. — Voy. *Détenu*.

EXCAVATIONS. Voy. *Voie publique.*

EXCUSES. Seuls cas dans lesquels les crimes ou délits peuvent être excusés, 65. — Emprisonnement auquel doivent être condamnés les individus déclarés coupables de crimes ou délits, mais excusables, 69. — Celui qui, ayant eu connaissance de crimes ou complots contre la sûreté de l'État, ne les a point révélés, n'est pas admis à excuse, 106. — Exceptions, 107. — Amendes et emprisonnement pour excuses de témoins et de jurés reconnues fausses, 236. — Réduction de peine pour crimes et délits à l'égard desquels l'auteur a été reconnu excusable, 326. — Voy. *Blessures, Castration, Meurtre, Parricide, Surveillance.*

EXÉCUTEURS DE JUGEMENTS. Voy. *Violences.*

EXÉCUTION JUDICIAIRE. Ce qui précède l'exécution d'un coupable condamné à mort pour parricide, 13. — Voy. *Condamnation, Place publique, Secours.*

EXHALAISONS. Amendes contre ceux qui auraient jeté ou exposé au-devant de leurs édifices des choses de nature à nuire par des exhalaisons insalubres, 471.

EXPERT. Celui qui a été condamné à la peine des travaux forcés à temps, au bannissement, à la réclusion ou au carcan, ne peut jamais être expert, 28. — Les tribunaux jugeant correctionnellement peuvent interdire pour un temps le droit d'être nommé expert, 42 et 43.

EXPOSITION D'ENFANT. Peine contre ceux qui auraient exposé et délaissé en un lieu solitaire un enfant au dessous de l'âge de sept ans, ou qui auraient donné l'ordre de l'exposer, 349 et 350. — Augmentation de la peine contre les tuteurs ou instituteurs de l'enfant exposé, 350. — Peines encourues dans le cas ou, par suite de l'exposition et du délaissement, l'enfant serait demeuré mutilé ou estropié, 351. — Peine de ceux qui auraient exposé et délaissé un enfant au dessous de sept ans dans un lieu non solitaire, 352. — Cas particuliers aux tuteurs et instituteurs, 353.

EXPOSITION PUBLIQUE. Les individus ayant moins de seize ans qui ont été condamnés à des peines afflictives ou infamantes, ne subissent pas l'exposition publique, 68. — Voy. *Carcan.*

EXPRESSIONS OUTRAGEANTES. Voy. *Injures.*

EXTORSION. Peines contre celui qui aurait extorqué par force, violence ou contrainte, la signature ou la remise d'un écrit, d'un acte, d'un titre, etc., contenant obligation ou opérant décharge, 400.

F.

FABRIQUES. Peines contre tout directeur, commis ou ouvrier de fabrique, qui aurait communiqué des secrets de cette fabrique à l'étranger, 418. — Peines contre celui qui, à l'aide d'une liqueur corrosive, ou par tout autre moyen, aurait volontairement gâté des marchandises ou matières servant à la fabrication, 443. — Voy. *Manufactures.*

FAILLITE. Voy. *Banqueroute.*

FALSIFICATION. Voy. *Contrefaçon, Effets publics.*

FAMILLE IMPÉRIALE. Voy. *Attentat.*

FARINES. Voy. *Commerce.*

FAUSSAIRE. Tout faussaire condamné à la réclusion ou aux travaux forcés à temps subit préalablement la peine de la marque, 165. — Celui qui a abusé d'un blanc-seing qu'on ne lui avait pas confié, est poursuivi comme faussaire, 407.

FAUSSE MONNAIE. Peine de mort, avec confiscation de
biens, pour contrefaçon ou altération des monnaies
d'or ou d'argent, et émission ou exposition des monnaies contrefaites, 132. — Pour contrefaçon, etc., des
monnaies de billon ou de cuivre, 133. — Pour contrefaçon des monnaies étrangères, 134. — Amendes
pour avoir fait usage de pièces fausses après en avoir
reconnu les vices 135. — Peines pour non-révélation
d'une fabrique ou d'un dépôt de monnaies contrefaites,
136. — Exceptions, 137. — Exemption des peines envers les coupables qui, avant la consommation des crimes, en auraient révélé les auteurs aux autorités constituées, 138. — Surveillance à laquelle ils peuvent
néanmoins être soumis, *ibib.* — Voy. *Contrefaçon.*

FAUSSES CLEFS. Peines contre les individus coupables de
vols commis à l'aide de fausses clefs dans une maison
habitée, 381 et 384. — Tous crochets, rossignols,
passe-partout, clefs imitées, contrefaites ou qui n'ont
pas été destinées aux serrures, cadenas et fermetures
auxquels le coupable les a employés, sont qualifiés
fausses clefs, 398. — Peine de quiconque a contrefait ou altéré des clefs, 399.

FAUSSSES SIGNATURES. Voy. *Faux.*

FAUX. Peines dont sont punis les fonctionnaires ou officiers publics qui ont commis un faux par fausses signatures, par altération des actes, écritures ou signatures, par supposition de personnes ou par des écritures faites ou intercalées sur des registres ou d'autres
actes publics depuis leur confection ou clôture, 145.
— Autre cas de faux en écriture publique, qui donne
lieu à la même peine, 146. — Quelle peine encourent
les autres personnes pour faux commis en écriture
authentique ou en écriture de commerce ou de

banque, 147.—Peine pour avoir fait usage des actes faux, 148. — Peine encourue par tout individu qui a commis un faux en écriture privée ou qui a fait usage de la pièce fausse, 150 et 151. — Faux commis dans les passe-ports, 153 et suiv. — Circonstances dans lesquelles le faux doit être puni des peines de simple police, 163. — L'application des peines portées contre ceux qui ont fait usage de monnaies, billets, sceaux, timbres, marteaux, poinçons, marques et écrits faux, contrefaits ou falsifiés, cesse quand le faux n'a pas été connu de la personne qui a fait usage de la chose fausse, 164. — Lorsque la peine du faux n'est point accompagnée de la confiscation des biens, il est prononcé une amende contre les coupables, *ibid.* — La marque est infligée à tout faussaire condamné à la réclusion ou aux travaux forcés à temps, 165.

FAUX TÉMOIGNAGE (peines contre les coupables de) en matière civile, correctionnelle et de police, 362 et suiv.

FEMME. Lorsqu'il est vérifié qu'une femme condamnée à mort est enceinte, elle ne subit la peine qu'après sa délivrance, 27.

FENÊTRES, (peines contre les expositions nuisibles sur les) 471.

FERMETURES. Voy. *Fausses clefs.*

FÊTES NATIONALES, Voy. *Condamnation.*

FÊTES RELIGIEUSES. Peines contre ceux qui auraient contraint ou empêché de célébrer certaines fêtes, ou d'observer certains jours de repos, 260. — Voy. *Condamnation.*

FEU. Voy. *Fours, Incendie.*

FEUILLES DE ROUTE. Peines encourues par les falsificateurs de feuilles de route, et par ceux qui font usage d'une feuille de route falsifiée, 156. — Les mêmes peines applicables aux personnes qui se sont fait délivrer une feuille de route sous un nom supposé, 156. — Peine contre l'officier public qui a délivré la feuille avec connaissance de la supposition de nom, 158. — Les peines des individus porteurs de fausses feuilles de route sont portées au *maximum* à l'égard des vagabonds ou mendiants, 281.

FEUILLES PÉRIODIQUES. Voy. *Ecrits.*

FIGURES. Voy. *Gravures.*

FILOUTERIES. Peine des larcins et des filouteries, 401. — Voy. *Interdiction, Surveillance de la haute police.*

FLAGRANT DÉLIT. Voy. *Adultère, Forfaiture, Mandat, Secours.*

FLÉTRISSURE des condamnés aux travaux forcés à perpétuité ou à temps, 20 ; des faussaires, *ibid.*

FONCTIONNAIRES PUBLICS. Peine contre les auteurs de fausses signatures de nom d'un fonctionnaire public, 118 ; et contre les fonctionnaires chargés de la police qui n'auraient point déféré à une réquisition tendant à constater les détentions illégales et arbitraires, 119. — Peines auxquelles peut donner lieu tout concert de mesures non autorisées par les lois, et pratiqué par une réunion d'individus ou de corps dépositaires de quelque partie de l'autorité publique, 123. — Augmentation de la peine si les mesures concertées étaient contraires à l'exécution des lois ou ordres du Gouvernement, 124 ; ou si elles avaient pour objet un complot attentatoire à la sûreté intérieure de l'Etat, 125. — Peines pour les délits de fonctionnaires qui se seraient ingérés dans des affaires ou commerces incompatibles avec leur qualité, 175 et suiv. — Peines contre les fonctionnaires publics qui se seraient

laissé corrompre, 177. — Peines encourues par les
coupables de voies de fait et de menaces envers les
fonctionnaires et agents publics, 179.—Peine contre
tout fonctionnaire public révoqué, destitué, suspendu
ou interdit légalement, qui aurait continué l'exercice
de ses fonctions, ou qui, étant électif ou temporaire,
les aurait exercées après avoir été remplacé. 197. —
Peines encourues par les fonctionnaires ou officiers
publics qui auraient participé à des crimes ou délits
qu'ils étaient chargés de surveiller ou de réprimer,
198. — Peines encourues par les fonctionnaires publics pour viol, 333 ; pour connivence avec les fournisseurs à l'effet de faire manquer le service des armées
de terre ou de mer, 432. — Voy. *Administrateurs,
Agents du Gouvernement, Arbres, Concussion,
Force publique, Grains, Lettres, Serment, Violences.*

Fonctions publiques. Les individus condamnés à la
dégradation civique sont destitués et exclus de
toutes fonctions ou emplois publics, 34. — Cas
dans lesquels les tribunaux peuvent interdire l'exercice du droit d'être appelé à des fonctions publiques,
42. — Peines contre celui qui, sans titre, aurait
rempli des fonctions publiques, 258. — Voy. *In-
diction.*

Force publique. Cas dans lesquels il y a lieu à la peine
de mort avec confiscation de biens, ou seulement à la
déportation, contre ceux qui auraient requis ou ordonné l'emploi de la force armée contre la levée des
gens de guerre légalement établie, 94. — Peine
contre tout fonctionnaire public, agent ou préposé du
Gouvernement, pour avoir requis ou ordonné l'emploi de la force publique contre l'exécution d'une loi,
contre la perception d'une contribution légale ou contre

l'exécution d'une ordonnance ou mandat de justice, ou tout autre ordre émané de l'autorité légitime, 188. —— Peine plus forte si la réquisition ou l'ordre ont été suivis de leur effet, 189. —— Cas dans lesquels ces peines ne sont point applicables aux fonctionnaires ou préposés inférieurs, 190. —— Circonstances propres à augmenter la peine des fonctionnaires supérieurs, 191. —— Peines contre tout commandant, tout officier ou sous-officier de la force publique qui, après en avoir été légalement requis par l'autorité civile, aurait refusé de faire agir la force à ses ordres, 234. —— Voy. *Autorité publique, Rébellion.*

Forêts. Voy. *Incendie.*

Forfaiture. Cas dans lesquels les officiers de police judiciaire, les procureurs généraux ou impériaux, leurs substituts, les juges, sont coupables de forfaiture, et punis de la dégradation civique, 121. —— Les fonctionnaires publics qui auraient arrêté de donner des démissions pour entraver un service quelconque, sont coupables de forfaiture et punis de la dégradation civique, 126. —— Divers cas de forfaiture pour les juges, les procureurs généraux ou impériaux, leurs substituts ou les officiers de police judiciaire, 127. —— Tout crime commis par un fonctionnaire public dans ses fonctions est une forfaiture, 166. —— Toute forfaiture pour laquelle la loi ne prononce pas de peine plus grave est punie de la dégradation civique, 167. —— Les simples délits ne constituent pas les fonctionnaires en forfaiture, 168. —— La forfaiture est encourue par tout juge ou administrateur qui se serait décidé par faveur ou par inimitié contre une partie, 181 et 183.

Forteresses. Voy. *Bandes armées, Places.*

FORTIFICATIONS. Voy. *Plans*.

FOSSÉS. Peines contre quiconque en aurait comblé, 456.

FOURNISSEURS. Peines contre tous individus qui, chargés, comme membres de compagnies ou individuellement, de fournitures pour le compte des armées de terre ou de mer, auraient, par leur faute personnelle, fait manquer le service, 430. —Peine plus forte en cas d'intelligence avec l'ennemi, *ibid.* — Cas dans lequel les agents des fournisseurs encourent la même peine, 431. — Peines encourues quand le service n'aurait été que retardé, s'il y a eu négligence ou fraude sur la nature ou la qualité des choses fournies, 433. — La poursuite des fournisseurs ne peut être faite que sur la dénonciation du Gouvernement, *ibid.*

FOURRAGES. Voy. *Grains*.

FOURS. Ceux qui ont négligé d'entretenir, réparer ou nettoyer les fours, cheminées ou usines où l'on fait usage du feu, sont punis d'une amende depuis un fr. jusqu'à cinq, 471. Voy. —*Incendie*.

FOUS. Amendes contre ceux qui en auraient laissé divaguer, 475; contre ceux qui auraient occasionné la mort ou la blessure d'animaux ou de bestiaux par l'effet de la divagation des fous ou furieux, 479. — Emprisonnement, suivant les circonstances, 480.

FRAIS. Les sommes provenant des paiements faits par les cautions d'un individu condamné pour crimes ou délits, sont affectées de préférence aux frais adjugés aux parties lésées, 46. — Durée de l'emprisonnement, pour acquit de frais prononcés au profit de l'État, après laquelle le condamné insolvable peut obtenir sa liberté provisoire, 53. — Voy. *Aubergistes, Condamnation, Responsabilité civile, Solidarité*.

FRANÇAIS. Voy. *Armes, Confiscation*.

FRAUDE. Voy. *Corruption, Marchandises.*

FRUITS. Amende pour avoir cueilli ou mangé sur le lieu les fruits d'autrui, 471.

FURIEUX. Amende contre ceux qui en auraient laissé divaguer, 475. — Voy. *Fous.*

G.

GAGES. Voy. *Maisons de prêt.*

GARDES CHAMPÊTRES. Délits de police correctionnelle qui donnent lieu à une peine plus grave lorsqu'ils ont été commis par des gardes champêtres ou forestiers, ou par des officiers de police, 434 à 462. — Voy. *Rébellion.*

GARDES FORESTIERS. Voy. *Gardes champêtres, Rebellion.*

GARDIENS DE PRISONS. Voy. *Concierges, Détenus.*

GARDIENS DE SCELLÉS. Voy. *Scellés.*

GENDARMERIE. Voy. *Détenus.*

GENS SANS AVEU. Voy. *Vagabondage.*

GEOLIERS. Voy. *Détenus.*

GESTES. Voy. *Outrages.*

GLANAGE. Amende contre ceux qui auraient glané, râtelé ou grapillé dans les champs non encore entièrement dépouillés et vidés de leurs récoltes, ou avant le moment du lever ou après celui du coucher du soleil, 471. — Peine d'emprisonnement pour les mêmes contraventions, 473.

GOUVERNEMENT. Délits qui donnent lieu d'en mettre les auteurs à la disposition du Gouvernement après avoir subi leur peine, 271 et 282. — Voy. *Attentats, Autorisation, Fournisseurs, Surveillance.*

GRAINS. Peines pour avoir coupé des grains ou des fourrages appartenant à autrui, 449. — Peine plus forte quand le crime a été commis en haine d'un fonctionnaire public et à raison de ses fonctions, ou pendant la nuit, 450. — Voy. *Commerce.*

GRAPILLAGE. Voy. *Glanage*.

GRAVURES. Peine encourue pour exposition ou distribution de chansons, pamphlets, figures ou images contraires aux bonnes mœurs, 287 et suiv. — Confiscation des écrits et gravures contraires aux mœurs, lesquels sont mis sous le pilon, 477. — Voy. *Contrefaçon*, *Crieurs*.

GREFFES. Voy. *Arbres*.

GREFFIERS. Voy. *Scellés*.

GROSSESSE. Voy. *Femmes*.

GUERRE. Peines contre ceux qui, par leurs manœuvres, auraient engagé des puissances étrangères à entreprendre la guerre contre la France, 76 ; et contre ceux dont les actions hostiles, non approuvées par le Gouvernement, auraient exposé l'Etat à une déclaration de guerre, 84 ; contre les complots tendant à exciter la guerre civile, 91.

GUET-APENS. Peines encourues pour violences envers des magistrats, lorsqu'il y a eu guet-apens, 332. — En quoi consiste le guet-apens, 298. Voy. *Assassinat*.

H.

HAIES. Peines pour avoir coupé ou arraché des haies vives ou sèches, 456.

HAUSSE. Voy. *Effets publics*.

HAUTE POLICE. Voy. *Surveillance de la haute police*.

HÉRITAGES. Voy. *Limites*.

HOMICIDE. Lorsque l'homicide a été commis volontairement, il est qualifié de meurtre, 295. — Peine de celui qui, par maladresse, imprudence, inattention, négligence ou inobservation des règlements, a commis involontairement un homicide, ou en a été involontairement la cause, 319. — Cas dans lesquels l'homi-

cide n'est qualifié ni de crime ni de délit, 327 et suiv.
— Voy. *Destruction.*

HOSPICES. Les réunions d'individus admis dans les hospices sont punies comme réunions de rebelles, 219.
— Voy. *Enfants.*

HÔTELIERS. Voy. *Aubergistes.*

HUISSIERS. Voy. *Arrêt, Détenus.*

I.

IDENTITÉ. Sur la seule preuve d'identité, le déporté rentré dans l'Empire est condamné aux travaux forcés à
perpétuité, 17. — Sur la même preuve, à l'égard du
banni, la peine de la déportation est contre lui prononcée, 33.—Voy. *Déportation.*

IMAGES. Voy. *Gravures.*

IMMONDICES. Amendes contre ceux qui imprudemment
auraient jeté des immondices sur quelques personnes,
471 ; et contre ceux qui en auraient jeté contre les maisons d'autrui ou volontairement sur quelqu'un, 474.
— Emprisonnement pour les mêmes causes, 476

IMPRESSION. Arrêts susceptibles d'être imprimés par extrait, 36. — Voy. *Contrefaçon.*

IMPRIMEURS. Voy. *Écrits, Gravures.*

IMPRUDENCE. Voy. *Homicide.*

IMPUTATION. Voy. *Calomnie.*

INATTENTION. Voy. *Homicide.*

INCENDIE. Peine de mort avec confiscation de biens contre
tout individu qui aurait incendié des édifices, des vaisseaux, des magasins ou autres propriétés appartenant à
l'État, 95. — Même peine contre quiconque aurait mis
le feu à des maisons, bâtiments, chantiers, forêts, bois
taillis, récoltes, 434. — Peine pour incendie causé
par vétusté, défaut de réparation ou nettoyage des

feurs, cheminées, maisons ou usines prochaines, ou par des feux allumés dans les champs à moins de cent mètres des maisons, édifices, forêts, bois, meules de grains, etc., ou par des feux ou lumières portés ou laissés sans précaution, et par des pièces d'artifice allumées ou tirées par négligence ou imprudence, 458. — Voy. *Menaces, Destruction, Secours.*

INDEMNITÉS. Celles auxquelles le coupable doit être condamné envers la partie quand il y a eu lieu à restitution, 51. — La cour ou le tribunal ne peut, même du consentement de la partie, appliquer ces indemnités à une œuvre quelconque, *ibid.* — Mode de règlement des indemnités dues aux auteurs d'ouvrages contrefaits ou de pièces représentées au mépris des règlements sur la propriété des auteurs, 429. — Voy. *Aubergistes, Responsabilité civile, Restitution.*

INDIGENCE. Voy. *Certificat d'indigence.*

INDUSTRIE. Peines contre ceux qui, dans la vue de nuire à l'industrie française, auraient fait passer en pays étranger des directeurs, commis ou ouvriers d'un établissement, 417. — Voy. *Manufacture.*

INFANTICIDE. C'est le meurtre d'un enfant nouveau-né, 300. — Ce crime est puni de mort, 302.

INFRACTIONS. Distinction entre celles que les lois punissent des peines correctionnelles ou afflictives, 1.

INHUMATION. La famille qui réclame le corps d'un supplicié doit le faire inhumer sans appareil, 14. — Peines contre ceux qui auraient fait inhumer un individu décédé sans l'autorisation préalable de l'officier public, ou auraient contrevenu d'une manière quelconque aux réglements concernant les inhumations précipitées, 358 et 359.

INJONCTIONS. Injures publiées dans les plaidoyers ou dans les écrits relatifs à la défense des parties qui donnent lieu aux juges d'en prononcer la suppression, de faire des injonctions aux auteurs du délit, de les suspendre de leurs fonctions, et de prononcer des dommages-intérêts, 377.

INJURES. Peines pour injures ou expressions outrageantes proférées dans des lieux publics ou insérées dans des écrits imprimés, 375; et pour celles qui n'auraient pas eu le double caractère de la gravité et de la publicité, 376. — Ce que doivent faire les juges saisis d'une contestation à l'égard des imputations et des injures contenues dans les écrits relatifs à la défense des parties ou dans les plaidoyers, 377. — Amende encourue par ceux qui auraient proféré des injures verbales contre quelqu'un, 471.

INONDATIONS causées par les propriétaires de moulins, étangs et usines; peines contre ceux qui les ont occasionnées, 457.

INSOLVABILITÉ. Durée de l'emprisonnement après laquelle le condamné insolvable peut obtenir sa liberté, 467. —Voy. *Amendes, Frais, Liberté provisoire.*

INSTIGATEURS. Voy. *Réunion armée, Sédition.*

INSTITUTEURS. Peines par eux encourues pour viol, 333. — Voy. *Exposition d'enfant.*

INSTRUMENTS D'AGRICULTURE. Amende contre ceux qui auraient laissé dans les champs des coutres de charrue, des pinces, des barres ou autres instruments dont pourraient abuser les voleurs et autres malfaiteurs, 471.— — Confiscation de ces instruments, 472. — Voy. *Champs.*

INSTRUMENTS DE CRIME. Cas dans lequel ceux qui ont procuré des armes, des instruments ou tout autre moyen pour faciliter l'exécution d'un crime, sont réputés complices, 60. — Peine pour avoir fourni des instruments de crime à des bandes armées, 96. — Voy. *Armes, Associations de malfaiteurs, Vagabondage.*

INTELLIGENCES. Peine de mort, avec confiscation de biens, encourue pour avoir pratiqué des machinations ou entretenu des intelligences avec les puissances étrangères à l'effet de les engager à commettre des hostilités contre la France, 76. — Pareille peine pour avoir pratiqué des manœuvres avec les ennemis de l'État à l'effet de leur faciliter l'entrée sur le territoire de l'Empire, de leur livrer des places ; de leur fournir des secours, etc., 77. — Peine du bannissement contre ceux dont la correspondance, sans avoir eu un but criminel, a procuré aux ennemis des instructions nuisibles à la France ou à ses alliés, 78. — Mêmes peines pour les manœuvres commises envers les alliés de la France agissant contre l'ennemi commun, 79 ; et pour avoir trahi le secret d'une négociation ou d'une expédition, ou livré des plans de ports et fortifications, 80 et 81. — Simple peine du bannissement si ces plans ont été livrés aux agents d'une puissance neutre ou alliée, 81. — Peine de mort et confiscation de biens pour intelligences pratiquées avec des bandes armées illégalement, 96. — Peines encourues par les fournisseurs qui auraient fait manquer le service des armées, et par les fonctionnaires qui auraient aidé les coupables, en cas d'intelligence avec l'ennemi, 430 et 432.

INTERCALATION D'ÉCRITURE. Voy. *Faux.*

INTERDICTION. L'interdiction à temps de certains droits civiques, civils ou de famille, est une des peines qui se prononcent en matière correctionnelle, 9. — Celui qui a été condamné à la peine des travaux forcés à temps, ou à la réclusion, est, pendant la durée de la peine, en état d'interdiction légale, 29. — Les tribunaux jugeant correctionnellement peuvent, dans certains cas, interdire temporairement, en tout ou en partie, l'exercice des droits civiques, civils et de famille, 42, 43 et 109. — Interdiction temporaire de l'exercice des fonctions publiques pour déni de justice et suppression ou ouverture de lettres, 185 et 187; pour exercice prolongé de ses fonctions de la part d'un fonctionnaire public revoqué, etc., 197. — Interdiction de tutelle et curatelle pour les pères, mères et autres personnes chargées de la surveillance de jeunes gens dont ils ont favorisé la prostitution, 334. — Droits dont les coupables de larcins et filouteries peuvent être interdits, 401. — Pareille interdiction contre les coupables d'escroqueries, 405; contre les coupables d'abus de confiance, 406; contre les individus qui auraient établi ou tenu des maisons de jeu, 410. — Voy. *Fonctionnaires publics*.

INTRODUCTION. Voy. *Contrefaçon*.

J.

JEUX DE HASARD. Amende contre ceux qui auraient établi dans les rues, chemins, places et lieux publics, des jeux de loterie ou d'autres jeux de hasard, 477. — Confiscation des tables, instruments et appareils de ces jeux ou loteries, ainsi que des enjeux et des lots proposés au jeu, 481. — Voy. *Maisons de jeux*.

JOURNALIERS. Voy. *Ouvriers*.

JOURNAUX. Voy. *Écrits*.

JOURS. Ceux d'emprisonnement sont de 24 heures, 465.

JOURS DE REPOS. Voy. *Fêtes religieuses*.

JUGEMENT. Voy. *Mandat*.

JUGES. Peine encourue par tout juge qui aurait détruit, supprimé, soustrait ou détourné les actes et titres dont il était dépositaire en cette qualité, ou qui lui auraient été communiqués en raison de ses fonctions,

173; qui se serait laissé corrompre ou qui se serait décidé par faveur ou par inimitié, 181 et 183. — Voy. *Corruption, Domicile, Déni de justice, Loi, Accusation, Autorité administrative, Pouvoir législatif, Forfaiture, Mandat, Dégradation civique, Revendication, Préposés du Gouvernement.*

JURÉS. Celui qui a été condamné à la peine des travaux forcés à temps, au bannissement, à la réclusion ou au carcan, ne peut jamais être juré, 28. — Les tribunaux peuvent, en certains cas, interdire pour un temps l'exercice du droit de remplir les fonctions de juré, 42 et 43. — Voy. *Corruption, Excuses.*

L.

LARCIN. Voy. *Filouterie.*

LÈSE-MAJESTÉ. Ce crime est puni comme parricide, et emporte de plus la confiscation des biens, 86. — Peine contre ceux qui, instruits de complots tendant au crime de lèse-majesté, ne les ont point révélés, 104.

LETTRES. Peines contre les fonctionnaires ou agents du Gouvernement ou de l'administration des postes, pour avoir commis ou facilité la suppression ou l'ouverture des lettres, 187.

LETTRES DE CHANGE. Voy. *Destruction.*

LIBERTÉ. Peine de la dégradation civique contre tout fonctionnaire, agent ou préposé du Gouvernement qui aurait ordonné ou fait quelque acte arbitraire et attentatoire à la liberté individuelle ou aux droits civiques des citoyens, 114. — Voy. *Insolvabilité, Restitution.*

LIBERTÉ PROVISOIRE. Durée de l'emprisonnement pour amendes et frais prononcés au profit de l'État, après laquelle le condamné insolvable peut l'obtenir, sauf à reprendre la contrainte par corps, s'il lui survient des moyens de solvabilité, 53.

LICENCIEMENT. Voy. *Commandement militaire.*

LIMITES. Peines pour avoir déplacé ou supprimé des bornes, pieds corniers, ou autres arbres plantés ou reconnus pour établir les limites entre différents héritages, 456.

LIQUEUR CORROSIVE. Voy. *Fabriques.*

LOGEMENT. Voy. *Aubergistes, Etat.*

LOGEMENT DE MALFAITEURS. Voy. *Association de malfaiteurs.*

LOGEURS. Voy. *Aubergistes.*

LOI. Peines contre les juges, les procureurs généraux, etc., qui auraient arrêté ou suspendu l'exécution des lois, ou délibéré sur leur publication ou exécution, 127. — Exécution des dispositions des lois et des règlements en vigueur en tout ce qui n'a pas été réglé par le Code, 484. — Voy. *Force publique, Bannissement, Déportation.*

LOTERIES. Peines contre ceux qui auraient établi ou tenu des loteries non autorisées par la loi, et contre les administrateurs et agents de ces établissements, 410. — Voy. *Jeux de hasard.*

M.

MACHINATIONS. Voy. *Dons, Intelligences.*

MACHINES. Voy. *Armes.*

MAGASINS. Voy. *Bandes armées, Boutiques, Incendie, Mine, Places.*

MAIRES. Voy. *Aubergistes, Epizootie, Préfets, Registres.*

MAISON. Quels bâtiments sont réputés maison habitée, 390. — Peines contre ceux qui auraient occasionné la mort d'animaux ou de bestiaux par le defaut de réparation ou d'entretien des maisons tombant en ruines, et par la négligence à placer les signaux d'usage, 479. —Voy. *Destruction*, *Incendie*, *Mine*.

MAISONS DE CORRECTION. On y renferme les individus condamnés à la peine d'emprisonnement, 40; et ceux qui, ayant moins de seize ans, ont commis avec discernement des crimes ou délits emportant des peines afflictives ou infamantes, 67.

MAISONS DE DÉPÔT. Voy. *Concierge*.

MAISONS DE FORCE. Voy. *Réclusion*, *Travaux forcés*.

MAISONS DE JEU. Peine contre ceux qui auraient établi et tenu une maison publique de jeux de hasard, 410.

MAISONS DE PRÊT. Peines contre ceux qui auraient établi ou tenu des maisons de prêt sur gages ou nantissement, sans autorisation légale, 411.

MAISONS GARNIES. Voy. *Aubergistes*.

MALADIE. Voy. *Violences*.

MALADIES CONTAGIEUSES. Voy. *Epizootie*.

MALADRESSE. Voy. *Blessures*, *Homicide*.

MALFAITEURS. Voy. *Assassinat*, *Association de malfaiteurs*.

MANDAT. Peine contre le concierge qui aurait reçu un prisonnier sans mandat ou jugement, 120.—Contre ceux qui ont provoqué, donné ou signé jugement, ordonnance ou mandat contre un membre du Sénat, du Conseil d'État ou du Corps législatif, sans les autorisations prescrites par les constitutions, et hors les cas de flagrant délit ou de clameur publique, 121. — Peines pour mandat décerné sans autorisation par des juges contre des préposés du Gouvernement prévenus de délits commis dans l'exercice de leurs fonctions, 129. — Voy. *Force publique*.

MANŒUVRES. Voy. *Intelligences.*

MANUFACTURES. Peines pour violation des règlements d'administration publique relatifs aux manufactures, au commerce et aux arts, 413 et suiv.—Voy. *Ouvriers.*

MARCHANDISES. Peines contre ceux qui, par des faits faux ou calomnieux semés dans le public, et par réunion ou coalition entre les principaux détenteurs d'une même marchandise ou denrée, auraient opéré la hausse ou la baisse du prix, 419.—Peine de celui qui aurait trompé l'acheteur sur la nature des marchandises, 423 — Voy. *Bandes armées, Fabriques, Marques particulières;*

MARIAGES. Peines contre celui qui aurait contracté un second mariage avant la dissolution du premier, 340; et contre l'officier public qui, connaissant l'existence du premier mariage, aurait prêté son ministère au second, *ibid.* — Voy. *Ministres des cultes.*

MARQUE. Elle peut être prononcée concurremment avec une peine afflictive, 7. — Crime dont la récidive entraîne la condamnation à la marque, 56. — Faussaires auxquels la marque est infligée, 165. — Elle l'est également aux vagabonds ou mendiants qui auraient commis un crime emportant la peine des travaux à temps, 280.—Voy. *Flétrissure.*

MARQUES DU GOUVERNEMENT. Peines pour contrefaçon des marques destinées à être apposées au nom du Gouvernement sur les denrées ou marchandises, 141.— Voy. *Marteaux de l'État.*

MARQUES PARTICULIÈRES.— Voy. *Sceaux particuliers.*

MARTEAUX DE L'ÉTAT. Peines pour contrefaçon ou falsification des marteaux de l'État servant aux marques forestières, 141 et 142.

MASSACRE. Voy. *Dévastation.*

MATÉRIAUX. Voy. *Voie publique.*

MATIÈRES D'OR OU D'ARGENT. Peines contre celui qui aurait trompé l'acheteur sur le titre de ces matières, 423.

MATRICES. Voy. *Contrefaçon.*

MÉDECINS. Voy. *Secrets.*

MÉDICAMENTS. Voy. *Avortement.*

MENACES. Cas dans lesquels les menaces par écrit, d'attentat contre les personnes, sont punies de la peine des travaux forcés à temps, ou d'un emprisonnement avec amende, 305 et 306. — Peines pour menaces verbales, 307. — Peines contre ceux qui auraient menacé de la mort des personnes par eux arrêtées, détenues ou séquestrées illégalement, 344. — Peines pour menaces d'incendie, 436. — Voy. *Attroupement, Dons, Outrage, Cultes, Violences.*

MENDICITÉ. Peine à infliger aux individus trouvés mendiant dans un lieu pour lequel il existerait un dépôt de mendicité, 274. — Peine contre les mendiants d'habitude valides trouvés dans les lieux où il n'existerait point encore de tels établissements, 275; et contre les mendiants même invalides qui auraient usé de menaces, seraient entrés sans permission dans une habitation, feindraient des infirmités ou formeraient des réunions, 276; contre les mendiants ou vagabonds qui auraient été saisis travestis ou munis d'armes et d'instruments propres à faciliter les vols, 277; autres dispositions communes aux vagabonds et aux mendiants, 278 et suiv.

MÈRES. Voy. *Pères, Mœurs.*

MESURES. Peines contre celui qui, par usage de faux poids ou de fausses mesures, aurait trompé sur la quantité des choses vendues, 423. — Confiscation des marchandises et des fausses mesures, 423; brisement de ces mesures, *ibid.*; ce qui a lieu lorsque le vendeur ou l'acheteur se sont servis d'autres mesures que celles qui ont été établies par les lois de l'État, 424. —— Amende pour emploi de faux poids ou de fausses mesures dans les magasins, boutiques, halles, foires ou marchés, 479; et pour emploi de poids ou de mesures différentes de celles qui sont établies par les lois en vigueur, *ibid.*—Emprisonnement, suivant les circonstances, des possesseurs de faux poids et de fausses mesures, et de ceux qui emploient des mesures différentes de celles que la loi a établies, 480.—Saisie et confiscation de ces différents poids et mesures, 481. —— Voy. *Action publique.*

MEULES DE GRAINS. Voy. *Champs.*

MEURTRE. Dans quel cas l'homicide est ainsi qualifié, 295. —— Le simple meurtre est puni de mort, 304. —— Cas dans lequel la peine est réduite aux travaux forcés à perpétuité, *ibid.*—Provocations qui rendent le meurtre excusable, 321. —— Autres circonstances qui produisent le même effet, 322. —— Seuls cas dans lesquels le meurtre commis par un époux sur l'autre soit excusable, 324. —— Lorsqu'un enfant délaissé est mort par suite de l'exposition, les coupables subissent la peine du meurtre, 351. —— Voy. *Assassinat.*

MINE. Peine de mort avec confiscation de biens contre tout individu qui aurait détruit par l'explosion d'une mine, des arsenaux, des vaisseaux, édifices, magasins ou autres propriétés de l'État, 95. —— Peine de mort contre ceux qui auraient détruit par l'effet d'une mine des bâtiments, maisons, édifices, navires ou bateaux, 435.

MINEUR. Peine contre quiconque aurait, par fraude ou

violence, enlever ou fait enlever, entraîner, détourner ou déplacer des mineurs des lieux où ils étaient mis par ceux à l'autorité desquels ils étaient confiés, 354 et suiv.—Seules poursuites qu'on puisse exercer contre le ravisseur qui aurait épousé la fille par lui enlevée, 357. — Voy. *Abus de confiance, Age, Enlevement.*

MINISTÈRE PUBLIC. Voyez *Accusation, Autorité administrative, Conflit, Dégradation civique, Loi, Mandat, Pouvoir législatif, Préposé du Gouvernement, Revendication.*

MINISTRE. Peine encourue par les ministres qui auraient ordonné ou fait des actes arbitraires, et qui, sur invitations légales, auraient refusé ou négligé de les faire réparer, 115. — Ce que les ministres doivent faire lorsque la signature à eux imputée leur a été surprise, 116. — Punitions des personnes qui auraient fait usage d'une fausse signature du nom d'un ministre, 118.

MINISTRES DES CULTES. Peines contre ceux qui procéderaient aux cérémonies religieuses d'un mariage sans justification d'un acte de mariage préalablement reçu par les officiers de l'état civil, 199 et 200. — Peines encourues par les ministres des cultes pour correspondance secrète avec des cours ou puissances étrangères sur des matières de religion, 207 et 208. — Peines encourues par les ministres des cultes pour viol, 333. — Voy. *Cultes.*

MINUTES. Voy. *Destruction.*

MŒURS. Peines contre toute personne qui aurait commis un outrage public à la pudeur, 330. — Peines contre quiconque se serait rendu coupable de viol ou de tout autre attentat à la pudeur, 331. —

Peine contre ceux qui auraient attenté aux mœurs en favorisant la débauche et la corruption de la jeunesse au dessous de l'âge vingt ans, 334. — Accroissement de peine lorsque la prostitution ou la corruption des jeunes gens a été facilitée par leurs pères, tuteurs et autres personnes chargées de leur surveillance, *ibid.* — Interdiction temporaire de tutelle, curatelle et de toute participation aux conseils de famille pour les personnes ci-dessus désignées, 335. — Privation particulière de droits et avantages accordés par le Code Napoléon aux père et mère sur la personne et les biens de l'enfant, *ibid.* — Voy. *Gravures, Surveillance de la haute police.*

MONNAIES. Amende contre ceux qui auraient refusé de recevoir les espèces ou monnaies nationales non-fausses ni altérées, selon la valeur du cours, 475. — Voy. *Fausse monnaie.*

MONUMENTS. Quelles peines sont infligées pour destruction, mutilation ou dégradation de monuments, statues et autres objets destinés à l'utilité ou à la décoration publique, 257.

MORT. Cette peine est tout à la fois afflictive et infamante, 7. — Tout condamné à mort doit avoir la tête tranchée, 12. — Crimes qui, commis par récidive, entraînent la condamnation à la peine de mort, 56. — L'accusé ayant moins de seize ans, qui, agissant avec discernement, a encouru la peine de mort, est condamné à la peine de dix à vingt ans d'emprisonnement dans une maison de correction, 67. — La peine de mort se prononce contre les recéleurs d'espions, 83; contre les coupables d'attentat ayant pour but de changer la forme du Gouvernement, 87; ou de troubler l'État par la guerre civile, le pillage public, etc. 91 et 125; contre les fabricateurs de fausse monnaie , 132 ; contre les contrefacteurs des sceaux de l'État , des billets de banque, des effets publics, 139. — Cas dans lesquels les violences commises envers les dépositaires de l'autorité publique donnent lieu à la peine de mort, 231 et 233. — L'assassinat, le parricide, l'infanticide, l'empoisonnement , le meurtre, sont punis de mort, 302. — Cas dans lequel le crime de castration est puni de mort, 316. — Circonstances dans lesquelles la peine de mort a lieu contre les coupables d'arrestations illégales et de séquestration, 344 ; contre les coupables de subornation de témoins, 365. — Cas dans lesquels les coupables de vols sont punis de mort, 381. — Destructions pour lesquelles la même peine est encourue, 434 , 435. et 437. — Voy. *Armes, Arrêt, Espionnage, Femme, Parricide.*

MORT CIVILE. Voy. *Condamnation.*

MOTEURS. Voy. *Travaux publics.*

MOULES. Voy. *Contrefaçon.*

MOULINS. Voy. *Usines.*

MOUTONS. Voy. *Empoisonnement.*

MUNICIPALITÉ. Déclaration à faire devant la municipalité par celui qui consent à se charger d'un enfant trouvé. 347.

MUNITIONS. Peine de mort, avec confiscation de biens, contre ceux qui auraient fourni ou procuré des armes ou munitions aux soldats par eux enrôlés, sans autorisation du pouvoir légitime, 94 ; ou à des bandes armées illégalement, 96. — Voy. *Association de malfaiteurs, Secours.*

MUR. Voy. *Escalade.*

MUSIQUE. Voy. *Contrefaçon.*

MUTILATION. Voy. *Monuments.*

N.

NANTISSEMENT. Voy. *Maisons de prêt.*

NAUFRAGE. Voy. *Secours.*

NAVIRE. Voy. *Incendie, Mine.*

NÉGLIGENCE. Peines auxquelles l'évasion des détenus donne lieu contre ceux à la négligence de qui elle peut être imputée, 237 et suiv. — Cessation de l'emprisonnement contre les conducteurs ou gardiens lorsque les évadés sont repris dans un délai de quatre mois, 246. — Voy. *Blessures, Gardiens de scellés, Homicide.*

NOM. Peine pour arrestation illégale faite sous un faux nom, 344.

NUIT. Peines pour vols commis la nuit, 382, 385 et 386

O.

OBLIGATION. Voy. *Extorsion.*

OFFICIERS DE JUSTICE. Voy. *Domicile.*

OFFICIERS DE L'ÉTAT CIVIL. Peines par eux encourues pour divers délits relatifs à leurs fonctions, 192 et suiv.; plus forte peine qui se prononce en cas de collusion, 195. Voy. *Inhumation, Mariage.*

OFFICIERS DE POLICE. Voy. *Aubergistes, Rébellion, Registres.*

OFFICIERS DE POLICE JUDICIAIRE. Les concierges des prisons sont tenus de leur représenter les détenus, 120. — Cas dans lesquels ces officiers sont coupables de forfaiture, 121. — Voy. *Accusation, Aubergistes, Dégradation civique, Domicile, Gardes champêtres, Préposés du Gouvernement, Rébellion, Registres.*

OFFICIERS DE SANTÉ. Voy. *Avortement, Certificat de maladie, Secrets.*

OFFICIERS DU MINISTÈRE PUBLIC. Voy. *Accusation, Autorité administrative, Conflit, Loi, Mandat, Préposés du Gouvernement, Pouvoir législatif, Revendication.*

OFFICIERS MINISTÉRIELS. Peines pour outrages et violences envers un officier ministériel ou agent de la force publique, 223 et 229. — Voy. *Rébellion.*

OFFICIERS PUBLICS. Voy. *Administrateurs, Concussion, Fonctionnaires publics, Soustraction, Violences.*

OFFRES. Voy. *Corruption.*

OR. Voy. *Matières d'or et d'argent.*

ORDONNANCE. Peines pour ordonnance rendue sans autorisation du Gouvernement contre ses agents ou préposés prévenus de délits commis dans l'exercice de leurs fonctions, 129. — Voy. *Force publique.*

ORDRE. Peine pour arrestation illégale faite sous un faux ordre de l'autorité publique, 344. — Peine pour vol commis en alléguant un faux ordre de l'autorité civile ou militaire, 381 et 384. — Voy. *Force publique, Préfets.*

OUTRAGES. Peines pour outrages par paroles envers des magistrats de l'ordre administratif ou judiciaire, 222; pour outrages faits par gestes ou menaces aux mêmes magistrats, 223 et 226; et pour outrages faits par paroles, gestes ou menaces à tous officiers ministériels ou agents dépositaires de la force publique, 224 à 227. — Voy. *Cultes.*

OUVERTURES SOUTERRAINES. L'entrée par une semblable ouverture autre que celle qui a été établie pour servir d'entrée, est une circonstance de vol de la même gravité que l'escalade, 397.

OUVRAGES. Voy. *Contrefaçon, Écrits.*

OUVRAGES DRAMATIQUES. Voy. *Théâtre.*

OUVRIERS. Les réunions des ouvriers ou journaliers dans les ateliers publics et manufactures sont punies comme réunions de rebelles, 219. — Peine pour vols commis par un ouvrier, compagnon ou apprenti dans la maison, l'atelier ou le magasin de son maître, 386. — Voy. *Coalition, Fabrique.*

P.

PAIX PUBLIQUE. Crimes et délits contre la paix, publique, 132 et suiv.

PAMPHLETS. Voy. *Gravures.*

PAPIERS. Voy. *Scellés.*

PAPIERS PUBLICS. Voy. *Effets publics, Journaux.*

PARC. Voy. *Champs, Enclos.*

PARI. Voy. *Effets publics.*

PAROLES. Voy. *Outrages.*

PARRICIDE. Comment le coupable condamné à mort pour parricide doit être conduit sur le lieu de l'exécution, 13. — Son exposition sur l'échafaud pendant la lecture de l'arrêt de condamnation, *ibid.* — On lui coupe le poing droit avant l'exécution, *ibid.* — Personnes dont le meurtre est qualifié de parricide, 299. — De quelle peine est puni le parricide, 302. — Le parricide n'est jamais excusable, 323. — Voy. *Lèse-majesté, Meurtre.*

PASSAGE. Voy. *Rues.*

PASSE-PARTOUT. Voy. *Fausses clefs.*

PASSE-PORTS. Les peines établies contre les porteurs de faux passe-ports sont portées au *maximum* à l'égard des vagabonds et des mendiants, 280.

PEINES. Les contraventions, les délits ou les crimes ne

peuvent être punis de peines que la loi n'avait pas
prononcées avant qu'ils fussent commis, 4. — Des
peines et de leurs effets, 6 à 11. — Nature des peines
en matière criminelle, 6. — Énumération des peines
afflictives et infamantes, 7. — Trois sortes de peines
seulement infamantes, 8. — En quoi consistent les
peines en matière correctionnelle, 9. — Sortes de
peines communes aux matières criminelle et correctionnelle, 11. — Des peines en matière criminelle,
12 à 39. — Des peines en matière correctionnelle,
40 à 43. — Des peines et des autres condamnations
qui peuvent être prononcées pour crimes ou délits,
44 à 55. — Des peines de la récidive, 56 à 58. —
Les peines ne peuvent être mitigées que dans les cas
où la loi permet d'en appliquer de moins rigoureuses,
65. — Cas d'aggravation dans les peines encourues
par les fonctionnaires ou officiers publics, 197. — Injures qui ne donnent lieu qu'à des peines de simple
police, 377. — Les peines de police sont l'emprisonnement, l'amende et la confiscation de certains objets
saisis, 464. — Voy. *Age, Condamnation, Mort.*

PEINTURE. Voy. *Contrefaçon.*

PERCEPTEURS DE TAXES ET CONTRIBUTIONS. V. *Rebellion.*

PÈRES ET MÈRES légitimes naturels et adoptifs : peines
contre les enfants qui les ont blessés, 312. — Voy.
Ascendants, Mœurs, Parricide.

PHARMACIENS. Voy. *Avortement, Secrets.*

PIEDS CORNIERS. Voy. *Limites.*

PIERRERIES. Peine contre celui qui aurait trompé l'acheteur sur la qualité d'une pierre fausse vendue pour
fine, 423.

PIERRES. Amende encourue pour ceux qui auraient jeté
des pierres ou d'autres corps durs contre les maisons
ou dans les jardins d'autrui, 475. — Emprisonne-

ment de ces individus, 476.——Peines contre ceux qui en jetant des pierres ou d'autres corps durs, auraient occasionné la mort ou la blessure d'animaux ou de bestiaux, 479. —— Voy. *Champs.*

PILLAGE. Voy. *Bandes armées, Dévastation, Réunion armée et séditieuse, Secours, Vols.*

PLACARDS. Voy. *Discours.*

PLACES DE GUERRE. Peine encourue pour des intelligences et manœuvres tendant à livrer aux ennemis des villes, forteresses, places, postes, ports, magasins, arsenaux, vaisseaux ou bâtiments de l'État, 77.——Voy. *Armes, Bandes armées, Commandement militaire.*

PLACE PUBLIQUE. C'est sur la place publique que les individus condamnés au carcan y sont attachés, 22.——Les arrêts de condamnation indiquent celles des places publiques sur lesquelles l'exécution doit avoir lieu, 26.

PLANCHES. Voy. *Contrefaçon.*

PLANS. Peine contre ceux qui auraient livré aux ennemis des plans de fortifications, arsenaux, ports ou rades, 81 et 82.

PLANTS. Voy. *Récoltes.*

POIDS. Voy. *Mesures.*

POINÇONS. Peines contre les contrefacteurs ou falsificateurs de poinçons servant à marquer les matières d'or ou d'argent, ou contre ceux qui auront fait usage de poinçons falsifiés, 140, 141, 142 et 143.

POING. Les coupables condamnés à mort pour parricide ont le poing droit coupé avant l'exécution, 13.

POISSONS. Voy. *Champs, Empoisonnement.*

POLICE. (haute) Voy. *Surveillance.*

PONTS. Voy. *Destruction.*

PORCS. Voy. *Empoisonnement.*

PORT D'ARMES. Celui qui a été condamné à la peine des

travaux forcés à temps, au bannissement, à la réclusion ou au carcan, est déchu du droit de port d'armes
et du droit de servir dans les armées de l'Empire, 28.
—— Les tribunaux correctionnels peuvent interdire
l'exercice du droit de port d'armes, 42.

PORTEURS DE CONTRAINTES. Voy. *Rébellion*.

PORTS. Voy. *Bandes armées, Commandement militaire,
Places, Plans*.

POSTE AUX LETTRES. Voy. *Lettres*.

POSTES DE GUERRE. Voy. *Bandes armées, Commandement militaire*.

POUVOIR LÉGISLATIF. Peines contre les juges, les procureurs généraux ou impériaux, leurs substituts et les
officiers de police judiciaire qui se seraient immiscés
dans l'exercice du pouvoir législatif, 127. —— Voy.
Préfets.

PRÉFETS. Peines contre les préfets, sous-préfets, maires et
autres administrateurs, qui se seraient immiscés dans
l'exercice du pouvoir législatif, ou qui auraient pris
des arrêtés généraux, ou des défenses quelconques à
des cours et tribunaux, 130. —— Voy. *Commerce*.

PRÉMÉDITATION. Peines contre les coupables de violences
exercées avec préméditation contre des magistrats,
232. —— En quoi consiste la préméditation, 297. ——
Lorsque les moyens employés volontairement pour détruire ou renverser des maisons, etc., ont occasionné
un homicide ou des blessures, le coupable est puni
comme les ayant commis avec préméditation, 437.
Voy. *Assassinat*.

PRÉPOSÉS DE LA POLICE. Voy. *Violences*.

PRÉPOSÉS DES DOUANES. Voy. *Rébellion*.

PRÉPOSÉS DU GOUVERNEMENT. Peines contre les juges ou officiers du ministère public ou de police qui auraient re-

quis ou rendu des ordonnances, ou décerné des mandats sans autorisation du Gouvernement contre ses agents ou préposés prévenus de crimes ou délits commis dans l'exercice de leurs fonctions, 129. — Voy. *Force publique, Soustraction, Violences.*

PRÉSENTS. Voy. *Corruption.*

PRISONNIER. Peine du concierge qui, sans mandat, aurait reçu un prisonnier, qui l'aurait retenu, ou refusé de le représenter à l'officier de police, et n'aurait pas exhibé son registre à cet officier, 120. — Les réunions de prisonniers, prévenus, accusés ou condamnés, sont punies comme réunions de rebelles, 219.

PROCÉDURES CRIMINELLES. Voy. *Scellés.*

PROCUREURS GÉNÉRAUX. Cas dans lesquels ils sont coupables de forfaiture, 121. — Voy. *Accusation, Autorité administrative, Conflit, Dégradation civique, Domicile, Loi, Mandat, Pouvoir législatif, Préposés du Gouvernement, Revendication.*

PROCUREURS IMPÉRIAUX. Voy. *Accusation, Autorité administrative, Conflit, Dégradation civique, Domicile, Forfaiture, Loi, Mandat, Pouvoir législatif, Préposés du Gouvernement, Revendication.*

PROMESSES. Voy. *Corruption, Dons, Récompenses.*

PRONOSTIQUEURS. Voy. *Songes.*

PROPRIÉTÉS. Crimes et délits contre les propriétés, 379 et suiv.

PROPRIÉTÉS PUBLIQUES. Voy. *Bandes armées.*

PROSTITUTION Voy. *Mœurs.*

PROVISION. Pendant la durée de la peine de la réclusion ou des travaux forcés à temps, il ne peut être remis au condamné aucune provision ni aucune portion de ses revenus, 21.

PROVOCATIONS. Voy. *Blessures, Censure de l'autorité publique, Meurtre, Réunion armée et séditieuse.*

PUBLICATION D'OUVRAGES. Voy. *Écrits.*

PUDEUR. Voy. *Mœurs.*

R.

RADES. Voy. *Plans.*

RAPT. Voy. *Mineurs.*

RATELAGE. Voy. *Glanage.*

REBELLION. Toute attaque, toute résistance avec violence et voies de fait envers les officiers ministériels, les gardes champêtres ou forestiers, la force publique, les préposés à la perception des taxes et contributions, leurs porteurs de contraintes, les préposés des douanes, les séquestres, les officiers ou agents de la police administrative ou judiciaire, etc., constitue un crime ou un délit de rebellion, 209. — Diverses peines encourues selon les différentes circonstances des rebellions, 210 et suiv. — Les provocateurs de rebellion sont punis comme les coupables mêmes de la rebellion, 217. — Réunions qui sont punies comme réunions de rebelles, 219. — Manière dont on fait subir la peine appliquée pour rebellion à des prisonniers prévenus, accusés ou condamnés relativement à d'autres crimes ou délits, 220. — Surveillance des chefs de rebellion après l'expiration de leurs peines, 221. — Voy. *Réunion armée et séditieuse.*

RECÈLEMENT. Peine de l'emprisonnement pour recèlement de criminels, 248. — Parents et alliés exceptés de cette peine, *ibid.* — Peine pour recèlement du cadavre d'une personne homicidée ou morte par suite de coups ou de blessures, 359.

RECÉLEURS. Ceux qui ont sciemment recélé des choses

enlevées, détournées ou obtenues à l'aide d'un crime ou d'un délit, sont punis comme complices, 62. — Seuls cas dans lesquels puisse leur être appliquée la peine de mort, des travaux forcés à perpétuité ou de la déportation. 63. —Peine de mort contre ceux qui auraient recélé les espions ou soldats ennemis envoyés à la découverte, 83. — Voy. *Espionnage.*

Recettes. Voy. *Confiscation.*

Récidive. Peines de la récidive pour crimes où délits, 56. — Cas dans lesquels il y a récidive pour contravention de police, 474. — Peine d'emprisonnement pour le cas de récidive de diverses contraventions de police, 478. — Voy. *Emprisonnement.*

Réclamation. Voy. *Vagabondage.*

Réclusion. Cette peine est afflictive et infamante, 7. Maison de force dans laquelle sont renfermés les individus condamnés à la réclusion, et travaux auxquels ils y sont employés, 21. —Durée de la peine de réclusion, *ibid.* — De quel jour se compte la durée de la réclusion, 23. — Les coupables condamnés à la réclusion sont pendant toute leur vie sous la surveillance de la haute police de l'Etat, 47. — Celui qui, ayant été condamné pour un crime, en a commis un second emportant la peine de la réclusion, doit être condamné aux travaux forcés à temps et à la marque, 56. — Temps pendant lequel doit être renfermé dans une maison de correction l'individu ayant moins de seize ans, qui, agissant avec discernement, a encouru la peine de la réclusion, 67. — La peine de la réclusion doit remplacer celle des travaux forcés et de la déportation à l'égard des individus âgés de soixante-dix ans, 71. — Tout condamné à la peine des travaux forcés en est relevé, dès qu'il a atteint l'âge de soixante-dix ans, pour être enfermé dans une maison de force pendant le temps qui reste à expirer de sa peine, 72. — Cas dans lequel la non-révélation d'un crime de lèse-majesté fait encourir la peine de la réclusion, 104. — Cette peine est encourue pour faux en écriture privée et pour usage de la pièce fausse, 148 et 149 ; pour concussion commise par des fonctionnaires publics, 174. — La peine de la réclusion est encourue par le juge prononçant en matière criminelle ou le juré qui s'est laissé corrompre, 181 ; par les fonctionnaires publics, agents ou préposés du Gouvernement, pour avoir requis ou ordonné l'emploi de la force publique contre l'exécution d'une loi, la perception d'une contribution légale, etc., 187.— Cas de rebellion qui donne lieu à cette peine, 209 et 210.— Cas dans lesquels la peine de réclusion a lieu pour raison de violences exercées envers des magistrats, 230 et 231.— Cas dans lesquels l'évasion de détenus donne lieu à la réclusion contre les préposés à leur garde ou conduite, 237 et 239. — Application de cette peine aux individus coupables de soustraction, enlevement ou destruction de pièces ou procédures criminelles, 254 ; aux individus chargés d'un service quelconque pour une association de malfaiteurs, 263 ; aux mendiants ou vagabonds qui auraient exercé quelque acte de violence, 279 ; aux individus qui ont porté des coups et fait des blessures volontaires, 309 et 312 ; aux individus qui ont procuré l'avortement et à la femme enceinte qui y a consenti, 317 ; aux coupables du crime de viol et de tout autre attentat aux mœurs, 331 ; aux coupables d'enlevement, de recélé ou de supposition d'un enfant, 345 ; d'onlevement de mineurs, 354 et suiv. ; de faux témoignage en matière civile, correctionnelle ou de police, 361 et suiv.; de vols commis dans diverses circonstances et par certains individus, 386 à 388 ; de contrefaçon ou altération de clefs par un serrurier de profession, 399 ; de communication des secrets d'une fabrique, 418 — La même peine contre les fournisseurs qui, sans contrainte par une force majeure, auraient fait manquer le service des armées de terre et de mer, 430. — Peine de réclusion pour avoir détruit ou renversé des édifices, des ponts, etc., avoir détruit des actes de l'autorité publique ou commis des pillages, 437, 439 et 441. — Voy. *Bannissement, Carcan, Condamné.*

RÉCOLTES. Peines pour dévastation des récoltes sur pied, ou des plants venus naturellement ou faits de main d'homme, 444. — Voy. *Champs, Incendie, Terrains.*

RÉCOMPENSES. Peines contre le faux témoin qui, en matière correctionnelle, de police ou civile, aurait reçu de l'argent, des récompenses ou des promesses, 364.

REGISTRE. Celui sur lequel les aubergistes et hôteliers doivent inscrire le nom, la profession et le domicile des personnes logées chez eux, 73. — Représentation à faire de ce registre sur la réquisition des maires, adjoints, officiers ou commissaires de police, ou des citoyens commis à cet effet, 475. — Voy. *Destitution, Scellés.*

RÉGLEMENTS D'ADMINISTRATION PUBLIQUE. Ceux qui doivent déterminer la manière d'employer les produits du travail des détenus dans les maisons de correction, 41. — Voy. *Loi.*

RÉGLEMENTS DE POLICE. Peine contre quiconque, par inobservation de ces règlements, aurait causé un homicide involontaire, 319. — Voy. *Loi.*

RENVOI. Cas dans lesquels les juges saisis d'une contestation où il a été allégué des injures portant le caractère d'une calomnie grave, doivent, outre la suspension provisoire, renvoyer les prévenus devant les juges compétents, 377. — Voy. *Surveillance de la haute police.*

RÉPARATION. Cas dans lesquels les outrages envers les dépositaires de l'autorité et de la force publique donnent lieu à une réparation, 226 et 227.

RÉPARATIONS CIVILES. Celles qui peuvent être dues par les dépositaires de la force publique pour avoir refusé de la faire agir sur la réquisition de l'autorité civile, 234. — Soustractions qui ne peuvent donner lieu qu'à des réparations civiles, 380. — Voy. *Dommages-intérêts, Restitution, Soustraction.*

REPRÉSAILLES. Peine du bannissement contre ceux qui, par des actes non approuvés par le Gouvernement, auraient exposé des Français à éprouver des représailles, 85.

RÉQUISITION. Voy. *Détention arbitraire, Force publique, Réparations civiles.*

RÉSERVOIRS. Voy. *Champs, Empoisonnement.*

RÉSISTANCE. Voy. *Rébellion.*

RESPONSABILITÉ CIVILE. Celle des logeurs et aubergistes, 73. — Dispositions du Code Napoléon auxquelles les cours et tribunaux doivent se conformer dans les autres cas de responsabilité civile qui peuvent se présenter dans les affaires criminelles, correctionnelles et de police, 74.

RESTITUTION. Les sommes provenant des paiements faits par les cautions d'un individu condamné pour crimes ou délits sont affectées de préférence aux restitutions envers les parties lésées, 46. — En cas de concurrence de l'amende ou de la confiscation avec les restitutions et les dommages-intérêts sur les biens insuffisants du condamné, ces dernières condamnations obtiennent la préférence, 54. — Restitutions et indemnités qui sont ordonnées dans le cas de soustractions commises par des dépositaires publics, 172. — Restitutions pour abus de confiance, 406. — Celui qui a trompé l'acheteur sur le titre, la qualité ou le poids des marchandises, est tenu, outre les restitutions et dommages-intérêts, d'une amende qui peut monter au quart de ces restitutions, 423. — Autres cas de restitution et de dommages-intérêts sur lesquels l'amende est réglée, 437, 439, 443, 455 et 457. — En cas d'insuffisance des biens d'un individu condamné pour contraventions de police, les restitutions et les indemnités dues à la partie lésée sont préférées à l'amende, 468. — Les restitutions, le paiement des indemnités et des frais, et généralement toutes les condamnations prononcées pour contraventions de police, entraînent la contrainte par corps, 469. — Lorsque ces condamnations sont prononcées au profit de l'État, les condamnés peuvent, après quinze jours d'emprisonnement, obtenir leur liberté, *ibid.* — Voy. *Aubergistes, Condamnation, Indemnité, Responsabilité civile, Solidarité.*

RETRAITE DE MALFAITEURS. Voy. *Association de malfaiteurs.*

RÉUNION ARMÉE ET SÉDITIEUSE. Dans quels cas est ainsi qualifiée une réunion d'individus pour un crime ou un délit, 213. — Comment sont punies les personnes munies d'armes cachées qui auraient fait partie d'une troupe ou réunion réputée non armée, 214. — Cas dans les-

quels les blessures et les coups sont imputables aux
chefs, auteurs, instigateurs et provocateurs des réunions séditieuses où il y a eu rebellion ou pillage,
313.

RÉUNIONS LITTÉRAIRES. Voy. *Sociétés.*

RÉVÉLATION. Dispositions relatives à la révélation ou
non-révélation de crimes contre la sûreté de l'État,
103 et suiv.; et de crimes relatifs à la fausse monnaie, 138; et à la contrefaçon des sceaux de l'État,
des effets émis par le trésor public, et des billets de
banque autorisés par la loi, 144.

REVENDICATION. Peines contre les juges qui, sur la revendication formellement faite par l'autorité administrative d'une affaire portée devant eux, auraient procédé au jugement avant la décision de l'autorité supérieure, 128. — Mêmes peines contre les officiers du
ministère public qui auraient fait des réquisitoires ou
donné des conclusions pour ce jugement, *ibid.*

REVENUS. Voy. *Provision.*

RÉVOCATION. Voy. *Fonctionnaires publics.*

ROSSIGNOLS. Voy. *Fausses clefs.*

ROULIERS. Voy. *Voituriers.*

ROUTES. Voy. *Arbres, Feuilles de route.*

RUES. Amendes contre ceux qui auraient négligé de nettoyer les rues et passages dans les communes où ce
soin est laissé à la charge des habitants, 471. —Voy.
Arbres, Jeux de hasard, Voituriers.

S.

SAGES-FEMMES. Voy. *Secrets.*

SAISIE. Voy. *Confiscation.*

SALAIRES. Voy. *Coalition.*

SANG. Voy. *Violences.*

Sceau de l'état. Ceux qui l'ont contrefait ou se sont servis du sceau contrefait, sont punis de mort, avec confiscation de biens, 139.

Sceaux particuliers. Peines pour contrefaçon du sceau, du timbre ou de la marque d'une autorité quelconque ou d'un établissement particulier de banque ou de commerce, 141, 142 et 143.

Scellés. Peine de l'emprisonnement contre le gardien des scellés brisés par sa négligence, 249. — Emprisonnement de plus longue durée si le bris de scellés s'applique aux papiers d'un prévenu de crimes emportant des peines afflictives, 250. — Peines contre ceux qui auraient brisé eux-mêmes les scellés ou auraient participé au bris, 250 et 251 ; pour vol commis à l'aide d'un bris de scellés, 252. — Peines contre les greffiers, archivistes ou autres dépositaires, dans le cas de soustraction de pièces, procédures criminelles ou registres, 253 ; et contre ceux qui se seraient rendus coupables de ces soustractions, 254.— Peine plus forte dans le cas où le bris des scellés, les soustractions, enlèvement ou destruction de pièces, auraient été commis avec violence, 255.

Secours. Peines contre ceux dont les manœuvres auraient eu pour objet de fournir aux ennemis de l'État des secours en soldats, hommes, argent, vivres, armes ou munitions, 77. — Amende contre ceux qui auraient refusé ou négligé de faire les travaux de service ou de prêter secours dont ils auraient été requis dans les circonstances d'accidents, tumulte, naufrages, inondations, incendie, et dans les cas de brigandage, pillage, flagrant délit, clameur publique ou exécution judiciaire, 475.

Secrets. Peines contre les officiers de santé, médecins,

chirurgiens ou pharmaciens, sages femmes, etc. qui, hors le cas où la loi les oblige à se porter dénonciateurs, auraient révélé les secrets à eux confiés, 378.

SECRETS DES ARTS ET MÉTIERS. — Peines contre ceux qui, auraient communiqué des secrets d'une fabrique dans laquelle ils étaient employés, 418.

SÉDITION. Peine contre ceux qui auraient fait partie d'une réunion séditieuse de bandes armées, 97 et 98. — Dispositions relatives aux individus qui, ayant fait partie des bandes, s'en seraient retirés sans opposer de résistance, 100. — Voy. *Bandes armées.*

SÉDUCTION. Voy. *Corruption.*

SÉNAT. Voy. *Forfaiture, Mandat.*

SÉPULTURE. Peines pour violation de tombeaux ou de sépultures, 360.

SÉQUESTRATION. Peines contre ceux qui, sans ordre des autorités constituées, et hors les cas où la loi ordonne de saisir des prévenus, auraient détenu ou séquestré des personnes quelconques, 341 et 342. — Circonstances atténuantes, 343. — Cas dans lesquels ce délit est puni de mort, 344.

SÉQUESTRE. Voy. *Rebellion.*

SERMENT. Poursuites et amendes encourues par les fonctionnaires publics qui seraient entrés dans l'exercice de leurs fonctions sans avoir prêté serment, 196. — Peines contre celui qui, sur le serment à lui déféré en matière civile, aurait fait un faux serment, 366.

SERRURES. Voy. *Effraction, Fausses clefs.*

SERRURIER. Peine contre celui qui a contrefait ou altéré des clefs, 399. — Plus forte peine en cas de complicité avec un voleur, *ibid.*

SERVICE. Peines pour refus d'un service dû légalement, 234 et suiv. — Voy. *Secours.*

SERVICE MILITAIRE. Voy. *Port d'armes.*

SERVITEURS A GAGES. Peines par eux encourues pour viol, 333. — Voy. *Domestiques.*

SIGNATURE. Voy. *Extorsion, Faux, Fonctionnaires publics, Ministres.*

SIGNAUX. Voy. *Maisons.*

SOCIÉTÉS. Nulle association de plus de vingt personnes dans le but de se réunir pour s'occuper d'objets religieux, littéraires, politiques, etc., ne peut se former qu'avec l'agrément du Gouvernement, sous peine d'être dissoute, 291 et suiv. — Peines encourues par les membres de ces assemblées dans le cas où il y aurait été fait des provocations à des crimes ou à des délits, 293. — L'autorisation du Gouvernement est nécessaire à tout individu pour accorder l'usage de sa maison aux membres d'une société même autorisée, 294.

SOLDATS. Voy. *Enrôlements.*

SOLIDARITÉ. Tous les individus condamnés pour le même crime ou pour le même délit, sont tenus solidairement des amendes, des restitutions, des dommages-intérêts et des frais, 55.

SONGES. Amende contre les gens qui font le métier de deviner et pronostiquer ou d'expliquer les songes, 479. — Emprisonnement qui peut être ordonné contre ces individus, 480. — Saisie et confiscation des instruments, ustensiles et costumes, 481.

SOUMISSIONS. Voy. *Enchères.*

SOUS-PRÉFETS. Voy. *Commerce, Préfets.*

SOUSTRACTION. Peines pour soustractions commises par des percepteurs, dépositaires ou comptables publics, 169 et suiv.; par des juges, administrateurs, fonctionnaires ou officiers publics pour soustraction, suppression ; destruction d'actes et de titres dont ils étaient dépositaires en cette qualité, 173. — Pareilles peines contre tous agents, préposés ou commis du Gouvernement, ou les dépositaires publics qui se seraient rendus coupables des mêmes soustractions, *ibid.* — Quiconque a soustrait frauduleusement une chose qui ne lui appartient pas est coupable de vol, 379. — Soustractions commises par des maris au préjudice de leurs femmes, et qui ne peuvent donner lieu qu'à des réparations civiles, 380. — Voy. *Abus de confiance ; Scellés.*

SPECTACLES. Voy. *Théâtres.*

STATUES. Voy. *Monuments.*

STILET. Voy. *Armes.*

SUBORNATION. Peine contre le coupable de subornation de témoins, 366.

SUBSTITUTION. Voy. *Enfant.*

SUBSTITUTS DES PROCUREURS GÉNÉRAUX ET IMPÉRIAUX. Voy. *Accusation, Autorité administrative, Conflit, Dégradation civique, Domicile, Forfaiture, Loi, Mandat, Pouvoir législatif, Préposés du Gouvernement, Revendication.*

SUFFRAGES. Tout citoyen qui, chargé de recueillir, dans un scrutin de dépouillement, des billets contenant les suffrages des citoyens, serait surpris falsifiant ces billets ou en soustrayant de la masse, serait condamné à la peine du carcan, 111. — Toutes autres personnes coupables des faits ci-dessus énoncés seraient punies de l'interdiction du droit de voter et d'être éligibles, 112. — Tous ceux qui auraient vendu leurs suffrages dans les élections seraient punis d'interdiction temporaire des droits de citoyen et de toutes fonctions ou emplois publics, 113. — Le vendeur et l'acheteur du suffrage condamnés chacun à une amende double de la valeur des choses reçues ou promises, *ibid.* Voy. *Délibérations de famille.*

SUPPLICE. La tête tranchée est le supplice de tout condamné à mort, 12.

SUPPLICIÉS. Délivrance du corps des suppliciés à leurs familles qui les réclament pour l'inhumation, 14.

SUPPOSITION DE PERSONNES. Voy. *Faux.*

SUPPRESSION D'ENFANT. Voy. *Enfant.*

SUPPRESSION D'ÉCRITS. Voy. *Injonction.*

SURETÉ PUBLIQUE. Voy. *État.*

SURPRISE. Dénonciation à faire par les ministres qui prétendent que la signature d'actes arbitraires leur a été surprise, 116.

SURVEILLANCE DE LA HAUTE POLICE. Le renvoi sous la surveillance spéciale ou à la disposition du Gouvernement est une peine commune aux matières criminelles et correctionnelles, 11. — Cautionnement de bonne conduite que la haute-police de l'État a droit d'exiger de l'individu placé sous sa surveillance s'il est en âge de minorité, 44. — Le même droit appartient à la partie intéressée, *ibid.* — Ce qui résulte du défaut de cautionnement, *ibid.* et 45. — Coupables que la nature de leur condamnation place de plein droit sous la surveillance de la haute police de l'État, 67 à 50. — Les coupables condamnés correctionnellement à un emprisonnement de plus d'une année, qui commettent un nouveau délit, doivent, outre les peines par eux encourues, être mis sous la surveillance spéciale du Gouvernement pendant cinq à dix années, 56.—Nombre d'années pendant lesquelles on peut mettre sous la surveillance de la haute police les individus ayant moins de seize ans qui ont encouru des peines afflictives et infamantes, 67. — Les individus déclarés coupables, mais excusables, peuvent être mis pendant cinq à dix ans sous la surveillance de la haute police, 69. — Ceux qui, ayant fait partie de bandes armées, s'en seraient retirés au premier avertissement des autorités civiles ou militaires, n'encourent aucune peine; mais ils peuvent être renvoyés sous la surveillance spéciale de la haute police, 100. — Pareille surveillance à l'égard des époux, des ascendants ou descendants, etc., qui n'auraient point révélé un complot contre la sûreté de l'État, 107; et à l'égard des coupables qui, avant l'exécution de ces complots, en auraient donné connaissance et auraient procuré l'arrestation des auteurs, 108. — Surveillance des individus qui, coupables de crimes relatifs à la fausse monnaie, auraient fait des révélations aux autorités constituées, 138. — Les chefs de rebellion peuvent, après l'expiration de leur peine, rester sous la surveillance spéciale de la haute police, 221. — Il en est de même d'individus convaincus d'avoir favorisé l'évasion de détenus, 246; et de ceux qui ont été condamnés pour menaces d'attentat contre les personnes, 308; pour blessures et coups volontaires dans des réunions séditieuses, et pour fabrique, débit ou port d'armes prohibées, 313 à 314; pour arrestations illégales et séquestration de personnes, 343; pour larcins et filouteries, 401; pour violation des réglements relatifs aux manufactures, au commerce et aux arts, 416, 419 et 420; pour destruction, dégradation ou dommages de différentes sortes, 444 et 452. — Voy. *Cautionnement, Mœurs.*

SUSPENSION. Voy. *Fonctionnaires publics, Injonction.*

T.

TAPAGE NOCTURNE. Voy. *Bruits nocturnes.*

TÉMOIN. Celui qui a été condamné à la peine des travaux forcés à temps, au bannissement, à la réclusion ou au carcan, ne peut être employé comme témoin dans les actes, ni déposer en justice, autrement que pour y donner de simples renseignements, 28. — Les tribunaux jugeant correctionnellement peuvent interdire pour un temps l'exercice du même droit, 42 et 43. — Voy. *Excuses, Faux témoignage, Subornation.*

TENTATIVE. Voy. *Crimes, Délits.*

TERRAINS. Amende pour avoir passé sur le terrain d'autrui préparé ou ensemencé, 471 ; et contre ceux qui y auraient laissé passer leurs bestiaux ou leurs bêtes de trait. de charge ou de monture avant l'enlèvement de la récolte, *ibid.* ; pour être entré sur le terrain d'autrui lorsqu'il était chargé de grains en tuyau, de raisins ou autres fruits mûrs ou voisins de la maturité, 475. — Pareille peine contre ceux qui auraient fait passer des bestiaux ou animaux de trait ou de monture sur des terrains ensemencés ou chargés d'une récolte, ou dans un bois taillis, *ibid.*

TÊTE TRANCHÉE. Supplice commun à tous les coupables condamnés à mort, 12.

THÉATRES. Peines contre tout directeur, tout entrepreneur de spectacles, toute association d'artistes, qui aurait fait représenter sur son théâtre des ouvrages dramatiques, au mépris des lois relatives à la propriété des auteurs, 428.

TIMBRES NATIONAUX. Peine des travaux forcés à temps pour contrefaçon ou falsification de timbres nationaux, 140, 141, 142 et 143.

TIMBRES PARTICULIERS. Voy. *Sceaux particuliers.*

TITRES. Peines contre celui qui se serait faussement attribué des titres impériaux, 259; et contre celui qui aurait commis un vol en prenant le titre d'un fonctionnaire public, ou d'un officier civil et militaire, 381 et 382. — Voy. *Destruction, Extorsion.*

TOMBEAU. Voy. *Sépulture.*

TORTURES. Peines contre ceux qui auraient fait supporter des tortures corporelles à des personnes par eux arrêtées et séquestrées illégalement, 344. Voy. —*Assasinat.*

TRAHISON. Voy. *Intelligences.*

TRANSPORTATION. Lorsque les individus déclarés vagabonds par jugement sont étrangers, le Gouvernement peut les faire conduire hors du territoire de l'Empire, 272.

TRAVAIL. Voy. *Boutiques.*

TRAVAUX CORRECTIONNELS. Les individus renfermés dans une maison de correction ont le choix de l'un des travaux qui y sont établis, 40. — Application des produits du travail des détenus pour délits correctionnels, 41.

TRAVAUX FORCÉS. Ces travaux à perpétuité ou à temps, sont au nombre des peines afflictives et infamantes, 7. — Travaux auxquels sont employés les individus condamnés à ces peines, 15. — Cas dans lesquels ils doivent trainer à leurs pieds un boulet, ou être attachés deux à deux avec une chaine, *ibid.* — Les femmes et les filles condamnées aux travaux forcés n'y sont employées que dans l'intérieur d'une maison de force, 16. — Le déporté qui rentre sur le territoire de l'Empire est condamné aux travaux forcés à perpétuité, 17. — La condamnation aux travaux forcés à perpétuité emporte mort civile, 18. — Nombre d'années pour lequel la condamnation aux travaux forcés à temps doit être prononcée, 19.—Flétrissure des condamnés aux travaux forcés à perpétuité, 20. — Application du produit des travaux des condamnés à la réclusion, 21. — De quel jour se compte la durée de la peine des travaux forcés à temps, 23. — Les individus condamnés à la peine d'emprisonnement ont le choix de l'un des travaux établis dans la maison de correction où ils sont renfermés, 40.—Les coupables condamnés aux travaux forcés à temps sont pendant toute leur vie sous la surveillance de la haute police de l'État, 47. — Celui qui, ayant été condamné pour crime, en a commis un second emportant la peine des travaux forcés à temps, doit être condamné à celle des travaux forcés à perpétuité, 56. — Si le second crime entraîne cette dernière peine, il y a lieu à la condamnation à mort, *ibid.* — L'accusé ayant moins de seize ans, qui, agissant avec discernement, a encouru la peine des travaux forcés à perpétuité, est condamné à la peine de dix à vingt ans d'emprisonnement dans une maison de correction, 67. — Durée de cette détention lorsqu'il n'avait encouru que la peine des travaux forcés à temps, *ibid.* — Peine des travaux forcés à temps contre les individus qui auraient fourni des logements ou lieux de retraite à des bandes illégalement armées, 99. — Même peine contre les auteurs de faux pour actes contraires aux constitutions, 118. — Peine des travaux forcés à perpétuité pour contrefaçon ou altération des monnaies françaises de billon ou de cuivre 133 ; et des travaux forcés à temps pour contrefaçon ou altération des monnaies étrangères. 134. — Même peine pour contrefaçon ou falsification de timbres, marteaux ou poinçons de l'État, 140.—Peine des travaux forcés à perpétuité pour faux commis par des fonctionnaires publics, 145. — Peine des travaux forcés à temps pour faux commis par d'autres personnes, en écriture authentique et publique, ou en écriture de commerce ou de banque, 146 et 147. — Cas dans lesquels la peine des travaux forcés à temps a lieu pour soustractions commises par les dépositaires publics, 168 et 170. — Cas dans lequel les violences envers des magistrats donnent lieu à la peine des travaux forcés à temps, 231. — Peine des travaux forcés à temps ou à perpétuité, dans le cas où l'évasion de prison a été favorisée par une transmission d'armes, 240. — Peine des travaux forcés à temps contre le dépositaire de pièces qui les a soustraites ou détruites, 251. — Même peine contre des malfaiteurs associés, 266. — Cas dans lequel la peine pour simple meurtre est celle des travaux forcés à perpétuité, 304. — Menaces qui font encourir la peine des travaux forcés à temps, 305. — Blessures et coups volontaires non qualifiés de meurtre qui donnent lieu à la peine des travaux forcés à temps ou à perpétuité, 310 et suiv. — La première de ces peines infligée pour crime de castration, et la seconde pour avoir procuré l'avortement d'une femme enceinte, 316 et 317. — Peine des travaux forcés à temps pour crime de viol commis sur la personne d'un enfant au dessous de quinze ans, 332. — Peine des travaux forcés à perpétuité pour crime de viol de la part de ceux qui ont autorité sur la personne envers laquelle l'attentat a été commis, 333. — Peine des travaux forcés à temps pour séquestration de personnes, et des travaux forcés à perpétuité, si cette séquestration a duré plus d'un mois,

341 et 342. — Travaux forcés à temps pour enlevement de filles au-dessous de seize ans, 355 et 356, pour divers cas de faux témoignages, 361 et suiv. — La peine des travaux forcés à temps ou à perpétuité, suivant les circonstances dans lesquelles on a suborné des témoins, 365. — Vols punis de l'une ou l'autre de ces deux sortes de travaux, 383 à 85. — Travaux forcés à temps pour avoir extorqué la signature ou la remise d'un écrit ou d'une pièce quelconque contenant obligation ou opérant décharge, 400. — Même peine contre les banqueroutiers frauduleux et les agents de change ou courtiers qui ont fait faillite, 402 à 404. — Travaux forcés à perpétuité contre les agents de change et courtiers convaincus de banqueroute frauduleuse, 404. — Travaux forcés à temps contre les fonctionnaires publics et agents du Gouvernement qui auraient aidé les fournisseurs à faire manquer le service des armées, 432. — Cas dans lesquels la peine des travaux forcés à temps, ou à perpétuité, est encourue pour menace d'incendie, 436. — Travaux forcés à temps pour pillage ou dégât de denrées et marchandises, commis en réunion et à force ouverte, 440 et 442. — Voy. *Age*, *Arrêt*, *Bannissement*, *Carcan*, *Condamnation*, *Condamnés*, *Marque*, *Réclusion*.

TRAVAUX PUBLICS. Peine de quiconque, par des voies de fait, se serait opposé à la confection de travaux autorisés par le Gouvernement, 438.

TRIBUNAUX DE POLICE. Voy. *Confiscation*.

TROMBLONS. Voy. *Armes*.

TROUBLES. Voy. *Cultes*.

TROUPES. Peine de mort, avec confiscation de biens, contre ceux qui auraient levé des troupes armées sans l'autorisation du pouvoir légitime, 92.

TUMULTE. Voy. *Secours.*

TUTELLE. Celui qui a été condamné à la peine des travaux forcés à temps, au bannissement, à la réclusion ou au carcan, est incapable de tutelle et de curatelle, si ce n'est de ses enfants, et sur l'avis seulement de sa famille, 28. — La même interdiction peut, dans certains cas, être prononcée par les tribunaux jugeant correctionnellement, 42 et 43. — Attentats aux mœurs qui font interdire les coupables de toute tutelle et curatelle, 335 — Voy. *Mœurs.*

TUTEUR. Voy. *Exposition d'enfant.*

U.

UNIFORME. Voy. *Costume.*

USINES. Peines contre les propriétaires ou fermiers jouissant de moulins ou usines qui, par l'élévation du déversoir de leurs eaux au dessus de la hauteur déterminée par l'autorité administrative, auraient inondé ou endommagé les chemins ou les propriétés d'autrui, 457. —Voy. *Fours, Incendie.*

USURPATION DE TITRES OU FONCTIONS. Peines auxquelles elle donne lieu, 258 et 259.

V.

VAGABONDAGE. Quels individus sont qualifiés de vagabonds ou gens sans aveu, 269 et 270. — Peines encourues par les vagabonds légalement déclarés tels, 271 et 272.—Les vagabonds nés en France peuvent être réclamés par délibération du conseil municipal de la commune où ils sont nés, ou cautionnés par un citoyen solvable, 273.— Dispositions communes aux vagabonds et mendiants, 277 et suiv.

VAISSEAUX, Voy. *Incendie, Mine, Places.*

VENDANGES. Voy. *Ban.*

VENDEURS. Voy. *Crieurs.*

VÉTUSTÉ. Voy. *Incendie, Villes, Bandes armées, Commandement militaire, Places.*

VIOL. Peine contre le coupable du crime du viol, 331. — Aggravation de la peine, si le crime a été commis sur un enfant au dessous de l'âge de quinze ans, 332. — Différentes sortes de peines suivant la classe des coupables, 333.

VIOLATION. Voy. *Sépulture.*

VIOLENCES. Peines pour violences exercées de la part d'un fonctionnaire ou officier public, d'un administrateur, d'un agent ou d'un préposé du Gouvernement ou de la police, d'un exécuteur des mandats de justice ou de jugement, d'un commandant en chef ou en sous-ordre de la force publique, 186. — Peines pour violences exercées contre un officier ministériel, ou agent de la force publique, ou un citoyen chargé d'un ministère de service public, 230. — Peines plus fortes en cas d'effusion de sang, blessures, maladie ou mort, 231 et 233. — Dans le cas même où ces violences n'auraient pas eu de pareilles suites, 232. — Peines contre les mendiants ou vagabonds qui auraient exercé quelque acte de violence, 279. — Peine pour vol avec violences ou menaces de faire usage d'armes, 382, 383 et 385. — Voy. *Avortement, Blessures, Bris de prison, Corruption. Etat, Meurtre, Rebellion, Scellés, Vols*

VIVIERS. Voy. *Champs, Empoisonnement.*

VIVRES. Peine pour avoir fourni des vivres à des bandes illégalement armées, 96.

VOIE PUBLIQUE. Amende contre ceux qui auraient em-

barrassé la voie publique en y déposant ou laissant
des matériaux propres à empêcher la liberté ou la sûreté du passage, 471. — Voy. *Arbres*.

VOIES DE FAIT. Cas dans lequel les voies de fait contre un magistrat donnent lieu contre le coupable à la peine du carcan, 228.—Voy. *Attroupement, Culte, Rebellion, Travaux publics, Violences*.

VOILE. Celui dont on couvre la tête du parricide conduit au supplice, 13.

VOIRIE. Peines contre ceux qui auraient négligé et refusé d'exécuter les règlements et arrêtés concernant la petite voirie, 471.

VOITURIERS. Amende contre les rouliers, charretiers, conducteurs de voitures ou de bêtes de charge, qui ne se seraient pas tenus constamment à portée de leurs chevaux et voitures, pour ne leur laisser occuper qu'un seul côté de chemin ou voie publique, 475; et contre ceux qui auraient laissé courir leurs chevaux dans l'intérieur d'un lieu habité, ou qui auraient violé les règlements contre le chargement, la rapidité ou la mauvaise direction des voitures, *ibid.* — Emprisonnement encouru pour les mêmes causes, 476. — Peine contre ceux qui auraient occasionné la mort ou la blessure de bestiaux par la rapidité, la mauvaise direction ou le chargement excessif des voitures, chevaux, bêtes de trait, de charge ou de monture, 479. — Emprisonnement suivant les circonstances, 480. —Voy. *Aubergistes*.

VOLS. Tout vol commis à l'aide d'un bris de scellés est puni comme le vol avec effraction, 253. — Il n'y a ni crime ni délit pour homicide commis en se défendant contre les vols et les pillages exécutés avec violences, 329. — Définition du vol, 379. — Vols accompagnés de circonstances qui rendent les coupables punissables de mort, 381. — Vols qui sont punis des travaux forcés à perpétuité ou à temps, 381 à 383. — Vols punis de la réclusion, 386 à 388, et 399. — Vols punis d'emprisonnement et d'amende, 400 et 401. — Voy. *Champs*.

VOTE. Cas dans lesquels les tribunaux peuvent interdire l'exercice des droits de vote et d'élection, 42; et de ceux de vote et de suffrage dans les délibérations de famille, *ibid.* — Interdiction temporaire du droit de vote contre ceux qui auraient empêché l'exercice des droits civiques, 109.

FIN DE LA TABLE ANALYTIQUE ET RAISONNÉE DES MATIÈRES.